IUFRO Research Series

The International Union of Forestry Research Organizations (IUFRO), with its 14,000 scientists from 700 member institutions in 100 countries, is organized into nearly 300 research units that annually hold approximately 60 conferences, workshops and other meetings. The individual papers, proceedings and other material arising from these units and meetings are often published but in a wide array of different journals and other publications. The object of the IUFRO Research Series is to offer a single, uniform outlet for high quality publications arising from major IUFRO meetings and other products of IUFRO's research units.

The editing, publishing and dissemination experience of CABI *Publishing* and the huge spread of scientific endeavours of IUFRO combine here to make information widely available that is of value to policy makers, resource managers, peer scientists and educators. The Executive Board of IUFRO forms the Editorial Advisory Board for the Series and provides the monitoring and uniformity that such a high quality series requires in addition to the editorial work of the conference organizers.

While adding a new body of information to the plethora currently dealing with forestry and related resources, this series seeks to provide a single, uniform forum and style that all forest scientists will turn to first as an outlet for their conference material and other products, and that the users of information will also see as a reliable and reputable source.

Although the official languages of IUFRO include English, French, German and Spanish, the majority of modern scientific papers are published in English. In this series, all books will be published in English as the main language, allowing papers occasionally to be in other languages. Guidelines for submitting and publishing material in this series are available from the Publisher, Books and Reference Works, CABI Publishing, CAB International, Wallingford, Oxon OX10 8DE, UK, and the IUFRO Secretariat, c/o Federal Forest Research Centre, Seckendorff-Gudent-Weg 8, A-1131, Vienna, Austria.

IUFRO Executive Board:

Titles Available:

1. *Forest Dynamics in Heavily Polluted Regions*
   Edited by J.L. Innes and J. Oleksyn

2. *Forest History: International Studies on Socioeconomic and Forest Ecosystem Change*
   Edited by M. Agnoletti and S. Anderson

3. *Methods and Approaches in Forest History*
   Edited by M. Agnoletti and S. Anderson

4. *Air Pollution and the Forests of Developing and Rapidly Industrializing Countries*
   Edited by J.L. Innes and A.H. Haron

5. *Forests in Sustainable Mountain Development*
   Edited by M. Price and N. Butt

Forest History: International Studies on Socio-economic and Forest Ecosystem Change

Report No. 2 of the IUFRO Task Force on Environmental Change

Edited by

M. Agnoletti
Department of Forestry and Environmental Sciences, University of Florence, Italy

and

S. Anderson
Forest History Society, Durham, North Carolina, USA

CABI *Publishing*
in association with
The International Union of Forestry Research Organizations (IUFRO)

CABI *Publishing* is a division of CAB *International*

CABI Publishing
CAB International
Wallingford
Oxon OX10 8DE
UK

Tel: +44 (0)1491 832111
Fax: +44 (0)1491 833508
Email: cabi@cabi.org
Web site: http://www.cabi.org

CABI Publishing
10E 40th Street
Suite 3203
New York, NY 10016
USA

Tel: +1 212 481 7018
Fax: +1 212 686 7993
Email: cabi-nao@cabi.org

A catalogue record for this book is available from the British Library, London, UK.

Library of Congress Cataloging-in-Publication Data

Forest History : international studies on socioeconomic and forest ecosystem change : report number 2 of the IUFRO Task Force on Environmental Change / edited by M. Agnoletti and S. Anderson

p. cm. -- (IUFRO research series ; 2)

Papers selected from a conference held in Florence in 1998.

Includes bibliographical references (p.).

ISBN 0-85199-419-9 (alk. paper)

1. Forests and forestry--Congresses. 2. Forests and forestry--Social aspects--Congresses. 3. Forest ecology--Congresses. I. Agnoletti, Mauro. II. Anderson, S. (Steven), 1956- III. IUFRO Task Force on Environmental Change. IV. Series.

SD118 .F53 000
634.9'09—dc21

99-088765

Published in Association with:

The International Union of Forestry Research Organizations (IUFRO)
c/o Federal Forest Research Centre
Seckendorff-Gudent-Weg 8
A-1131 Vienna
Austria

ISBN 0 85199 419 9

Typeset by AMA DataSet Ltd, UK
Printed and bound in the UK by Biddles Ltd, Guildford and King's Lynn

Contents

Contributors

M. Agnoletti, Department of Forestry and Environmental Sciences, Via San Bonaventura 13, I-50145 Florence, Italy
J.A.S. Amarilla, Universidad Complutense, Madrid, Spain
E. Amodio, Escuela di Antropología, Universidad Central di Venezuela, Caracas, Venezuela
S. Anderson, Forest History Society, 701 Vickers Avenue, Durham, NC 27701, USA
M. Armiero, Department of History, University of Naples, Via Giuseppe Bonito 29, 80129 Naples, Italy
S.C. Artacho, Departamento Territorio y Patrimonio Histórico, Universidad de Jaén, Paraje de las Lagunillas, s/n CP 23071, Jaén, Spain
J. Björklund, Department of Economic History, Umeå University, S–90187 Umeå, Sweden
J.I.J. Blanco, Facultad de Ciencias Económicas y Empresariales, Universidad Complutense, Campus de Somosaguas, 28223 Madrid, Spain
S.Z. Blanco, Extremadura University, Spain
A. Corvol-Dessert, CNRS, 45, rue d'Ulm, 75005 Paris, France
B. Cowell, Department for Culture, Media and Sport, 2–4 Cockspur Street, London SW16 1TL, UK
J. Dargavel, Centre for Resource and Environmental Studies, Australian National University, Canberra, ACT 0200, Australia
M.N. Delgado, Departamento Historia Contemporánea, Universidad de Granada Campus de Cartuja s/n, 18071 Granada, Spain
A.L. Freschi, Department of Forestry and Environmental Sciences, University of Florence, Via San Bonaventura 13, 50100 Florence, Italy

M.A. Gallo, Liceo Scientifico Statale Decollatura (CZ), viale Stazione 70, 88100 Gallina di Reggio Calabria, Italy
M. Gangemi, Facoltà di Economica, Istituto di Storia Economica, Università degli Studi di Bari, Bari, Italy
E. Garnier, Laboratory of Historical Science, University of Besancon, 12, rue des Ejols, 88120 Vagney, France
M. González de Molina, Departamento Historia Contemporánea, Universidad de Granada Campus de Cartuja s/n, 18071 Granada, Spain
B.-S. Grewe, Universitaet Trier, Neuere u. Neueste Geschichte, D-54286 Trier, Germany
F. Hatzenberger, Université Pierre et Marie Curie (Paris VI), 10 rue Laugier, 75017 Paris, France
L. Hermanin, Department of Forestry and Environmental Sciences, University of Florence, Via San Bonaventura 13, 50100 Florence, Italy
F. Iovino, Dipartimento di Agrochimica & Agrobiologia, Università di Reggio Calabria, Piazza San Francesco di Sales, 89061 Gallina di Reggio Calabria, Italy
E. Johann, Österreichischer Forstverein, Arbeitsgruppe Forstgeschichte, Bertoldstrasse 17, D-79085 Freiburg, Austria
M. Jones, Countryside Research Unit, School of Leisure and Food Management, Sheffield Hallam University, City Campus, Sheffield S1 1WB, UK
C.M. Manuel, Departamento de Silvopascicultura, Escuela Técnica Superior de Ingenieros de Montes de Madrid, Ciudad Universitaria s/n 28040 – Madrid, Spain
D.G. Mártinez, Zaragoza University, Spain
P. Mitchell-Banks, Central Coast Consulting, 4439 West 7th Avenue, Vancouver, BC, V6R 1X1, Canada
G. Modert, Institut für Neuere Geschichte, Ludwig-Maximilians-Universität München, Trautenwolfstraße 3/I, D-80802 Munich, Germany
V. Moriniaux, Institut de Géographie, Université Paris IV-Sorbonne, 191, rue Saint-Jacques, 75005 Paris, France
P. Nanni, Rivista di Storia dell'Agricoltura, Accademia dei Georgofili, Logge degli Uffizi Corti, 50122 Florence, Italy
A. Ortega, Departamento Historia Contemporánea, Universidad de Granada Campus de Cartuja s/n, 18071 Granada, Spain
J.F.Z. Pineda, Malaga University, Spain
T. Raffaelli, Dipartimento di Economia dell' Impresa, della Tecnologia, dell' Ambiente, viale Sant' Ignazio 17, 09123, Cagliari, Italy
A.S. Rawat, 2B Sleepy Hollow, Naini Tal, UP 263001, India
E. Rico Boquete, Universidad de Santiago de Compostela, Francisco Mosqueda 10, 4C, 28043 Madrid, Spain

I.D. Rotherham, Countryside Research Unit, School of Leisure and Food Management, Sheffield Hallam University, City Campus, Sheffield S1 1WB, UK
F. Salvestrini, Department of History, University of Florence, Piazza San Remigio, 1, 50122 Florence, Italy
A.O. Santos, Departmento Historia Contemporánea, Universidad de Granada, Campus de Cartuja S/N CP 18071, Granada, Spain
L.F. Sedrez, History Department, Stanford University, Stanford, CA 94305-2024, USA
J.J. Serovayskay, University of Karaganda, University Str. 28, 470012 Karaganda 12, Kazakhstan
M. Suttor, Université d'Artois, Centre de Recherches des Anciens Pays-Bas à l'Eurorégion, 9 rue du Temple – BP 665, F-62030 ARRAS Cedex, France
J.P. Torrente, Instituto Feijoo de Estudios del Siglo XVIII, Universidad de Oviedo, Campus del Milán, s/n E-33011 Oviedo, Spain
J. Tsouvalis, School of Geography, University of Nottingham, University Park, Nottingham NG7 2RD, UK
C. Watkins, School of Geography, University of Nottingham, University Park, Nottingham NG7 2RD, UK

Preface

Most of the chapters presented in the book are based on presentations given at the International IUFRO conference 'History and Forest Resources' held in Florence (Italy) in May 1998. We would like to thank the Ente Cassa di Risparmio di Firenze and the Forest Service of the Tuscan Regional Government for sponsoring the meeting, the Italian Academy of Forestry Science for holding and promoting it, the staff for secretarial and organizational assistance and the members of the scientific and organizing committee, Florence City Council for the reception at the Palazzo della Signoria, Roberto Guicciardini for the hospitality offered at the Gargonza castle and Luigi Hermanin for leading the post-conference excursion to the Vallombrosa forest. We would also like to thank IUFRO, especially the Leader of Division 6, Niels Eler Koch and the President of IUFRO, Jeff Burley, for promoting the meeting and supporting the publication of this volume. We give special thanks to Elisabeth Johann, leader of the forest history subject group, for her effective collaboration and all her efforts to make the conference successful, and Lars Ostlund and John Dargavel for their collaboration and suggestions for the publication. Special thanks are also extended to the International Programs of the United States Forestry Service for support of the early editing efforts required for this volume, Gary Blank and Andrea Anderson for their assistance in editing, and Tim Hardwick and the staff of CABI *Publishing* for their assistance.

One of the most important achievements of the conference was to bring together scientists from different research fields and different cultural backgrounds, stimulating an interdisciplinary approach to forest history. In order to preserve this value, the editors and CABI *Publishing* decided to allow the authors to keep their own style when reporting references. Thus, both reference styles used in scientific and historical papers are found, according to the professional background of each author. Another book based on papers from the conference (*Methods and Approaches in Forest History*) has also been published by CABI *Publishing*.

M. Agnoletti
Chairman, IUFRO S6.07.02

S. Anderson
President, Forest History Society

Introduction: Factors and Processes in the History of Forest Resources

Mauro Agnoletti

Department of Forestry and Environmental Sciences, Via San Bonaventura 13, I-50145 Florence, Italy

1 Introduction

The beginning of the 21st century imposes a need for profound reflection on the way modern society is dealing with the issue of integrating socio-economic development and environmental conservation. Forest history investigations have shown that most of the forest ecosystems on this planet are affected by human activity. Unfortunately, environmental policies, especially forest policy, is still taking the results and suggestions of these studies into but little account. The limited awareness of forest history research, among both the general public and the scientific community, leads to a lack of knowledge of forest ecosystem dynamics and to a reduced capability for understanding the effects of today's political decisions and economic trends on the forest environments. The chapters in this book analyse some of the main issues concerning the interrelationships between human society and the forest environment, also presenting current debates in related disciplines such as environmental history and historical ecology.

Some of the most debated issues concern the extent to which the forest environment has been modified by man. The real effects of human activity seem alternately to be either over- or underestimated, according to the approaches and findings of the various studies (Balée, 1998). Increasing uncertainty about forest history in Europe has recently been cited by Watkins (1998) who reported on the results of investigations presented at a major conference held on the subject of forest and woodland history in Nottingham in 1996. Although we must agree on the variability of the effects of interference on local forest ecosystems, forest resources were often affected by

(eds M. Agnoletti and S. Anderson)

widespread economic, social and political changes causing similar effects on a large scale. According to this approach most of the authors in this book present studies related to large geographical areas, in order to define the factors and processes affecting forest changes using different temporal scales. However, the analysis of the evolution of man/forest relationships over time should also be supported by local investigations to clarify the extent to which general historical patterns can be applied to regions or countries. In the same way, the definition of a local model needs to be provided at a general level to evaluate its relevance and to present comprehensive pictures, which cannot be replaced by a number of microhistories. This approach suggests integrating methods in forest history as proposed by an increasing number of researchers, matching problems such as the traditional opposition between historians and those with backgrounds in the natural sciences (Agnoletti and Anderson, 2000).

2 Fuel wood and timber from the Middle Ages to the Enlightenment

The importance of woodlands as a strategic resource in the Middle Ages has been already assessed by forest historians, but the real impact of medieval societies on forest resources is still unclear, even when considering only timber or fuel wood. Suttor's description of the growth of the timber trade along the Meuse River from the 11th to the 17th century reminds us that the need for fuel wood and charcoal for mining and timber for shipbuilding affected the history of many European forests. The need for charcoal seems the reason for the heavy exploitation of many forests, especially in countries like Germany, the leading country for metal production in the 16th century (Devèze, 1962). However, the results of historical research on this matter are quite controversial. Together with interpretations such as those by Dornic (1984) or Gabbrielli (1982), stressing the generalized destruction of forests for iron smelting, books like the one edited by Woronoff (1990) on France seem to demonstrate a lesser impact of metallurgy on the reduction of forests. This approach, which tends to diminish the oft-claimed destructive impact of industrial activities on forest resources, also appears in other chapters in this volume, but the metal industry was only one of the many activities that used large amounts of wood. The organization of forest utilization to supply fuel wood for mining and glassmaking as well as timber in the mountain region of the Vosges, is described by Garnier. However, to understand the effects of these activities on forest changes it is important to know what kind of wood was utilized. The preference for broad-leaved trees such as beech, oak or chestnut, rather than conifers, for producing charcoal had a great impact on forest successions. A change in the main species populating forests is reported for the 14th–16th centuries in the Vosges (Boithias and Brignon, 1985) and also in the Italian Dolomite alps (Agnoletti, 1998) as a result of forest succession, alternating beech and spruce or fir. An economic interpretation of this succession, assessing the major role played by mining activities preferring beech to produce charcoal, as opposed to the timber industry favouring conifers, must take the ecology of beech and spruce into account. Management practices favouring the utilization of one species, progressively turning the forest into an ecosystem dominated by a single species, ends by promoting the regeneration of the other. Alternation of species dominating the same forest, beech and fir for instance, is a well known process in ecological studies, although rarely linked to human influence.

The presence of mines also led to significant social and political problems, partly because the forests surrounding the mines were often reserved, reducing the customary rights of local communities, and partly because of the workers engaged in this activity. The rise of mining activities in Trentino (north-eastern Italian Alps) around the 13th century brought thousands of miners from southern Germany who settled there, creating an important German community, strongly affecting the local forest ecosystems and the life of the mountain communities.

Another major issue affecting not only the forests in Europe, but also those in several tropical countries is definitely shipbuilding, to which Suttor's chapter also refers. Generations of historians have dedicated their time to studying the development of maritime powers such as Venice, Holland and England. However, very few studies focus on the effects shipbuilding had on the forest of the countries and regions from which timber was collected. As reported by several authors, timber played an important strategic role for countries like Holland or Venice (Lane, 1973; North, 1996), but very little is known about the organization of forest utilization and transportation, or the role played by different wood species in shipbuilding. The decrease in Venetian power and the rise of Holland in the 16th and 17th centuries had much to do with the availability and the price of oaks, while the supply of conifers did not seem to be a major problem. In the 14th century Venice had already established an efficient timber industry for collecting conifer wood from the Alps for building galleys, while a severe shortage of oak wood was already being reported on the plains. From the 15th century, Venice reserved all the oaks in its dominions for shipyards, and drew up forest cadastres where each oak tree was marked and measured. This policy led to the development of conflicts with local populations that opposed the Republic's control of oak forests; furthermore, the Republic imposed the duty of carrying out utilization on these people. Although the demand for nearly 50,000–55,000 m^3 of oak wood every year created a major problem for Venice, it was little indeed compared with Holland's wood requirements: that country, in 1697 had 10,000 ships, almost half the entire world's fleet (Silvy-Leligois, 1962). The building of a Dutch galleon required at least five to six times more wood than a Venetian galley, but the amount of forest cut for this purpose could differ according to the species composition, ecological conditions and management system. In

Venice, in order to build one galley almost 300 ha of oak forest and 122 ha of conifer forest, managed with selective cuttings were cut, while 2–3 ha and 2 ha, respectively were needed if the clear cutting method was adopted (Agnoletti, 1999). This information gives us a rough idea of the problem related to supplying wood for all the major European maritime powers, and how the impact of cuts in different types of forests could vary. Holland imported wood, not only from the Baltic countries, but also from all the regions crossed by the Rhine River as far as Switzerland. The forest history of these areas from the 15th to the 18th century was affected by the timber trade. This was a lucrative business but often created a reason for conflict since the local populations needed oak woods to support their own farming and livestock-raising activities.

Human pressure on wood resources was already heavy in the old continent, but only began in the New World at the time of the Spanish colonization. As reported by Amodio and Hatzenberger, the environment that the first European settlers found was already modified by indigenous populations, so the idea of a ‘natural environment’ cannot be correlated with South America’s forests. The approach of a very traditional Catholic culture that applied European categories of knowledge to a completely new environment, alternately considered a paradise or an inferno, posed many problems including lack of knowledge of the quality of different wood species. However, the initial feeling of uncertainty and wonder regarding the natural resources of the area rapidly changed as the Spanish quickly began exploiting wood resources, first using dye wood for the clothing industry and later timber.

According to several authors, pressure on European forest resources increased until the late 17th and early 18th century when the area of woodlands seemed to reach its lowest level (Kirby and Watkins, 1998). However, this time scale can probably only be applied to some parts of the continent, as the factors causing the reduction of forests did not appear everywhere at the same time. In Italy the deforestation rate increased rapidly during the 18th century, not as the result of the farmers’ need to extend grain cultivation for their own consumption, but rather due to the actions of other forces, such as the landowners’ desire for increased profits from expanding agriculture (Vecchio, 1974). As reported by Nanni, in Tuscany deforestation took place mostly at the end of the 18th and the beginning of the 19th century as a result of the expansion of agriculture. This process originated from the Grand Duke’s free trade policy liberalizing the grain and wood markets, and eliminating former restrictions on wood cutting. However, with regard to the 18th century, several authors criticize the interpretation of terms such as tillage and deforestation that are often used as synonyms, and they also question previous analyses of official documents concerning the reduction of woodlands (Corvol, 1995).

If the old continents offer a great variety of interpretations regarding the late 18th century, the situation described in South America is quite different. A new cultural view of nature that matured during the Enlightenment and the growing need for timber favoured a more rational approach to the tropical forest environments which, in turn, led to dramatic changes in the landscape. The colonial impact on those territories increased steadily, but it would be accelerated even more by the North American approach in the 19th century. A typical expression of the 'manifest destiny' doctrine, also applied to South America, was reported by Sedrez. As a result, the native forest in the south and south-east sections of Brazil, covering 75% of the land in the 16th century, would be reduced to 13% by 1980, and this is just one example among many for that area (Dargavel and Kengen, 1992).

This progressive increase of human pressure on the forest resources of the New World was backed by the rise of scientific forestry in Europe, due to growing concerns about the wood supply and its rising costs. The development of forestry was initially expressed in France by the works of Duhamel du Monceau and then later in Germany, at the beginning of the 19th century, through the foundation of forest schools that replaced customary forest practices based on traditional local knowledge. This shift from the old to the new forest culture may well be illustrated by the situation of the Venetian shipyards in 1797, on the advent of the French army. The French were shocked by the poor conditions of the shipyards that once had dominated the world. Venetian silvicultural knowledge also appeared disappointing as it was based only on the practical experience of the shipwrights who had no scientific background. The real issue was that Venice, like many other Italian states, had lost its political and economic power, especially in sea trade.

3 The scarcity of wood: a key issue

The relationships between industrial development and forest resources present different patterns according to the ecological features of the forests and the socioeconomic environment. Although in countries like England from the early 18th century on, nearly 90% of fuel requirements were met by coal (Collins, 1996), wood was still the most important source of energy and the need for timber was increasing. In poor regions, such as southern Italy or Spain (Armiero, Torrente), the presence of industry in areas with limited amounts of forests required an organization that also took the needs of local populations into account in order to avoid deforestation and social conflicts. Armiero suggests that industrialization promoted the development of silviculture as a science supporting production, but for Italy very little evidence can be found to confirm this assumption until the second half of the 19th century. The situation in the Vosges mountains or the Alps could be considered completely different from that of the southern regions. Here the organization of forest utilization to produce fuel wood, charcoal and timber was more effective in supporting industrial development. This was due to the availability of large forests – permitting important technological developments rarely described in either forest history or the history of technology – and to reduced demographic pressure.

The issue of wood scarcity seems to attract the attention of recent approaches, arguing overestimation of forest destruction often ascribed to industrial needs versus traditional interpretations that present the destruction of forests as a general feature of our history. This reductive interpretation of negative human influence should be clearly separated from the approach proposed by historical ecologists suggesting that human actions did not necessarily lead to the degradation of the forest environment (Balée, 1998; Agnoletti and Anderson, 2000). A different appraisal of wood shortages is probably the result of greater attention given to investigative methods that try to avoid the spread of what Oliver Rackham called pseudo-histories having no connection with real facts. Nevertheless, it probably also reflects a cultural tendency that tried to attenuate the idea of the 'destructiveness of man', in the words of George Perkins Marsh (1985), reducing the sense of guilt, but also promoting some kind of reevaluation of the role economics played in its relations with the environment. This sort of historical revisionism, that comes mainly from historians, seems to be a crucial issue in the current debate on forest policy: some foresters assess the opposition between economics and ecology as one of the main problems in managing environmental resources (Ciancio and Nocentini, 2000).

In her chapter, Corvol-Dessert strongly criticizes the interpretations of wood shortages at the beginning of the 19th century, which are due to misleading interpretations of sources and a lack of consideration of the features of the wood market. This same approach is presented from a different angle in the chapters by Armiero and Grewe, which link France, Italy and Germany to opinions already expressed by Collins in 1996 regarding England. In his chapter Grewe deals with this topic by focusing his attention on who is complaining about wood shortages and who benefits from these claims, concluding that the matter was often an excuse for excluding local people from using the forest resources. Sometimes the state forest administrations seemed to promote this public feeling in order to maintain their power, as Corvol describes for France. These actions probably prevented a real shortage of timber and fuel wood, but induced a crisis in the availability of resources for the populations as reported by Modert and Grewe. From a general standpoint the situation seems to reflect the advantages the state could obtain from producing and selling timber for financial gain, thereby reducing or even impeding local uses according to an economic and silvicultural system devoted to maximizing wood production that was typical of Germany in the mid-19th century. Johann presents a different perspective for the Austrian Alps. There, population growth in the 18th–19th century reduced the amount of land available for each family, causing a profound crisis since survival was not possible without integrating wood resources with agriculture and livestock raising. According to the author, this integration was more important than timber production as a local resource, however heavy grazing and litter utilization could permanently reduce soil fertility. This is also the reason for degradation in the Italian Apennine mountains and in many other Mediterranean countries, where grazing and agriculture caused deforestation leading to progressive reduction of soil fertility through erosion and dry climate.

4 The role of technology

The purported scarcity of wood was often an effect of the increasing price of timber assortments, due to higher transportation and labour costs, making cuts too expensive. The issue of transportation is quite crucial when considering the availability of the wood supply. The establishment of effective forest utilization systems was often linked to the possibility of floating timber. In the Baltic Sea, special ships were designed to transport timber, while on the Rhine River 350-m-long rafts travelled to Holland, while much smaller ones were used along the alpine streams (Agnoletti, 1995). Here, the need to harvest timber not too far from waterways forced dealers to make cuttings almost in the same areas thus degrading the forests. The technical systems based on water as a source of energy and means of transportation lasted until the development of efficient road systems. Nevertheless, until the end of the 18th century in northern Italy, horse-drawn transportation could still cost 30% more than water-borne means. In the second half of the 19th century the railways brought about a great innovation and opened entire regions to forest exploitation, facilitating foreign competition and the development of a market economy as reported by Björklund. This pattern of development can be seen in many areas according to the periods when the roads and railways were built. In the case reported by Gangemi for Calabria in southern Italy, the lack of suitable roads and rivers made wood transportation from the mountains to the coastal ports nearly impossible, although local people were forced to provide labour for this purpose. In the early decades of the 20th century when a railway was built connecting northern and southern Italy, it was still more convenient for northern Italian dealers to purchase timber in Austria because the railway tariffs were lower.

The technological revolution affected not only transportation, but also saw-milling. Water-powered sawmills were replaced by steam power, but the change took place at very different rates. In countries like the United States, steam-powered saws were already widespread in the mid-19th century (Stanger, 1967) and they were also common in Great Britain, Scandinavia and Central Europe. At the same time, water-powered mills were still popular in the Alps, the Black Forest, the Vosges and Eastern Europe. During the 19th century Venetian sawmills, or the two- or three-stage transmission mills so popular in France and Germany could still compete with steam saws, but the development of more productive mills with circular or band saws reduced their numbers (Agnoletti, 1994). In the long run, there was no way to compete with the North American or Scandinavian production that already dominated the timber market in the beginning of the 20th century. Nevertheless, particular economic and market conditions – in Italy, for example – allowed small hydraulic mills to continue to exist until 1960. Since the beginning of the century the country had been importing most of its lumber from Austria, reducing saw-milling and developing a furniture industry that achieved a leading market position. This underlined the destiny of a country with no raw materials that was compelled to turn to processing industries.

5 Social conflicts

Many chapters (Armiero, Cowell, Cruz, Garnier, Grewe, Modert, Ortega, Rawat and Serovayskay) report on how changes in management forms due to property variations, legislation or the imposition of different management regimes often led to conflicts with local populations. Conflicts cannot be regarded simply as political and social matters, they are also evidence of disturbances at the ecosystem level, deeply affecting forest evolution. Traditional cultures and early agrarian societies often established complex systems of practices that were frequently described as 'activation practices' (Moreno *et al.*, 1998) to enhance the availability of environmental resources as reported by many anthropologists (Netting, 1981). If not affected by excessive population growth, these agroforestry systems could maintain a certain stability in forest ecosystems, but disturbances generated by the imposition of different political and economic systems could break this often uncertain equilibrium, provoking the local populations to react. The cases presented in the chapters by Ortega and Cruz very clearly describe how a rural society, based on the traditional agricultural/forestry/grazing systems, was modified by the commercial forestry policy that destroyed these. These dramatic changes generated a crisis in the local ecosystems, interrupting energy flows and food chains, and also stimulated the reactions of local populations expressed through the spread of forest fires and other forest crimes, leading to further upheavals in the forest ecosystems. As reported by Cowell, citing Freeman (1997), the conflicts among different social groups over forest resources did not automatically lead to ecological instability as long as there was a body of laws and regulations to settle claims and provide sustainable management. This could be the case represented by several alpine communities, which freely established a body of rules on how to use forest resources, regulating their relationship with the environment and often leading to fights with external forces that threatened their rights. In other contexts, the ability of outside authorities to achieve control of forest resources seems questionable, especially when legislative acts tended to exclude people from using the forests, as described also by Grewe. The wide range of reactions of local communities to the creation of royal reserves, and the imposition of noble authorities on forest lands in England from the 11th to the 19th century, as discussed by Serovayskay and Cowell, is an example that can also be applied to other countries. In this case it was not industrial utilization that caused the conflicts, but rather, the reduction of common rights in forests reserved for hunting, considered a key element of aristocratic power and prestige (Watkins, 1998).

An interesting comparison with the European situation, taking colonialism and the development of industrial silviculture into account, is presented in the chapters by Rawat and Dargavel. Rawat stresses the paradoxes of English policy in India, promoting agriculture at the beginning of the 18th century and stimulating an economy of self-sufficiency for the local populations. Towards the end of the century, with the introduction of modern silviculture and the development of an imperial model of forestry which would also be applied in other parts of the Empire, limitations were imposed on the use of forests and grazing, provoking violent reactions on the part of the population. An issue linked to the same problem and quite typical of the approach of European settlers colonizing new lands, is presented by Dargavel for the Australian continent. The application of modern management forms by the Australian colonists often resulted in a destructive impact on any previous forest patterns established by pre-colonial societies creating what Dargavel calls the 'wood of neglect', a very degraded forest territory. The legal reservation of the best forests for the new settlers, the dispossession of the original inhabitants, the decrease in their traditional rights and management by professional foresters seem to be the main elements of the imperial model of forestry as applied in Australia. According to the author, the contrast between well planned forestry conducted on the basis of European forestry and the wood of neglect, symbolizes the current situation of many forests in developing countries when compared with temperate forests. This problem has not been solved in either forest use or social terms and it also represents the main reason for the development of modern environmentalism. More generally, chapters such as that by Rawat confirm a tendency to politicize forest history in India, whose features and dangers have been clearly reported in Grove's introduction to *Nature and the Orient* (Grove *et al.*, 1998).

6 The legislative aspects

Some of the chapters seem to confirm the opinion expressed by Trifone in 1957, that the rise of forest laws expressed mostly the need of the dominant classes to keep forest resources under their control. However, the extent to which the promulgation of a law was effective in preserving a forest is not clear, as no law was able to achieve full control of forest resources, as all the chapters on social conflicts show. The establishment of the royal forests in England was meant to reserve their use for the Crown, excluding the common rights of the local inhabitants, but the history of these forests shows that it was not sufficient to preserve them. The officials paid to protect the trees of Sherwood Forest were the first to make deals to cut them (Watkins, 1998), while the forest crimes listed by Serovayskay show how little the institutions could do to prevent local populations from using the forest. This history is very close to that of the oak forests reserved for the shipyards by the Venetian Republic from the 15th to the 18th century: no threats of the death penalty or other punishment were successful, either in stopping abuses or even in saving the forests from destruction. Not all the forests reserved for the Venetian shipyards were destroyed, but those that were left were located in the mountains where human pressure was much lower than on the plains, and these forests comprised conifers and beech, not oak.

The mountain community of the Alps actually developed a more balanced approach to the forest environment and maintained the forests in fairly good condition through the ages. However, it is worth noting that in the 16th century, in the Dolomite Alps, in order to achieve at least a minimal repression of forest crimes, the law stated that one-third of the income from forest fines should be kept by the wardens charging the fine. A similar situation can be found in the Medici laws in Tuscany, as described in Salvestrini's chapter, which extended state control over almost all the forest resources, but did not prevent the Apennines from being deforested in the 16th century. According to Gallo, writing about the Calabria mountains in southern Italy, when the forest came under state administration this limited the application of common rights and reduced deforestation conducted for extending agriculture. But the conservation of the pine forests of the Sila and Aspromonte ranges was the result of the impossibility of carrying out forest utilization at economic prices, as illustrated by Gangemi.

Although the possibility of applying restrictions to forest use also depended on the state's ability to achieve effective control of the territory, forest laws seemed to be much more effective when they liberalized forest cuts. The application of the free trade policy in late 18th century Tuscany, as well as permission for private owners to enclose their lands, caused the rapid expansion of cultivated lands and the rapid reduction of forest. This led the authorities to worry about forest conservation, a growing concern for nearly all Italian regions, where the forest only covered 17% of the land as opposed to the estimated 75% in pre-Roman times. The foundation of the Forest School of Vallombrosa near Florence soon after the unification of Italy marked the beginning of a national forest policy. This policy aimed at promoting forestry to control erosion and prevent flooding, and to increase timber availability for the birth of Italian industry, as reported by Raffaelli. The rise of state forestry, as in England, Spain and the British Empire was supported mostly by the knowledge expressed by the German rather than the French school which also had a long tradition. In Italy, the years straddling the two centuries marked the shift from an economic culture based on free trade hegemony protecting private property, to a new wave of interventionist and protectionist ideas which led to the creation of a state forest agency in 1910. The aims were to achieve self-sufficiency in timber production, to provide an example to private owners as to how a correctly managed forest could become a profitable business, and mainly – though questioned by some economic historians – to improve soil conservation through afforestation and watershed management. Initially, state-owned forests were good business, but the idea of becoming self-sufficient in timber production was merely wishful thinking. This goal was never achieved, not even under the autarchy of the fascist regime, despite forest plantations and acquisitions of forest territory at the end of the First World War. These disappointing results seem quite similar to the policies in force in Spain during the same period as reported in the chapters by Blanco and Rico Boquete where the state's efforts to increase forests were devoted more to protection than to production. From this standpoint, Italian and Spanish policies differ markedly from the example given by Mitchell-Banks who discusses a timber-exporting country with a strong timber industry that is quite reluctant to accept limitations on forest exploitation in order to enhance environmental protection.

The drop in fuel wood and charcoal production after the end of the Second World War due to the use of new energy sources affected the history of most Italian forests, as well as the destiny of the state forest administration under the impact of growing economic problems. In general, the turn-of-the-century positivist culture which made the forest an object of scientific management, had to face the fact that timber production and land preservation could not always coexist. The latter involved expenditures that differed markedly from the economic investments that would bring a direct return sooner or later. The history of forest policy as described by Freschi shows increasing bureaucracy and inefficiencies affecting Italian forest policy after the Second World War, when forests progressively reduced their economic value. The transfer of administrative functions to the regional governments in the 1970s marked the end of a concept of state intervention. This led to uncertainties and ineffectiveness in environmental policy that was incapable of preventing conflicts concerning the utilization of forest lands.

7 Changes in forest ecosystems

There is still little attention dedicated to the changes in the forest ecosystems due to human action, compared with the amount of studies considering human society and its relations with the forest environment. Even when taking into account the effects of different socio-economic and climatic conditions and the different species involved, it becomes progressively clear that a reduction in broad-leaved species, especially oaks, and a growth of conifers has affected the history of many European regions from the Middle

Ages to the present. Contrasting forces were involved in this process, speeding it up or slowing it down, and creating different patterns, if we compare Great Britain (Rackham, 1995) with some areas of France (Du Bois, 1996), Sweden (Nilsson, 1997), Germany (Brandl, 1992) or Italy (Susmel, 1994). However, the significant role played by man is unquestionable. On one hand, oaks were one of the most important trees in the Middle Ages, a crucial resource in agriculture as farmers used their wood and also leaves and acorns to feed livestock, mainly pigs. On the other hand, the large amounts of fuel wood and charcoal needed for cooking, heating and most proto-industrial activities such as mining, metallurgy, brick and lime kilns burned oak in addition to other broad-leaved species such as beech. Oak forests grew mostly on plains and hilly areas and, as more and more territory was converted into arable land and pastures, most of the oak woods were destroyed. The species and their respective distribution also made a great difference. *Quercus sessilflora* and *Quercus peduncolata* which were very good for shipbuilding and producing acorns for livestock, were heavily exploited, but they were much more widespread in northern countries. These species were less common in the drier, southern countries where xerophytic oaks, such as *Quercus ilex* were quite widespread but of little use for feeding livestock or shipbuilding, and were mostly used for fuel wood. As long as ships were made of wood (at least 50–60% of the wood needed to build a galley was oak), the need for these trees by the Italian maritime republics, Holland, France, Spain and England, from the 15th to 19th century marked the history of many regions. The reduction in oak forests concerned not only the countries where the ships were built, but also the countries that supplied the shipyards. Having no forests of its own, Holland had to acquire timber wherever possible, but Venice, Great Britain, France and Spain did the same, and even resorted to importing it from the New World.

Dutch timber dealers traded in oak with farmers of the Black Forest, near Switzerland and would float it down the Rhine (Carle *et al.*, 1988) or they bought it in the Baltic countries as did England; Venice did the same, acquiring oak from the Dalmatian coast, and France sought oak even in Albania and Italy as did Spain (Bamford, 1956). Some of the chapters discuss the matter of price, but there was no price question for Venice, which simply did not have enough oak and bought it from Dalmatia, or even had its ships built there. Corvol questions the assertions about reductions of timber for the navy, giving increasing prices and transportation problems as the reason, but she also admits to a scarcity of curved pieces that were always obtained from oaks. Nevertheless, the price of Italian oak in the port of Toulon in 1770 was higher than French products and the amount of timber imported from Italy and Albania exceeded the domestic output. The oak forests suitable for shipbuilding along the Tuscan coast, already sparse around the mid-18th century, were all destroyed during the following decades as were those in the Venetian dominions and the south. Although shipbuilding was not the primary cause, since potash production, for instance, seems to have been another industry that consumed hardwoods in the Baltic regions, as was fuel wood production and later railway construction in Italy, the reduction of oak forests seems certain. The beech forests were also reduced in some areas like the Vosges, Baden-Württemberg and the Alps, but this has to be related also to the growing importance of conifers.

The chapter by Moriniaux offers an interesting account of the rise of conifers in France at the beginning of the 19th century, but similar evidence can be found regarding the forest history of other European countries in the other chapters. Part of the growing interest in conifers may be linked to the spread of some exotic species as reported by Moriniaux, but there is good evidence that their increased importance must also be related to the growing timber industry. Conifers such as spruce, fir and pines grow faster than oak, yielding straight logs that are easy to saw and process by sawmills and used by carpenters, while bark and litter could be used by farmers, and the branches used for firewood and charcoal. The growth of the European population also meant the growth of the construction industry, which preferred beams and planks made from conifers, while the development of scientific forestry promptly favoured the spread of these species. The leading theories of the German Forest School, based on economic and financial issues promoted geometrically shaped monospecific plantations, often using artificial regeneration. Even-aged forests replaced uneven-aged stands, while selective cuttings, not even considered a proper silvicultural method, were superseded by clear cuts or shelter wood felling. Conifers began to be planted by public and private owners, but the reason was not only economic. In fact, they were considered particularly suited for restoring soil fertility in degraded areas, a great problem in Mediterranean regions. The use of conifers for afforestation was also one of the most common silvicultural activities conducted by state forest services in all European countries, although this policy was implemented earlier in Germany, and later in countries such as England (see Tsouvalis and Watkins' chapter), Italy and Spain (Rico Boquete, Blanco, Manuel). In areas like the Alps and the Vosges, this process was already clear in the 17th–18th centuries, but the tendency seemed to accelerate at the turn of the 19th century. In the north-eastern Italian Alps, the percentage of conifers rose from 54% to 74% in almost 80 years from the end of the 19th century, and today, in high stands, the presence of broad-leaved species is just 2%, with a dramatic fall in beech and the disappearance of chestnut. The importance of chestnut from the Middle Ages and its reduction in the 19th and 20th centuries was related to the importance of chestnut flour as a food for mountain populations, while its reduction is being linked to the abandonment of mountain areas and decreases in agriculture. Along with afforestation, another reason for the expansion of conifers was the spread of these species deriving from the decrease in agriculture and successions taking place in former pastures and degraded areas, especially after fires. The expansion of larch in the Alps and of pine at lower altitudes is also due to the conifers' ability to colonize poor soils, and this is a common trend from Scandinavia to the Mediterranean.

The technical changes in agriculture and reductions in cultivated land reported in recent years introduce another aspect of forest changes, which is the reduction of coppice woods. Coppice woods have been one of the most common forms of forestry practice associated with agriculture since Roman times (Di Berenger, 1859–1863), yielding a large number of assortments such as poles, branches, timber, leaves and fuel wood to support farming. The preference given to coppices was due to the higher number of assortments produced and to the short rotation systems adopted (as little as 6 years), allowing more frequent harvesting compared with the much longer time needed for a high stand. The flexibility of this management form was achieved through different techniques, such as coppice with standards, simple coppice and selective coppice. Roman civilization diffused coppice technique in many European countries and coppice woods became an important feature of the landscape, especially in wine growing areas (Ostermann and Reif, 2000). Coppice woods, especially coppice with standards, seems an increasing management form in medieval England (Rackham, 1995), while its growth seems to be associated with the progressive reduction of wooded pastures, as reported by Rotherham and Jones. Tsouvalis and Watkins and Rotherham and Jones report on the reduction in coppices in England, associated with the rising importance of coal in place of charcoal, and the change in farming techniques at the end of the 19th century. As the value of the timber assortments produced by high stands increased, there was a gradual tendency to turn coppice woods into high forests, a process stimulated also by the development of scientific forestry. The consideration that the long-term investments needed for a high stand could not be supported by private owners was at the basis of the rise of the state forest agency in Italy, and this policy is still being conducted by many public administrations. In this way, a valuable historic form of landscape will be lost and economic interest in fuel wood production seems to be the only reason for the existence of these woods.

If the development of the timber industry can be questioned as the reason for deforestation, one common effect of the growth of sawmill activities was the reduction in the average size of trees. As related by Björklund, this problem has been affecting the Swedish timber industry since 1880, but in the Italian Alps it was already noted in the early decades of the 19th century when sawmills reduced the size of the frames, adapting them to smaller dimensions (Agnoletti, 1994). The harvesting of big trees was due to the higher prices paid for large saw logs, but the reduction slowly changed the technological features of sawmills and affected market conditions. The elimination of big trees changed the features of most uneven-aged forests, reducing age classes and slowly turning them into almost even-aged stands. Today, the pulp and paper industries, and the mechanization of forest utilization favour small trees, even less than 30 cm. Although this certainly satisfies the current needs of production, it is important to remember that this technological change was brought about by the heavy forest exploitation of the past. Different growing rates in northern and southern countries affect this as well, as the shorter growing season in Scandinavia means that it takes longer for a tree to attain large size, but certainly modern industrial plantations have shortened rotation times to meet industrial needs.

Compared with Europe and North America where the effects of industrial forestry profoundly modified the features of many forests (Langston, 1995), developing countries present a different situation. As reported by Amodio and Hatzenberger, the forest environments the European settlers found was already affected by the local populations, but the amount of change brought about by colonialism had a much more powerful impact. The export of Brazilian wood (*Caesalpinia echinata*) to Europe, not as timber, but because of its red dye, reduced the coastal forests of this species to a very small area by the 18th century although the inland forests were mostly untouched. Logwood (*Haematoxylon campechianum*), used for the same purpose met the same fate; it was exploited until the early 20th century when chemical dyes came to dominate the European market (Tucker, 1992). The 17th century mahogany trade also affected the Caribbean, which was compensating for the shipbuilding timber shortage that already existed in Europe. By 1800, mahogany was being used for furniture and became a more important commodity than dye wood; other species such as Spanish cedar were also heavily exploited. For Spain, and also for England, and later the USA, South America was merely an inexhaustible resource to exploit. The only ‘problem’ was how to do this since utilization was very difficult. Forests at great distances from the water remained largely untouched and only technological innovations in sawmills and transportation changed the situation in the early decades of the 20th century. Later, forest plantations of eucalyptus and pine were created in south-eastern Brazil to overcome wood shortages.

In the Caribbean, Cuba’s mahogany forests were totally cleared to make room for sugar cane plantations, and the same destiny touched Haiti, where coffee plantations were created on previously forested territories and lemon and orange trees were introduced by settlers. The detailed description of forest changes presented by Hatzenberger shows how the effects of exploitation caused the destruction of certain tree species, modifying the features of local ecosystems and inducing secondary successions favouring xero-mesophilous shrubby species. Still in Haiti, from the 19th century on, the reduction of broad-leaved species was followed by the expansion of pines, partly as a result of the colonization of degraded soils and partly because of afforestation.

Regarding Australia, along with the destruction of native forest ecosystems Dargavel reports an extension of afforestation with pine commencing in the mid-19th century. The population expanded rapidly and domestic production could not meet the demand for softwood, which was imported mostly from Europe and North America. Most plantations were established using *Pinus radiata* and other species such as *Pinus pinaster*, but *Eucalyptus* spp. were also planted since the commercially usable hardwood timber from native eucalyptus forests was quite limited. Half of these forests were cleared or heavily modified for farming and much of the native wood destroyed.

A very brief account of forest changes is presented in Rawat's chapter. He reports on a progressive reduction of oak forests in the Central Himalayan region due to the spread of agriculture and the need for fuel wood, timber, leaves and litter. In this case commercial timber exploitation cannot be considered a reason for deforestation; if no alternative energy resources are found, the need for fuel wood will probably lead to deforestation.

8 Conclusion

Although a reduction in forested land over time seems certain, there is a growing number of historians who question the true extent to which deforestation or wood shortages actually occurred; they argue that the problem has been overestimated, especially when related to industrial needs. This trend is probably the result of the greater attention paid to the analysis of sources, trying to reduce the effects of misleading interpretations and misunderstandings that are quite widespread in environmental history, but it may also be due to a cultural trend that is trying to diminish the concept of human destructiveness. Even if the real meanings of terms such as 'forest' or 'deforestation' could be debated, and thus data about surface changes questioned, at issue is actually a better understanding of changes in species composition, structure and density as determined by forest practices carried out at different spatial and temporal scales. The reduction of broad-leaved species and the growth of conifers as a general trend affecting European forest history seem certain, due both to the effects of industrial needs and to secondary successions in areas subjected to fire or heavy grazing.

The growth of conifers is also the result of the application of silvicultural theories supporting the concept of maximizing forest production, which developed at the beginning of the 19th century and affected the actions of the national forest services in many countries throughout the world. From this viewpoint, countries with large forests, and leading wood markets, as in North America or Scandinavia, induced changes on a very large scale. This imposes a reflection on the meaning of terms such as 'sustainable management', taken into consideration also in the recent matters of ecocertification. It seems questionable to consider 'sustainable' management forms preserving forest extension, but reducing their composition to one or two conifer species, especially in respect to biodiversity issues. However, conifers were also used in afforestation due to their ability to grow on degraded soil, favouring the improvement of site conditions.

According to evidence from research, local communities in traditional, pre-industrial agrarian contexts often seem able to establish a more balanced relationship with the forest environment, improving biodiversity and maintaining more stable ecosystems. Although this model seems to be affected by demographic trends and the availability of sufficient forest resources, sudden political, economic and technological changes often trigger social conflicts and ecological crises. In wealthy societies, the decrease in farming and the general reduction in pressure on wood resources leads to an extension of forests as well as a decrease in certain management forms such as coppicing. However, the loss of the traditional features of the agro-forestry landscape is felt as a loss of an important part of the cultural heritage and it also reduces biodiversity. In a very broad sense, in rich countries forest management keeps forests in acceptable but often very artificial conditions, while in poor countries deforestation and degradation caused by industrial exploitation and demographic pressure are still continuing. One of the most important findings of the studies presented in this volume, is that the forest is not just a background against which human history developed, but an actor in this history, playing an active role in all the many different ecological environments. The debate on the relationships between society and forest resources is one of the crucial issues of our times. Hopefully, historical research will be able to make a substantial contribution to clarify these relations and to the definition of strategies for environmental policies at the beginning of the new millennium.

References

Agnoletti, M. (1994) Technology, economics and forestry: water powered sawmills in Italy's Cadore region. *Forest and Conservation History* 38, 24–32.

Agnoletti, M. (1995) From the Dolomite to Venice: raft and river driving along the Piave river in Italy. *Industrial Archaeology* 21, 15–36.

Agnoletti, M. (1998) *Segherie e Foreste nel Trentino dal Medioevo ai Giorni Nostri.* Museo degli Usi e Costumi delle Genti Trentine, San Michele All'Adige.

Agnoletti, M. (1999) Organisation technique de l'approvisionnement en bois pour la république de Venise XIVe–XVIIIe siècle. In: Corvol, A. (ed.) *Forêt et Marine.* L'Harmattan, Paris, p. 226.

Agnoletti, M. and Anderson, S. (eds) (2000) *Methods and Approaches in Forest History.* CAB International, Wallingford, UK.

Balée, W. (1998) Historical ecology: premises and postulates. In: Balée, W. (ed.) *Advances in Historical Ecology.* Columbia University Press, New York, pp. 13–29.

Bamford, P.W. (1956) *Forests and French Sea Power.* University of Toronto Press, Toronto, Ontario, 240 pp.

Boithias, J.L. and Brignon, M. (1985) *Les Scieries et les Anciennes Sagards des Vosges.* CREER, Saint Diè, France.

Brandl, H. (1992) *Entwicklungen und Tendenzen in der Forstgeschichte seit ende des 18. Jahrhunderts.* Mitteilungen der Forstlichen Versuchs und Forschungsanstalt Baden-Württemberg, Freiburg.

Braudel, F. (1966) *Le Méditerranée et le Monde Méditerranéen à l'Époque de Philippe II.* Colin, Paris.

Carle, U., Delfs, Ebeling, Eimmermacher, Fuchs, Keweloh, Metzen, Pich, Piot, Prooje, Radkau, Scheifele, Schweizer and Wiemer (1988) *Auf den Spuren der Floesser.* Konrad Theiss Verlag, Stuttgart.

Ciancio, O. and Nocentini, S. (2000) Forest management from positivism to the culture of complexity. In: Agnoletti, M. and Anderson, S. (eds) *Methods and Approaches in Forest History*. CAB International, Wallingford, UK.

Collins, J.T. (1996) The wood-fuel economy of eighteenth century England. In: Istituto Internazionale di Storia Economica 'F.Datini,' Atti della XXVII Settimana di Studi: *L'Uomo e la Foresta, secc. XIII–XVIII*, Prato 8–13 Maggio 1995, a cura di S. Cavaciocchi, Collana Atti delle settimane di studi ed altri convegni n. 27, Florence, pp. 1097–1122.

Corvol, A. (1995) Connaître la forêt: problèmes et documents. In: Instituto Internazionale di Storia Economica 'F. Datini,' Atti della XXVII Settimana di Studi: *L'Uomo e la Foresta, secc. XIII–XVIII*, Prato 8–13 Maggio 1995, a cura di S. Cavaciocchi, Collana Atti delle settimane di studi ed altri convegni n. 27, Florence, pp. 45–77.

Dargavel, J. and Kengen, S. (1992) Promise and performance of industrial plantations in two regions of Australia and Brazil. In: Steen, H.K. and Tucker, R.P. (eds) *Changing Tropical Forests*, Forest History Society, Durham, North Carolina, pp. 217–232.

Devèze, M. (1962) Les forêts de l'Allemagne au seizième siècle. *Revue Forestière Française* 6, 479–493.

Di Berenger, A. (1859–1863) *Dell'Antica Storia e Giurisprudenza Forestale in Italia.* Longo, Treviso-Venice, pp. 139–142.

Dornic, F. (1984) *Le Fer contre la Forêt*. Editions de l'Ouest, Angers, France.

Du Bois, J.J. (1996) Espaces et paysages forestiers du Nord-Ouest de la France du XIII au XVIII siècles: l'apport de la biogéographie historique. In: Istituto Internazionale di Storia Economica 'F.Datini,' Atti della XXVII Settimana di Studi: *L'uomo e la Foresta, secc. XIII-XVIII*, Prato 8–13 Maggio 1995, a cura di S. Cavaciocchi, Collana Atti delle settimane di studi ed altri convegni n. 27, Florence, pp. 252–296.

Duby, G. (1962) *L'Économie Rurale et la Vie des Campagnes dans l'Occident Médiéval (France, Angleterre, Empire, IX–XV Siècles)*. Montaigne, Paris.

Freeman, M. (1997) Wychwood Forest, Oxfordshire: an episode in its recent environmental history. *Agricultural History Review* 45, 137–148.

Gabbrielli, A. (1982) Boschi e Magona. *Rivista di Storia dell'Agricoltura* 1, 107–154.

Grove, R.H., Damodaran, V. and Sangwan, S. (1998) *Nature and the Orient.* Oxford University Press, Delhi, India.

Kirby, K.J. and Watkins, C. (1998) Ecology and European woodland. In: Kirby, K.J. and Watkins, C. (eds) *The Ecological History of European Forests.* CAB International, Wallingford.

Lane, F. (1973) *Venice. A Maritime Republic.* Johns Hopkins University Press, Baltimore, Maryland.

Langston, N. (1995) *Forest Dreams, Forest Nightmares: the Paradox of Old Growth in the Inland West.* University of Washington Press, Seattle, 368 pp.

Marsh, G.P. (1885) *The Earth as Modified by Human Action.* Scribner, New York.

Moreno, D., Cevasco, R., Bertolotto, S. and Poggi, G. (1998) Historical ecology and post-medieval management practices in alderwoods (*Alnus incana* (L.) Moench) in the northern Apennines, Italy. In: Kirby, K.J. and Watkins, C. (eds) *The Ecological History of European Forests*. CAB International, Wallingford, pp. 185–201.

Netting, R.McC. (1981) *Balancing on an Alp. Ecological Change and Continuity in a Swiss Mountain Community*. Cambridge University Press, Cambridge.

Nilsson, G. (1997) Forests in the temperate–boreal transition: natural and man-made features. *Ecological Bulletins* 46, 61–71.

North, M. (1996) Trade and production of timber and timber by-products in the Baltic region, 1575–1775. In: Istituto Internazionale di Storia Economica 'F. Datini', Atti della XXVII Settimana di Studi: *L'Uomo e la Foresta, secc. XIII–XVIII*, Prato 8–13 Maggio 1995, a cura di S. Cavaciocchi, Collana Atti delle Settimane di Studi ed Altri Convegni n. 27, Florence, pp. 883–895.

Ostermann, R. and Reif, A. (2000) Socioeconomic and ecological aspects of coppice woods history in the lower Vosges and Black Forest. In: Agnoletti, M. and Anderson, S. (eds) *Methods and Approaches in Forest History*. CAB International, Wallingford, UK.

Rackham, O. (1995) *Trees and Woodland in the British Landscape*. Weidenfeld & Nicolson, London.

Silvy-Leligois, P. (1962) Origine et importance des bois acheminés par eau vers la Hollande aux XVII et XVIII siècles. *Revue Forestière Française* 6, 511–531.

Stanger, F.M. (1967) *Sawmills in the Redwoods: Logging on the San Francisco Peninsula, 1849–1967*. San Mateo County Historical Association, San Mateo, California, 160 pp.

Susmel, L. (1994) *I Rovereti di Pianura della Serenissima*. Cleup, Padua.

Trifone, R. (1957) *Storia del diritto forestale in Italia*. Accademia Italiana di Scienze Forestali, Florence.

Tucker, R.P. (1992) Foreign investors, timber extraction and forest depletion in central America before 1941. In: Steen, H.K. (ed.) *Changing Tropical Forests*. Durham.

Vecchio, B. (1974) *Il Bosco negli Scritori Italiani del Settecento e dell'Età Napoleonica*. Einaudi, Turin.

Watkins, C. (1998a) Themes in the history of European woods and forests. In: Watkins, C. (ed.) *European Woods and Forests: Studies in Cultural History*. CAB International, Wallingford, UK, pp. 1–10.

Watkins, C. (1998b) 'A solemn and gloomy umbrage': changing interpretations of the ancient oaks of Sherwood forest. In: Watkins, C. (ed.) *European Woods and Forests: Studies in Cultural History*. CAB International, Wallingford, UK, pp. 93–113.

Woronoff, D. (1990) *Forges et Forêts: Recherches sur la Consommation Proto-industrielle de Bois*. EHESS, Paris.

Les Ressources Forestières et le Développement Economique de la Vallée Mosane du 11e au 17e Siècle d'après l'Étude du Trafic Fluvial

1

Marc Suttor

Maître de Conférences, Université d'Artois, Centre de Recherches 'Des Anciens Pays-Bas à l'Eurorégion', 9 rue du Temple – BP 665, F-62030 ARRAS Cedex, France

The relationship between the evolution of forest exploitation and the economic development of a region is well illustrated by the study of the traffic on the Meuse River from the 11th to the 17th century.

The Meuse River carries important wood resources from the Ardennes forest to each city of the valley and to the regions downstream which are almost without woodland, such as Zeeland, Holland and Gelderland. Timber, firewood and charcoal meet the needs of building and shipyards, as well as those of the early metallurgical industry, along the river or in the Low Countries.

During the 16th century, the increasing needs of Dutch shipbuilding for the most important European merchant fleet and of more numerous smelting furnaces led to intensive wood exploitation and to a very important traffic on the river.

In the 16th and 17th centuries, the first problems concerning forest exploitation appear. The answers proposed to reduce them included successive orders for the preservation of the forests and the use of coal for the industry in the basin of Liège.

So the evolution of the wood traffic on the river, the easiest means of transport for this material, demonstrates the link between the intensity of forest exploitation and economic development in the Meuse valley.

(eds M. Agnoletti and S. Anderson)

L'histoire des bois et forêts de Belgique a déjà fait l'objet d'une étude assez fouillée, réalisée en 1927–1930 par le comte Goblet d'Alviella.[1] Toutefois, cette publication semble quelque peu vieillie et, surtout, elle n'aborde pas les questions qui prennent place 'en aval' de l'exploitation forestière, à savoir le commerce et le trafic du bois. Or ceux-ci semblent d'excellents révélateurs pour préciser l'évolution des ressources ligneuses.

Nous nous proposons d'illustrer ici cette relation en ce qui concerne le bassin mosan, qui draine la majorité des étendues boisées de l''espace belge', par une analyse détaillée des transports fluviaux en rapport avec le développement économique de la vallée, et ce du IIe au 17e siècle.[2]

Il s'agira tout d'abord de noter l'origine de ce matériau ou combustible et d'examiner la manière dont s'opère ce trafic par eau. On pourra ensuite répertorier les premières mentions de ce commerce au 11e siècle et en définir les caractéristiques, puis analyser la forte croissance de ces échanges au début du 16e et au 17e siècle, avant d'observer les premiers indices d'une difficulté d'approvisionnement en bois. Il faudra enfin évoquer les différents problèmes documentaires suscités par cette étude.

De vastes étendues de forêts, des feuillus, surtout chênes et hêtres, avoisinent la rivière et ses affluents.[3] En fait, à la fin du 18e siècle, la couverture boisée de l''espace belge' se concentre pour plus de 80% dans le bassin mosan.[4] Au milieu de celui-ci, les domaines forestiers du comté de Namur s'étendent, en 1265, sur près de 10,000 ha, soit environ le douzième de la superficie totale de cette principauté.[5] Les principales zones boisées se trouvent en Ardenne. La Lesse et l'Ourthe-Amblève, mais aussi la Semois, traversent cette région et 'emportent' le matériau jusqu'au fleuve. En outre, la couverture forestière de ce bassin comprend aussi la Marlagne (sur la rive gauche de la Meuse, peu en amont de Namur), qui comporte près de 5700 ha, et le Condroz namurois près de 1000 ha en 1294. Plus en aval, on citera les bois autour de Clermont-sous-Huy et au-dessus de Seraing, non loin de Liège. On mentionnera enfin une vaste zone de forêts (l'Hertogenwald) dans le duché de Limbourg, en bordure du plateau des Hautes Fagnes.[6]

D'un autre côté, on constate que l'exploitation des ressources ligneuses, malgré le peu de renseignements dont l'on dispose à ce propos, ne présente guère de particularités. Notons seulement qu'au moment de la coupe, on sépare le merrain (*mairin*, terme qui, dans cette région, désigne toujours du bois d'œuvre), les *leingnes* (ou bois de chauffage) et les écorces. En outre, le matériau peut déjà subir une transformation importante sur le lieu de l'abattage, telle que la fabrication de douves ou de roues de chariots.[7]

Le transport du combustible et du matériau s'effectue soit par bateau, soit par flottage. Bois travaillés, bois de chauffage, charbon de bois sont embarqués à bord des bâtiments et on emmène en remorque des radeaux de poutres et de fagots.[8] Le flottage du matériau, permis par l'existence de nombreuses rivières drainant les ressources des forêts d'Ardenne vers le fleuve, s'opère selon deux procédés: 'à bûches perdues' ou par trains de radeaux constitués de troncs assemblés, les *givées*.[9] Ces radeaux comprennent 30–45 pièces de bois, disposées en longueur, par rangées de 12 ou 15, et maintenues ensemble par un treillis de perches horizontales assujetties à l'aide de liens de fagots. Les *givées*, composées de dix ou 12 radeaux enchaînés les uns aux autres, sont conduites par deux hommes, l'un à la tête et l'autre à la queue du train de bois, pour éviter les bancs de gravier et les rochers. Les dimensions des *givées*, formées de 300 à 540 grumes, peuvent être estimées à une longueur de 150 à 200 m, une largeur de 6 à 7.5 m et une hauteur de 1 à 1.5 m.[10] Notons encore qu'elles permettent le transport d'un volume non négligeable de bois semi-ouvrés et de nombreuses autres marchandises.[11] Quant à la durée du transport, on observe que le flottage d'un train de radeaux depuis Namur jusqu'à Liège s'effectue en 2 ou 3 jours, et jusqu'à Maastricht en 3 jours.[12]

Il convient enfin de rappeler que la navigabilité mosane se révèle très variable: les conditions de navigation se modifient en effet tout au long de l'année et tout au long du cours d'eau, en fonction de son profil, de son tracé et des variations saisonnières. Le fleuve présente d'importantes inégalités de pente. Il offre en fait l'aspect d'un immense escalier, qui voit se succéder des biefs aux eaux calmes et des bancs de gravier parcourus par un flot rapide. Ces barres de roches produisent des courants très puissants, parfois étroits et pas toujours assez profonds. Ainsi, la tâche des conducteurs de *givées* s'avère délicate quand ils doivent diriger cette masse de plusieurs tonnes et lui faire passer ces rapides. Elle se complique encore et devient dangereuse lors des hautes eaux, lorsque des courants irréguliers se forment et rendent périlleuse la descente. C'est une des raisons pour lesquelles le flottage s'effectue surtout au printemps et au début de l'été pendant les eaux moyennes qui offrent les conditions de navigabilité les plus favourables, mais aussi à l'époque de la circulation maximale sur la rivière. Bien entendu, cela contrarie et entraîne de grands dangers pour la navigation. En outre, la longueur des *givées* peut provoquer leur arrêt dans les méandres prononcés: il faut alors détacher certaines pièces du train pour franchir l'obstacle, ce qui prend du temps et produit l'encombrement du cours d'eau.[13]

La Meuse apparaît dès le milieu du 11e siècle comme une véritable 'route du bois'.[14] A cette époque déjà, l'évêché de Liège est réputé à Bruges pour ses exportations de 'batterie' et de *grant marrien*.[15]

Le fleuve se voit assez tôt ponctué de marchés pour ce matériau.[16] Liège constitue ainsi un point important pour le commerce des grumes, et ce au moins dès le début du 14e siècle. On recueille des *mairins* au rivage de la Boverie en 1316 et, en 1340, on rencontre *Sur-Meuse* (à l'amont du pont des Arches) les ateliers des *mairniers*. Ces marchands de poutres encombrent la berge, empêchant le halage, et leurs activités gênent la navigation jusqu'au 17e siècle. Il existe encore un débarcadère à Avroy pour les bois de construction semi-ouvrés, transportés par bateau, et ce à partir de 1350. A Maastricht, on cite déjà le *forum lignorum* au milieu du 13e siècle. On entreprend en outre des travaux sur le cours d'eau à la fin du 15e siècle pour améliorer le flottage.[17] Ces deux centres de commerce des grumes contribuent aussi bien à l'approvisionnement des régions avoisinantes qu'à l'exportation. Tout d'abord, la fourniture régulière des villes mosanes ne pose pas de problème, même si la consommation de ce matériau dans la vallée s'avère importante.[18] Ensuite, il existe des débouchés considérables pour les *mairins* plus en aval, en Gueldre et en Hollande, régions pauvres en forêts.[19]

Le flux d'exportation de ce matériau vers le delta du fleuve semble en effet considérable. En 1243, l'abbaye des Dunes (sur le littoral flamand, près de Furnes) obtient de Guillaume II de Hollande une franchise de tonlieu dans son comté pour les arbres coupés dans des bois aux environs de Givet (à la frontière belge). Il est question de *givées* dans un traité, conclu en janvier 1251, concernant le péage de Heusden (en Brabant septentrional). En 1333, le comte Guillaume III de Hollande et de Hainaut vend à un marchand liégeois une coupe de 300–450 chênes provenant de la forêt de Smuid sur la Lomme (non loin de Saint-Hubert), qui seront livrés soit à Liège, soit à Dordrecht. A cette occasion, le comte exempte ce transport du paiement de tout tonlieu en basse Meuse. A partir de cette époque, une documentation plus abondante vient préciser l'importance de ces échanges et on trouve trace du transport de ce matériau à de nombreux postes de péage sur le cours inférieur de la rivière.[20]

Dordrecht constitue alors une très importante étape du bois, article le plus courant, après le vin, dans le commerce de la ville. Ce matériau provient aussi bien du Rhin que de la Meuse et l'étape alimente tant les chantiers navals qu'un marché de redistribution pour toute la région littorale. C'est à Dordrecht, en effet, que les Brugeois achètent de grandes quantités de 'bois liégeois', qui provient, en fait, par le fleuve, de différentes forêts ardennaises. À Bruges, le bois originaire du bassin mosan concurrence sérieusement celui de la Baltique et du Rhin: entre 1359 et 1384, il représente un minimum de 33.6% des livraisons, avec un maximum de 79.5% en 1368–1369.[21]

Au 15e siècle, de nombreux trains de radeaux continuent à descendre le cours d'eau vers la Gueldre et la Hollande. Parmi de multiples exemples, on note en 1470 le passage au tonlieu du pont des Arches à Liège de 19 *givées*, flottées depuis Fumay (à l'amont de Givet), en direction de Ruremonde (peu après Maastricht).[22] En 1491, plusieurs trains descendent encore depuis la haute Meuse vers Namur, Liège et la Hollande.[23] Toutefois, seules les séries chiffrées des comptes de péage peuvent procurer une indication relativement précise sur l'importance de ces échanges fluviaux. Dans la seconde moitié du 15e siècle, il passe en moyenne 139 *givées* chaque année entre 1463 et 1493 à Vireux-Wallerand (en amont de Givet) mais plus de 700 trains à Hastière (non loin de Dinant) de 1472 à 1474, ce qui laisse supposer qu'un nombre assez considérable de ces radeaux sont formés après Vireux. Cela constitue un volume considérable, lorsque l'on sait que chaque train de radeaux comporte de 300–540 troncs d'arbres, ce qui représenterait pour les chiffres extrêmes, tout en insistant sur le caractère artificiel de cette 'équivalence', entre 60,000 et 380,000 grumes par année.[24]

Ajoutons que ces échanges ne portent pas exclusivement sur les pièces de charpente. Le bois 'travaillé' approvisionne encore diverses industries, comme l'armurerie, où il représente un 'matériau stratégique' dans la fabrication de fûts, de manches et de poignées d'armes, de hampes de traits, de machines et d'instruments de jet, de boucliers, d'affûts et de chariots d'artillerie. Dans l'industrie minière, ce matériau sert à l'étançonnement des fosses et des galeries.[25] Mais le commerce porte aussi sur des bois semi-ouvrés tels que les 'ouvrages de charlerie' ou roues pour les charrons, et surtout les *clippe-clappes* ou *fustaille*, c'est-à-dire des douves pour les tonneliers. Ces artisans doivent être nombreux à Dordrecht, point de départ d'un très important trafic des harengs sur la Meuse et le Rhin.[26]

Tout comme le matériau, le bois de chauffage ou *leingne*, abondant partout, descend la rivière, transporté par bateau.[27] La mention la plus ancienne de ce trafic figure dans le record des droits du comte de Namur à Dinant au milieu du 11e siècle.[28]

On signale le passage de nombreuses cargaisons ou *navées* de 'bois à brûler' sur le cours d'eau aux winages de Fumay, Vireux-Wallerand et Givet au 15e siècle.[29] Ainsi, on compte le passage annuel de 26 cargaisons de *leingnes* dans la seconde moitié du 15e siècle à Vireux-Wallerand.[30] Namur et peut-être Liège constituent des marchés pour le commerce du bois de chauffage. À Namur, l'étape obligatoire apparaît en juin 1424. A Liège, un tonlieu perçu sur le bras d'Avroy au moins dès 1350 touche les *leingnes*.[31] Mais ces villes alimentent essentiellement des échanges locaux.[32]

Toutefois, les bateliers namurois mènent aussi des cargaisons de 'bois à brûler' en direction du pays de Liège, et des *naiveurs* liégeois, à leur tour, chargent des *leingnes* pour les transporter à Maastricht.[33] On peut parler d'exportation de bois de chauffage vers les régions presque dépourvues de forêts situées en aval de cette ville, puisque ce combustible apparaît dans les tarifs des différents tonlieux de la basse Meuse à partir du 14e siècle.[34]

Le charbon de bois reste très longtemps le combustible 'industriel' le plus employé, même après la découverte de la houille à la fin du 12e siècle. La production du charbon de bois (obtenu par carbonisation à feu couvert) se fait dans les grandes forêts d'Ardenne, ce qui explique la prolifération des forges et fourneaux dans cette région.[35] Cependant, certaines activités industrielles se concentrent très tôt dans les villes. Ainsi, les travailleurs du métal doivent se procurer de ce combustible pour alimenter leurs fours. L'approvisionnement de ceux-ci suppose qu'il existe un trafic du charbon de bois, dont une partie s'effectue très probablement par la voie fluviale.

Il faut cependant attendre le 15e siècle pour rencontrer l'attestation la plus ancienne de tels échanges sur la rivière.[36] On lève alors des droits sur les cargaisons de ce combustible, qui descendent toujours la Meuse depuis Fumay jusqu'à Liège. Dans la seconde moitié du siècle, on note le passage annuel de 550 t de charbon de bois, en moyenne, au winage de Vireux-Wallerand.[37] En ce commerce, Namur et Liège occupent une place importante, vraisemblablement en raison des grandes quantités de combustible nécessaires à l'approvisionnement de leurs 'industries'. À Liège, on exige un tonlieu pour les cargaisons amenées aux débarcadères d'Avroy. Il semble aussi qu'une part non négligeable de ces échanges est destinée aux régions situées plus en aval et parvient, entre autres, jusqu'à Dordrecht.[38]

Mais au 16e siècle, ces divers trafics se développent. Les besoins d'agglomérations telles que Liège s'accroissent de manière significative, en raison, notamment, de l'expansion de l'armurerie et de l'industrie minière. D'autres indices révèlent l'importance de ce commerce: les recettes que procure l'exploitation des forêts, la perception de droits sur le transport du matériau à la plupart des péages ou encore les conflits concernant ces échanges à plusieurs tonlieux dans la partie inférieure de la vallée.[39]

De manière plus particulière, on note la forte croissance de l'exportation de bois de marine à destination des chantiers navals de Hollande, qui commencent alors à construire ce qui deviendra la flotte commerciale la plus puissante d'Europe.[40] En 1533, ce sont 400 pièces de charpente qui descendent le fleuve de Sedan (en amont de Mézières) à Liège.[41] On compte ainsi, en moyenne, 204 *givées* ainsi que 34 *navées* de *mairins* enregistrées sur le cours d'eau à Vireux-Wallerand chaque année, entre 1532 et 1561.[42]

Les données concernant le trafic du bois de chauffage au 16e siècle proviennent seulement de quelques rares comptes de tonlieux. A Vireux-Wallerand, on paie en moyenne chaque année entre 1532 et 1561 des droits pour le passage de 82 *navées* de ce combustible, et pour la descente de 90 cargaisons à Givet de 1540 à 1561. Cela représente un trafic non négligeable, en nette augmentation par rapport au siècle précédent. On peut croire aussi qu'une quantité assez importante de bois de chauffage est exportée vers le cours inférieur de la rivière.[43]

Les besoins en charbon de bois augmentent de façon très sensible à cette époque. En effet, la métallurgie se développe alors de manière considérable dans le Namurois et à Liège. Cette industrie, qui engloutit un volume énorme de ce combustible, nécessite l'acheminement de très grandes quantités de charbon de bois. Ce trafic s'intensifie malgré le fait que l'emploi de la houille comme combustible 'industriel' commence à se répandre. En effet, la sidérurgie namuroise continue, très longtemps, à consommer plutôt du charbon de bois. Il existe à Namur une étape obligatoire pour ce produit, signalée en 1568.[44] À Liège aussi, les besoins de la métallurgie poussent, dans le dernier quart du siècle, des marchands à passer d'importantes commandes – plusieurs centaines de tonnes – près de Huy et dans l'Entre-Sambre-et-Meuse (en amont de Namur).[45] Mais on a besoin aussi de ce combustible plus en aval: des bourgeois de Zaltbommel (non loin de Dordrecht) remontent en 1518 le fleuve pour en acheter à Venlo, entre cette ville et Maastricht.[46] Entre 1532 et 1561, ce sont plus de 2400 t de charbon de bois qui descendent chaque année le cours d'eau à Vireux-Wallerand, ce qui atteste l'intensité accrue de ces échanges.[47]

Au 17e siècle, ces trafics semblent continuer à se développer. De nombreux actes notariés en témoignent. Ainsi, en février 1624, un contrat porte sur l'acheminement de 800 chênes depuis la Lesse jusqu'à Liège et on note encore une très importante commande passée le 16 mars 1643 à Deville (en aval de Mézières).[48] La rivière se voit sillonnée de marchands mosans et hollandais, les conflits d'intérêts s'exacerbent. Même si nos dépouillements pour cette période restent fort incomplets, les coups de sonde réalisés dans les archives montrent que les volumes de bois descendant alors le fleuve s'avèrent encore plus considérables qu'auparavant.[49]

C'est aussi à cette époque que l'exploitation intensive des forêts de la vallée commence à poser problème. En effet, les besoins d'activités telles que la construction navale, mais surtout la consommation accrue de charbon de bois par la sidérurgie, véritable 'dévoreuse de forêts', poussent à pratiquer des coupes fort importantes, comme dans la région comprise entre Revin (peu après Deville) et Givet.[50] Cette situation inquiète les autorités, qui tentent de prendre des mesures, et ce tout d'abord pour les zones boisées proches des forges et fourneaux, le pays de Franchimont (non loin de Liège) dès 1507, où le prince-évêque Erard de la Marck ordonne 'que chacun charbonnier [. . .] lasseroient sur chacun journal de bois diex stallons chaisnes [. . .] ou fawes', le comté de Namur en 1535 et le ban de Seraing en 1545.[51] Ces ordonnances visent de plus en plus les maîtres de forges et leurs fournaises, mais, à vrai dire, réitérées à de multiples reprises, elles montrent leur inefficacité et, partant, la persistance du problème.[52] Celui-ci s'accentue: au début du 17e siècle, ferrons, forgerons, charbonniers étendent leur prospection au duché de Luxembourg, 'en prenant toutes sortes de bois, chênes, faulx et autres, . . . pour réparer et entretenir leurs forges, fourneaux, platineries, fenderies', au point d'entraîner en 1617 la promulgation d'un édit fort répressif, défendant, entre autres, 'de faire plus d'une etre ou place à cuire charbon en un arpent' et d'en installer de nouvelles.[53] Par contre, dans la région liégeoise, l'emploi croissant de la houille pour de nombreuses industries permet de préserver quelque peu les ressources forestières.[54]

Il reste à examiner les problèmes documentaires rencontrés dans cet exposé. En effet, la documentation s'avère tout à fait déficiente si l'on désire quantifier ces échanges. On le sait, toute étude quantitative de ce type se heurte, au Moyen Âge, à d'incontournables difficultés. En général, on s'appuie sur les comptes de péage. Bien sûr, ils comportent des erreurs, des lacunes. Nombre d'entre eux ne mentionnent pas les multiples exemptions, notamment en ce qui concerne des échanges locaux certainement importants, ce qui fausse la signification des recettes. Enfin, les fraudes viennent encore altérer les chiffres. Toutefois, on estime que le volume de ces inexactitudes reste à peu près constant et que ces documents, malgré leurs défauts, demeurent les plus représentatifs de l'intensité du trafic. Ils nous livrent des données chiffrées qui expriment des ordres de grandeur vraisemblables et comparables entre eux. Mais il importe que ces comptes présentent un caractère de continuité suffisant pour que l'on puisse observer une évolution à plus ou moins long terme. C'est là que se pose le problème lorsque l'on étudie le commerce mosan. Tout d'abord, les premiers documents conservés apparaissent très tard, dans la seconde moitié du 15e siècle. Ensuite, leur nombre trop limité et des lacunes trop importantes rendent malaisée leur utilisation et empêchent tout traitement statistique. En outre, les quelques registres tenus de 1470 à 1476 pour l'un de ces péages – celui du pont des Arches à Liège, situé à un endroit 'stratégique' pour ces échanges – ont été dressés dans de telles conditions que les chiffres enregistrés ne révèlent en aucune manière le volume habituel des transports fluviaux à hauteur de cette ville à cette époque. Dès lors, si l'on tire de ces comptes des renseignements certes précieux, car ils constituent l'essentiel de nos informations chiffrées, ils se révèlent toutefois fort incomplets et peu satisfaisants.[55]

Au vrai, il faut attendre le 17e siècle pour pouvoir disposer de sources d'archives diversifiées, en quantité et en qualité suffisantes pour se prêter à une analyse quantitative sérieuse. Alors se pose un autre problème: celui du dépouillement systématique de centaines de registres, à la recherche de renseignements très épars. Nous venons à peine d'entreprendre ce travail de longue haleine.

Enfin, un état de guerre latent et l'apparition puis le renforcement progressif dans ces régions au 17e siècle d'une politique économique protectionniste (le mercantilisme) gênent toute étude approfondie de l'évolution des échanges sur la rivière.[56]

Au terme de cet exposé, il convient de souligner la grande richesse des étendues boisées drainées par le fleuve et ses affluents. De vastes forêts de feuillus permettent une exploitation importante et régulière des ressources ligneuses. Le transport de ces matériau et combustible abondants s'opère, de manière très commode, par bateau ou, le plus souvent, par flottage de trains de dix à 12 radeaux (comportant de 300 à plus de 540 grumes), ce qui entraîne quelques difficultés pour la navigation.

Ce trafic apparaît fort tôt sur la Meuse et satisfait non seulement les besoins locaux, mais encore l'approvisionnement du delta, région pauvre en forêts. Toutefois, c'est surtout aux 16e et 17e siècles que se développent de façon extraordinaire ces échanges, essentiels tout autant à la construction de la flotte des Provinces-Unies qu'à l'industrie métallurgique de la vallée. Le cours d'eau se voit sillonné de nombreux marchands, les conflits d'intérêts s'enveniment. On peut affirmer, pour autant qu'une base documentaire soudain plus abondante ne produise un effet de loupe, que les volumes de bois descendant alors le fleuve s'avèrent considérables. D'un autre côté, dès le 16e siècle, l'exploitation intensive du combustible au profit de l'industrie sidérurgique inquiète et provoque la réaction des autorités, au demeurant peu efficace.

Malheureusement, la documentation disponible, lacunaire, n'autorise pas une quantification précise avant le 17e siècle, et se révèle l'objet de recherches longues et aléatoires par après.

Il n'en demeure pas moins que les capacités considérables des forêts mosanes, associées aux facilités remarquables offertes par la voie d'eau pour l'acheminement du bois, ont contribué au développement d'un trafic d'une grande importance au sein de la vallée, et même d'une portée assez extraordinaire au niveau international.

Notes

1 Comte Goblet d'Alviella, *Histoire des bois et forêts de Belgique. Des origines à la fin du régime autrichien*, 4 t., Paris-Bruxelles, 1927–1930.

2 En général, voir Suttor, M. (1986) *La Navigation sur la Meuse Moyenne des Origines à 1650*, Liège-Louvain, Id., 'Un 'scandale géographique :' le pays mosan au Moyen Age,' *Cahiers de Clio*, fasc. 105, 1991, p. 33–54 et notre thèse *La Meuse entre Mézières et Maastricht des Origines à 1600. Trafic Fluvial, Pouvoir et Société*, à paraître dans la *Nouvelle Bibliothèque du Moyen Age* en 2000.

3 Essences: comte Goblet d'Alviella 1965, t. 1, p. 400–402 et 422–426 ainsi que la *Carte de cabinet des Pays-Bas autrichiens levée à l'initiative du comte de Ferraris*, Bruxelles.

4 80%: Goblet d'Alviella, t. 1, p. 82–90, 145–158, 200–206 et 392–426 ainsi que la carte 9 publiée par Sosson, J.-P. (1977), *Les Travaux Publics de la Ville de Bruges, XIVe–XVe siècles. Les Matériaux. Les Hommes*, Bruxelles, p. 321. – Bien entendu, la superficie des forêts a considérablement varié au cours du dernier millénaire, essentiellement sous l'action de l'homme: Suttor, M. (1999) 'Prospection, commerce et transport du bassin mosan aux Pays-Bas XIIIe–XVIIe siècle,' *Forêt et Marine*, L'Harmattan, Paris, pp. 163–164. Sur l'évolution de la couverture boisée, cfr encore Goblet d'Alviella, t. 1, p. 400–402 et 422–426 ainsi que A. Noirfalize (1985), 'La forêt wallonne et ses types écologiques,' *Le grand livre de la Forêt Wallonne*, Liège, p. 65–74.

5 Fanchamps, M.-L. (1966) Transport et commerce du bois sur la Meuse au Moyen Âge, *Le Moyen Age*, t. 72, p. 60 et n. 4.

6 La forêt est très dense le long de la Lesse (*ibid.*, p. 63–67), de l'Ourthe et de l'Amblève (*ibid.*, p. 72–76 et Marquet, L. (1971) 'Transports d'autrefois. La navigation sur l'Ourthe moyenne,' *Ardenne et Famenne*, t. 11, p. 211–212), de la Semois 1973 (Id., 'Le flottage du bois,' *Enquêtes du Musée de la Vie Wallonne*, t. 13, pp. 200–204). Marlagne et Condroz namurois en 1294: Brouwers, D.D. (1926) *Cens et Rentes du Comté de Namur au XIIIe siècle*, t. 2 (1), Namur, p. CXLVI). Clermont-sous-Huy et Seraing: Fanchamps, *op. cit.*, p. 60. Hertogenwald: *ibid.*, p. 61, Goblet d'Alviella (1927), t. 2, Paris-Bruxelles, p. 197–225, Buis, J. (1985), *Historia Forestis. Nederlandse Bosgeschiedenis*, Wageningen, p. 235 et F. Letocart (1988–1989), *Les Domaines Forestiers dans le Duché de Limbourg: Limites, Évolution et Gestion des Origines au XVe Siècle*, Liège.

7 Bois de construction, de chauffage et écorces sont déjà distingués dans le record des droits du comte de Namur à Dinant au milieu du 16e siècle: *infra*, n. 14. Transformation

sur le lieu de l'abattage: Fanchamps (1972), 'Le commerce sur la Meuse moyenne dans la seconde moitié du 15e siècle et dans la première moitié du 16e siècle d'après des comptes de tonlieux,' *Histoire Économique de la Belgique. Traitement des Sources et État des Questions*, Bruxelles, p. 292 ainsi que *infra* et n. 26. Exploitation: Goblet d'Alviella, t. 1, p. 90–99, 155–158, 273–319, t. 2, p. 103–119, 197–244 et 255–331.

8 Transport par bateau: Walch, M.-S. (1932) *Les Relations Économiques du Pays Mosan aux XIIIe, XIVe et XVe siècles*, Liège, p. 93 (acte de 1562), Fanchamps, 'Transport et commerce du bois,' p. 64 et Id., 'Le commerce sur la Meuse moyenne,' p. 292–293. – Radeaux remorqués: Thimister, O. (1878) *Cartulaire ou Recueil de Chartes et Documents Inédits de l'Église Collégiale de Saint-Paul, Actuellement Cathédrale de Liège*, Liège, p. 144 (acte du [12] décembre 1317), Archives Générales du Royaume (désormais A.G.R.), *Chambre des Comptes*, reg. 1004, f°21 (première moitié du 16e siècle), Philippe de Hurges, *Voyage à Liège et à Maastricht en 1615*, éd. Michelant, H. (1872) *Société des Bibliophiles Liégeois*, Liège, p. 128 et planche p. 126 ainsi qu'une vue de Namur par M. Merian (deuxième quart du 17e siècle: Suttor, *La Navigation*, p. 155).

9 Questienne, P. (1978) *La Navigation en Wallonie*, p. 21. – A propos du flottage 'à bûches perdues,' voir Marquet, 'Le flottage du bois,' p. 166–170.

10 Sur cette technique: Marquet, *op. cit.*, p. 171, 173–174, 178–179, 184, 192 et Buis, *Historia Forestis*, p. 157–158. Le mot *givée* est attesté dès le 14e siècle [avant 1393]: Borgnet, J. (1851–1859) *Promenades dans Namur*, Namur, p. 158, n. 1. – Description et dimensions des trains de radeaux: Philippe de Hurges, *Voyage à Liège et à Maastricht*, p. 126–128 et planche p. 126; voir aussi Archives de l'Etat à Liège (désormais A.E.L.), *Notaire T. Pauwéa*, 1664, f° 317 et une peinture de Dinant en 1870 par H. Boulenger (Suttor, *La Navigation*, p. 155). – Sur le mode d'assemblage des *givées*, cfr A.E.L., *Notaire G. Lien*, 1637–1649, f° 144–147 v° et Marquet, p. 192. – Maniement des radeaux: *infra* et n. 13.

11 Transport de diverses marchandises sur les *givées*: de Hurges, p. 126. – On rencontre ces trains de bois sur d'autres fleuves et rivières. Sur le haut Danube, voir Neweklowsky, E. (1958) *Die Schiffahrt auf der oberen Donau und ihren Nebenflüssen*, Munich, pp. 38–46. Pour le Rhin, voir Buis, pp. 505–509 (au moins dès la fin du 13e siècle) ainsi que Silvy-Leligois, P. (1962) 'Origine et importance des bois acheminés par eau vers la Hollande aux 17e et 18e siècles,' *Revue Forestière Française*, t. 14, p. 525 et planche p. 524; sur ce fleuve, on trouve des trains de radeaux de 300 à 320 m de long, sur plusieurs dizaines de mètres de largeur (jusqu'à 55 m au 18e siècle), pour une hauteur de 1.3 m, composés de 110 à 900 troncs: Buis, p. 507 et 953, n. 111 ainsi que Silvy-Leligois, planche p. 524.

12 La durée du flottage est estimée à partir d'une enquête de janvier 1699: Archives de l'Etat à Namur (désormais A.E.N.), *Conseil de Namur*, enquête n°8333. – Les coupes se pratiquent à la fin de l'automne et en hiver: l'absence de sève favorise la conservation des grumes en évitant la formation de champignons, surtout pour les hêtres; elle assure la qualité et la flottabilité des *mairins*. Au printemps, en effet, ceux-ci ont perdu une part suffisante de leur humidité et leur masse volumique avoisine alors 800 kg m^{-3}. Les données de la dendrochronologie confirment les informations des documents écrits à propos des saisons auxquelles on pratique les coupes: Hoffsummer, P. (1995) *Les Charpentes de Toitures en Wallonie. Typologie et Dendrochronologie (XIe–XIXe siècle)*, Namur, p. 60. J. Hébert, Maître de Conférences aux Facultés agronomiques de Gembloux, a confirmé et complété toutes les informations techniques qui précèdent. Nous l'en remercions vivement.

13 Notes sur la navigabilité et les techniques de navigation: Suttor, M. *La Navigation*, p. 48–59 et 108–123. – Maniement des radeaux: A.E.L., *Echevins de Liège. Obligations*, reg. 56, f° 60 v° (acte du 15 octobre 1576), de Hurges, p. 128 et A.E.L., *Notaire T. Pauwéa*, 1664, f° 317. – A propos des risques et des difficultés engendrés sur le Rhin par le flottage des trains de bois, cfr Buis, p. 508–509.

14 La première mention du commerce du bois sur la Meuse, et tout particulièrement du flottage, date de [1056–1064], à Dinant (éd. G. Despy, *Les Chartes de l'Abbaye de Waulsort. Etude Diplomatique et Édition Critique*, t. 1, Bruxelles, 1957, p. 90; à propos de cet acte et de sa datation, corrigée par J.-L. Kupper, 'Une 'conventio' inédite entre l'évêque de Liège Théoduin et le comte Albert II de Namur (1056–1064),' *Bulletin de la Commission Royale d'Histoire*, t. 145, 1979, p. 8–9), voir Suttor, M. *La Meuse entre Mézières et Maastricht*.

15 L. Gilliodts-Van Severen, 1904 *Cartulaire de l'ancienne estaple de Bruges. Recueil de documents concernant le commerce intérieur et maritime, les relations internationales et l'histoire économique de cette ville*, Bruges, p. 19; Pirenne, H. (1951) 'Esquisse d'un programme d'études sur l'histoire économique du Pays de Liège,' *Histoire économique de l'Occident médiéval*, Bruges, p. 493. Cfr encore *infra*.

16 Huy, et Namur constituent des marchés régionaux du bois de chauffage: Suttor, M. (1994) 'Prospection, commerce et transport', p. 167 et 'Le contrôle du trafic fluvial: la Meuse, des origines à 1600,' *Revue du Nord*, t. 76, p 13 et n. 25 ainsi que *infra*.

17 Liège, Boverie, 1316: Poncelet, E. (1898) *Le Livre des Fiefs de l'Eglise de Liège sous Adolphe de la Marck*, Bruxelles, p. 178. *Sur-Meuse*, 1340: Id., *Les Domaines Urbains de Liège*, Liège, 1947, p. 195; problèmes pour la navigation: Suttor, *La Navigation*, p. 115–116 et les dessins dressés lors d'un procès en 1553 (A.E.L., *Cartes et Plans*, 220 bis et salle d'exposition, n°6). Avroy, 1350: Fairon, E. *Régestes de la cité de Liège*, t. 1, Liège, 1933, n° 449, p. 381–382. – Maastricht, *forum lignorum*: Questienne, *La Navigation*, p. 23, ainsi que des actes de 1347 (Willemsen, M. 'Inventaire chronologique des chartes et documents de l'église de Saint-Servais, à Maastricht,' *Publications de la Société Historique et Archéologique dans le Limbourg*, t. 3, n° 115, p. 421) et de 1592 (Hankart, R. 'Notes sur la vie quotidienne à Avroy. Les forêts et le commerce du bois,' *Bulletin de la Société Royale 'Le Vieux-Liège,'* t. 5, 1960, p. 466, n. 30). Travaux: Questienne, p. 23.

18 Approvisionnement en bois en général: Suttor, 'Prospection', p. 167, et *La Meuse entre Mézières et Maastricht*, ainsi que Hoffsummer, *L'Évolution des Toits à deux Versants dans le Bassin Mosan: l'Apport de la Dendrochronologie (XIe–XIXe siècle)*, Liège, 1989–1990.

19 Régions dépourvues de forêts: Buis, p. 487–488. Le marché hollandais, mentionné au moins depuis le 14e siècle, alimente surtout la construction navale (Fanchamps, 'Transport,' p. 79, Niermeyer, J.F. (1942) 'Dordrecht als handelsstad in de tweede helft van de veertiende eeuw,' *Bijdragen voor Vaderlandsche Geschiedenis en Oudheidkunde*, t. 3, p. 31 et Buis, p. 480–484). Il devient très important aux 16e–17e siècles (voir *ibid.* et *infra*).

20 Acte de 1243: Niermeyer, J. F. (1968) *Bronnen voor de Economische Geschiedenis van het Beneden-Maasgebied, I, 1104–1399*, t. 1, La Haye, n° 36, p. 15. – Traité de 1251: *ibid.*, n° 49, p. 21. 1333: Bruwier, M. 1958 'Note sur l'exploitation des bois de Mirwart par le comte de Hainaut en 1333,' *Etudes sur l'Histoire du Pays Mosan au Moyen Âge. Mélanges Félix Rousseau*, Bruxelles, p. 149–150 et Fanchamps, 'Transport,' p. 65, n. 24. – Comptes des tonlieux de Mook-Heerewaarden, 1394: Niermeyer, *Bronnen*, t. 1, n° 679, p. 375 et Thurlings, T.L.M. (1945) *De Maashandel van Venlo en Roermond in*

de XVIde eeuw (1473–1572), Amsterdam, p. 50; compte, 1348–1349 et tarif, [1355–1361] du tonlieu de Venlo: Niermeyer, J.F. 'Dordrecht als handelsstad,' p. 3, n. 1 et Id., *Bronnen*, t. 1, n° 473, p. 249; tarif du tonlieu de Woudrichem (confluent Waal–Meuse), [1355–1358]: *ibid.*, n° 441, p. 223–224; tarif du tonlieu de Dordrecht, [1355–1358]: *ibid.*, n° 441, p. 226.

21 Dordrecht, étape du bois et appellation 'bois [du pays] de Liège :' Niermeyer, 'Dordrecht,' p. 21 et 35; règlement pour le métier des marchands de bois de Dordrecht, 1367: Id., *Bronnen*, t. 1, n° 499, p. 277. – A propos des achats de bois mosans par les Brugeois à Dordrecht entre 1334 et 1386, cfr Sosson, *Les Travaux Publics*, p. 102–104 et 111–112 ainsi que tableaux XII et XVII. Notons que Bruges forme aussi un marché de redistribution de ce matériau pour la Flandre (*ibid.*, p. 103–104 ainsi que Gaier, C. 1973 *L'Industrie et le Commerce des Armes dans les Anciennes Principautés Belges du XIIIe à la Fin du XVe Siècle*, Paris, p. 211 et n. 140).

22 1470: Fairon (1912) 'Note sur la domination bourguignonne dans la principauté de Liège (1468–1476),' *Bulletin de l'Institut Archéologique Liégeois*, t. 42, p. 77.

23 1491: Walch, *Les relations économiques du pays mosan*, p. 8.

24 Moyennes calculées à partir des chiffres notés par Fanchamps, 'Commerce,' annexe. Voir encore *supra* et n. 10 ainsi que Suttor, *La Meuse entre Mézières et Maastricht*. – On ne dispose pas de chiffres comparables pour d'autres péages dans cette partie de la vallée: *infra* et Suttor, *op. cit.* – De manière indirecte, on peut déduire l'intensité du commerce et, partant, du trafic de ce matériau à partir de l'estimation des domaines forestiers (Suttor, 'Prospection', p.169).

25 Industrie armurière: Gaier, *op. cit.*, p. 208–209. Industrie minière: Buis, p. 185.

26 Ce commerce apparaît dès le XIVe siècle, mais se développera surtout au XVIe siècle: Fanchamps, 'Commerce,' p. 292, 'Transport,' p. 61–62 et Suttor, M., *La Meuse entre Mézières et Maastricht*. – Dordrecht: Niermeyer, 'Dordrecht,' p. 35, 177 et 184.

27 Le tri s'opère sur la coupe (*supra* et n. 7), mais les documents ne précisent pas toujours s'il s'agit de bois de chauffage ou de construction. – Transport par bateau: *supra* et n. 8 ainsi que d'autres mentions dans de nombreux textes (15 septembre 1350: Fairon, *Régestes de la cité de Liège*, t. 1, n° 449, p.381, etc.).

28 *Supra*, n. 14.

29 Fumay, 1431: Archives Départementales du Nord (désormais A.D.N.), *Série B* 7835; Vireux-Wallerand, 1463–1561 et Givet, 1540–1561: Fanchamps, 'Commerce,' annexe. – *Navée*: unité de mesure pour plusieurs marchandises pondéreuses (cfr Suttor, *La Meuse entre Mézières et Maastricht*, glossaire).

30 1463–1493: Fanchamps, 'Commerce,' annexe.

31 Namur: le bois de chauffage constitue une des marchandises sur lesquelles on impose un contrôle dès l'apparition de l'étape (Suttor, 'Le contrôle du trafic fluvial,' p. 13). – Liège: acte du 15 septembre 1350 (*supra*, n. 27); ce tonlieu apparaît à plusieurs reprises par la suite et l'emplacement de ce péage, de part et d'autre du pont d'Avroy, suggère qu'il existe là un marché du bois de chauffage (Suttor, *La Meuse entre Mézières et Maastricht*).

32 Parmi de nombreux exemples, pour Namur: Génicot, L. (1970) *La Crise Agricole du Bas Moyen Âge dans le Namurois*, Louvain-Gand, p. 148; pour Liège: comptes de 1458–1464 (A.E.L., *Collégiale Saint-Denis*, reg. 596, f° 23; reg. 597, f° 17 et 42). Malheureusement, le trafic de ce combustible au niveau local laisse peu de traces dans la documentation, notamment en raison de nombreuses exemptions: *infra* et n. 55.

33 Namurois: règlement de l'étape, 5 janvier 1465 (éd. Brouwers (1920), *Cartulaire de la Commune de Namur*, t. 4, Namur, n° 395, p. 244). Liégeois: acte du 9 septembre 1483 (A.E.L., *Métiers de Liège*, reg. 216, f° 7).
34 Comptes du tonlieu de Mook, 1394–1395 (Thurlings, *De Maashandel van Venlo en Roermond*, p. 50); tarifs des tonlieux de Woudrichem, [1355–1358] (Niermeyer, *Bronnen*, t. 1, p. 224) et Dordrecht, [1355–1358] (*ibid.*, p. 226; voir aussi Id., 'Dordrecht,' p. 11 et Fanchamps, 'Transport,' p. 79).
35 A propos de la production du charbon de bois, cfr Gaier, *L'Industrie et le Commerce des Armes*, p. 215. – Sur le rôle de ce combustible dans la localisation de l'industrie métallurgique, voir Joris, A. (1969) 'Probleme der mittelalterlichen Metallindustrie im Maasgebiet,' *Hansische Geschichtsblätter*, t. 87, p. 59 et Gillard, A. (1971) *L'Industrie du Fer dans les Localités du Comté de Namur et de l'Entre-Sambre-et-Meuse de 1345 à 1600*, Bruxelles, p. 50–51. En outre, le charbon de bois entre dans la composition de la poudre noire (Gaier, p. 216) et est utilisé dans les fours à chaux (compte de la collégiale Notre-Dame de Dinant en 1474–1475: éd. Brouwers, D.D. (1908) *Cartulaire de la Commune de Dinant*, t. 8, Namur, p. 138).
36 Approvisionnement des fours dans les villes: Gaier, p. 215 et n. 154. Première mention du transport de charbon de bois sur la Meuse, dans le tarif du tonlieu perçu par l'abbaye de Saint-Laurent, de part et d'autre du pont d'Avroy à Liège en 1426 (transcrit par Jean de Stavelot, *Chronique*, éd. Borgnet, A. (1861) *Commission Royale d'Histoire. Publications in-4°*, Bruxelles, p. 235).
37 Fumay, 1431: A.D.N., *Série B* 7835; Hastière, 1472–1474: Fanchamps, 'Commerce,' p. 293 et annexe; Dinant et Namur, [1477–1479]: éd. Borgnet, J. et Bormans, S. (1873) *Cartulaire de la Commune de Namur*, t. 2, Namur, p. 348; Liège, 1426: *supra*, n. 36. A Vireux-Wallerand, de 1463 à 1493, ce trafic représente 9.5% de la recette totale: Fanchamps, 'Commerce,' p. 293 et annexe; les estimations sont réalisées grâce à des équivalences notées par Gaier, *L'Industrie*, p. 207.
38 *Supra*, n. 36. En dehors de l'approvisionnement des industries, il faut aussi satisfaire la demande des diverses communautés de la ville et des particuliers (par exemple, pour Liège: A.E.L., *Collégiale Saint-Denis*, reg. 596, f° 48 et 86; reg. 597, f° 20 et 46 v°; 1459–1464). – Le charbon de bois est taxé à Dordrecht dès la seconde moitié du 14e siècle (Niermeyer, 'Dordrecht,' p. 11). Mais il s'agit là de la seule mention relevée et l'on ignore d'où provient ce combustible (du Rhin ou de la Meuse?). En 1420, la ville de Zwolle, dans l'Overijssel, commande 200 tonneaux de charbon de bois, peut-être produit dans nos régions (Gaier, p. 216).
39 Expansion de l'industrie: Suttor, M. *La Meuse entre Mézières et Maastricht* et 'Un 'scandale géographique': le pays mosan,' p. 34–35, 37–39, 42–45 et 48–50. – Importance du domaine forestier namurois en 1551: Baelde, M. (1971) *De Domeingoederen van de Vorst in de Nederlanden omstreeks het Midden van de Zestiende Eeuw (1551–1559)*, Bruxelles, p. 313–314. Conflits en basse Meuse au 16e siècle: Thurlings, *De Maashandel*, p. 46. Conflits entre Huy, Liège, Maastricht, Ruremonde et Venlo: *ibid.*, p. 68–71 et Suttor, *L'Exercice du Pouvoir dans la Vallée de la Meuse*, à paraître en 2000.
40 Bruijn, J.R. (1980) 'Scheepvaart in de noordelijke Nederlanden,' *Algemene Geschiedenis der Nederlanden*, t. 7, Haarlem, p. 138.
41 Hankart, 'Notes sur la vie quotidienne à Avroy,' p. 466 et n. 31.
42 Moyennes calculées à partir des chiffres notés par Fanchamps, 'Commerce,' annexe. – On ne dispose pas de chiffres comparables pour d'autres péages dans cette partie de la vallée: *infra*.

43 Vireux-Wallerand, 1532–1561; à Givet, entre 1540 et 1561, on note le passage de 90 *navées* chaque année, en moyenne, ce qui signifie que l'on charge encore des bateaux de bois de chauffage entre ces deux péages (Fanchamps, 'Commerce,' annexe et *supra*, n. 42).
44 Augmentation du trafic au 16e siècle: cfr *infra* et n. 47. Le 23 juin 1576, les Liégeois se plaignent de la hausse des droits de tonlieu qui touchent les cargaisons de charbon de bois descendant la Meuse à Namur (Bormans, S. (1871) 'Table des registres aux recès de la Cité de Liège,' *Bulletin de la Société Scientifique et Littéraire du Limbourg*, t. 11, p. 279). – On reconnaît des qualités différentes aux deux combustibles: Gaier, p. 206–207. A propos de la grande consommation de charbon de bois par l'industrie du fer namuroise, voir Gillard, *L'Industrie du fer*, p. 50–51. – Etape à Namur: acte du 14 novembre 1568 (éd. Brouwers, *Cartulaire de la Commune de Namur*, t. 4, n° 304, p. 58).
45 Il s'agit là d'un véritable commerce en gros: Lejeune, J. (1939) *La Formation du Capitalisme Moderne dans la Principauté de Liège au XVIe Siècle*, Liège, p. 152–153 ainsi que p. 207 et n. 3.
46 Thurlings, *De Maashandel*, p. 125.
47 Ce trafic représente, à Vireux-Wallerand, 14.5% de la recette (Fanchamps, 'Commerce,' annexe), soit une augmentation considérable par rapport au siècle précédent; pour les estimations, cfr *supra*, n. 37. – A Givet, en 1540–1561, il passe en moyenne par an 228 *navées* de charbon de bois (Fanchamps, p. 293 et annexe) et à Château-Regnault, on compte 18 bateaux chargés de ce combustible en 1581 (pour 36 bâtiments avec une cargaison de gueuses de fer, par exemple: Rigault, J. (1974) 'La navigation de la Meuse au XVIe siècle vue du péage de Château-Regnault,' *Actes du 95e Congrès National des Sociétés Savantes. Philologie et Histoire*, t. 2, Paris, p. 136).
48 Contrats de 1624: A.E.L., *Notaire H. Oupie*, août 1623–juillet 1624, f° 131 v°, du 16 mars 1643: *Notaire G. Lien*, 1637–1649, f° 144–145 et de nombreux actes conservés aux A.E.L. Voir aussi le témoignage de Philippe de Hurges, p. 127.
49 Thurlings, p. 75 et p. 122 ainsi que *supra*, n. 39.
50 Consommation de charbon de bois: Lejeune, *La formation du capitalisme*, p. 152–153, 207 et n. 3, Gillard, *L'Industrie du Fer*, p. 50–51 ainsi que Gaier, *L'industrie*, p. 207 et 215–216. Au 17e siècle, il faut 100 ha de forêt par an pour alimenter un fourneau, 20 kg de charbon de bois, c'est-à-dire 100 kg de chêne, pour produire 4 ou 5 kg de fer: Jacquet, P. et Jacquet-Ladrier, F. (1987) *Albums de Croÿ*, t. 15, Bruxelles, p. 22. – Coupes réalisées entre Fumay-Revin et Givet: Fanchamps, 'Transport,' p. 60.
51 Pays de Franchimont: mentionné édit de 1507, dans un acte du 9 mai 1519 (éd. Polain, L. (1869) *Recueil des Ordonnances de la Principauté de Liège*, 2ème série, *1507–1684*, t. 1, Bruxelles, p. 30); comté de Namur: ordonnance du 18 janvier 1535, éd. Lameere, J. (1902) *Recueil des Ordonnances des Pays-Bas*, 2e série, *1506–1700*, t. 3, Bruxelles, p. 456–457 (cfr aussi Goblet d'Alviella, t. 1, p. 263–264); ban de Seraing: acte du 12 décembre 1545, éd. L. Polain et S. Bormans., *op. cit.*, p. 139.
52 Mandements du 15 mai 1557 et du 20 juillet 1559 pour le marquisat de Franchimont: éd. *ibid.*, p. 255–256 et 257–258; édits du 11 juillet 1541 et du 24 novembre 1559 pour le comté de Namur: éd. Lameere, J. et Simont, H. (1907) *op. cit.*, t. 4, Bruxelles, p. 304 ainsi que Vicomte Terlinden et Bolsée, J. (1978) t. 8, Bruxelles, p. 35–36. Sur l'inefficacité de cette législation, voir Goblet d'Alviella, t. 1, p. 286–288.

53 Brandts, V. (1912) *Recueil des Ordonnances des Pays-Bas. Règne d'Albert et Isabelle. 1597–1621*, t. 2, Bruxelles, p. 344, 346 et 352–354. Cfr encore Goblet d'Alviella, t. 1, p. 255–256, 335–336 et 448–452.
54 Emploi de la houille: Lejeune, *La formation du capitalisme*, p. 160–162.
55 Absence du trafic local dans la documentation: *supra*, n. 32. Exemption des établissements ecclésiastiques, grands consommateurs de ces matériau et combustible: Fanchamps, 'Transport,' p. 81. Qualités requises pour l'utilisation informatisée des comptes: Mollat, M. (1977) 'Comptes portuaires et informatique,' *Informatique et Histoire Médiévale*, Rome, p. 151 et 153 ainsi que Suttor, *La Meuse entre Mézières et Maastricht*; on ne dispose d'aucune série suffisante pour les péages depuis Dinant jusqu'à Maastricht; les comptes de la basse Meuse, dont ceux de Lith, très intéressants, ne sont pas publiés et ne paraissent pas complémentaires avec ceux de la Meuse moyenne: *ibid.*
56 Harsin, P. (1928) 'Etudes sur l'histoire économique de la principauté de Liège particulièrement au 17e siècle,' *Bulletin de l'Institut Archéologique Liégeois*, t. 52, p. 69 et 118 et Van Houtte, J. A. (1972) 'Le tonlieu de Lith et le commerce sur la Meuse de 1551 à 1701. Résultats d'une recherche collective,' *Histoire Économique de la Belgique*, Bruxelles, p. 301.

'The Coveted Tree': the Industrial Threat to the Vosges Forest in the 16th and 18th Centuries

2

Emmanuel Garnier

Laboratory of Historical Science, University of Besancon, 12, rue des Ejols, 88120 Vagney, France

At any epoch, and depending on the existing technical means, the investments allowed and the expectations expressed, the forest keeps changing in its phytosociological organization, its floristic level, its layers and its landscapes. In that regard, the pre-industrial growth of the 16th and 18th centuries, which involved the whole of western Europe, definitely marks a basic break with the forest ecosystem. In France this issue generates a lively debate between the diverging views of historians, geographers and foresters.[1] This chapter aims at studying the impact generated by the industrial pressure on the Vosgian forest populations, following a multi-scale approach. Its scope of investigation covers the southern part of the Vosges mountains. It is partly the result of historical studies carried out for the National Forestry Commission and the Regional Natural Reserve of the 'Ballons des Vosges'.

1 The Vosgian forest facing pre-industrial growth: sources and stakes

To carry out scientific and historic research upon this topic, historians rely on several sources of information. The old archives form the main documentation that researchers can use. The pieces of information are numerous, geographically scattered and sometimes written in Old German.

Early maps are an excellent resource to learn about, assess or even appropriate an area. The map of 'Hautes Chaumes' designed in 1579 is the first

(eds M. Agnoletti and S. Anderson)

chronological example of cartography concerning the Lorraine part of the Vosges mountains. None the less, maps in different scales drawn by land surveyors and geographers did not appear before the beginning of the 18th century. Within the scope of our theme they provide a precise understanding of forestry evolution. Despite their weaknesses and their lack of precision, the old maps are necessary to the evolutionary understanding of woodlands exposed to industrial exploitation. Historians are then able to draw up synthetic maps, which simply originate in the cross-checking of old ones (Fig. 2.1).

The texts by forestry administration as well as by the great abbeys, such as Murbach in Alsace, Lure in Franche-Comté and Remiremont in Lorraine, are undoubtedly at the heart of research documentation. Very rich quantitatively and qualitatively, they include numerous reports on visits to woods and hammering (marking of trees before felling). They record the state, nature, quality and limits of the forests. The papers written by foresters describe the different technical operations used in silviculture with precision and often record the names of tree species.

Another significant source of information is the iconography of the time, which can sometimes give detailed information concerning the techniques of forestry used by manufacturers. In this regard the drawings by the painter Heinrich Gross of Saint-Die in 1529 are a useful source of information. His collection of drawings recounts the activity of the mine in Sainte-Marie-aux-Mines. He focused on the geography of the forest where the rails, planks and pit props necessary to the stability of the galleries were made.

Toponymy is of great help in localizing the cantons exploited for industrial purposes ('glass-making canton', for example). However, what we learn from it is limited because it requires comparisons with old maps using lots of toponyms which have now disappeared from modern ones.

Remote sensing is a very recent tool which, paradoxically, can also provide historians with interesting data. Through the combination of several channels of SPOT satellite images, we can get a picture which shows clear colouring differences in the midst of woodlands. Within an archaeological approach, this process allows us to localize precisely the old pioneering populations of glass-makers, nowadays victims of the forestry reconquest.

Right away we should insist upon the fact that the Vosges mountains in the 16th century still corresponded to a territory with strategical confines and, as such, one open to colonization. The latter led to clearings in the Vosgian forest, a process which reached its highest point at the dawn of the 17th century. The discovery of ore deposits in the heart of the forest obviously contributed to the move towards human settlements. The competing exploitation of the mines of Château-Lambert in Franche-Comté and Le Thillot in Lorraine, from the end of the 15th century onward, was in keeping with a policy of sovereignty assertion of those two states. Indeed, conflicts could not but occur, since miners of both villages – a mere 2 km apart – worked on the same veins and relationships were not always easy. On several occasions shootings

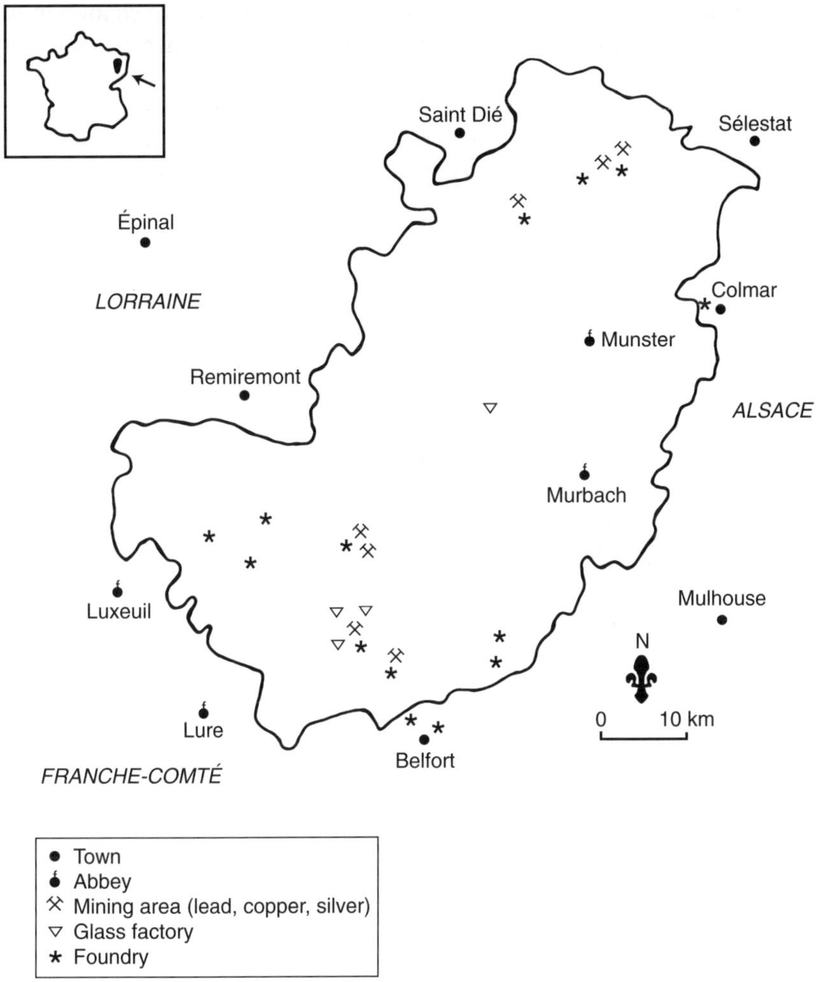

Fig. 2.1. The industrial threat to the Vosges forest (16th and 18th centuries).

occurred between the coalmen of the two states. Consequently, the forest quickly became a strategic raw material, receiving full attention from the 'grueries' (forestry divisions) specially set up to provide the local mines with wood. In 1612 the man in charge of the Lorraine division, who was also called the 'ferweser' (a technical Old German word referring to the mine manager), denounced the damage caused in the woods by the miners of Franche-Comté.[2] These degradations were perpetrated by organized groups of 10–15 men, who operated at night. The dispute was finally taken before the highest state representative and the Duke of Lorraine decided to intercede with the parliament of Franche-Comté to secure reparation. Only in 1668 was a final sharing of the mines decided. However, strained relationships continued throughout the

17th century, the region experienced the inexorable rise of French monarchy, which reached its peak under the reign of King Louis XIV. Concretely, this evolution led in 1681 to the occupation by the French troops of the Haute-Moselle valley, which belonged to the Lorraine dukedom. The greed caused by the mining and forestry richness of this region played a major part in this annexation.

In addition to these constituted states, the wealth of the lands, and therefore the forests, owned by the great Vosgian abbeys must not be omitted. Bolstered by the success of the first clearings made by the Irish monks, with Saint Colomban at their head, the monasteries of Murbach in Alsace, Remiremont in Lorraine and Lure and Luxeuil in Franche-Comté quickly set up real principalities, benefiting from the decline of the Carolingian monarchy. Today, these religious communities own huge properties made up of several thousand hectares of forests. For example, the Lure Abbey administered as its own property nearly 2207 ha of forest in 1790. Therefore the monks could not but oppose the Duke of Burgundy as early as the 15th century, since their aim was to control the veins of Plancher-les-Mines situated on their land. This dispute saw a favourable outcome for the monks, as a ducal adjudication dated 1472 granted them the tenure of the mines in exchange for royalties.[3]

The Vosges uplands correspond to the southern part of the primary mountain range, which was raised in the Tertiary period because of the alpine orogenesis. As shown on the map drawn in 1579 by an official of the Duke of Lorraine, the forest covered the entire landscape. However, the valley, cleared and developed since the Middle Ages, as well as the mountain-top moorlands, called 'chaumes', presents an alternative limited by the supremacy of trees. They trace a bluish line on the balloon-shaped mountains. The map also shows a recent mining development near Gerardmer lake. This document, however unreliable it may seem, has the merit of highlighting the imposing mass of the forest, represented by a dark blue-green, cloud-like shape. This mountain is home to beech–fir forests, which cover huge areas on both sides of the U-shaped valleys of glacial origin. Fir trees above an altitude of 800 m have the best conditions for growing and nowadays the state-owned forests of the region are known for the beauty of their fir plantations. Beech–fir forests create a typical scenery, with old firs and scattered beeches, whose density is higher on the north-facing slopes and in damp areas. Above 800 m, beeches tend to outnumber the other species, whereas spruces, rarely native to the region, grow around peat bogs.

It is within such a geographical frame that the Vosgian forest has been exploited for industrial purposes, as a forest which seems to offer inexhaustible fuel supplies.

2 A multifaceted industrialization

Sawing is a traditional activity in the Vosges. In the archives, sawmills are called 'haut-fer' (or saw). They were all situated along a river, since these machines were operated by a paddle-wheel. The machinery consisted of a dam situated upstream, which could be 50 m wide on the Moselotte River. There was also a canal to supply the wheel with water and a lock. These mills were a direct threat to woodlands, as confirmed by the programme of regional development called 'Revers aux Chiens'. No fewer than four mills appeared, all situated on the edge of forests. Most of them belonged to a sovereign or an abbot who hired them to private individuals. In the Vosges the tenants often came from Germany or German-speaking Switzerland, an example of this being the sawmill built in the heart of the Saint-Antoine Forest in 1705. The latter was hired by the Franckenhauser brothers, who came from Lucerne in Switzerland.[4]

The metallurgy adventure that the Vosges mountains experienced between the 16th and 17th centuries originated in Central Europe. It was due to the existence of metals in the fractures, or veins, which cross the old hercynian mountainous massifs of Europe. These veins are found mostly in Central Europe: Austrian Alps (Schwaz), Saxony, Bohemia, Harz (Central Germany). To the west, the Black Forest and Vosges mountains border the 'German mining region'. This area possesses an outstanding historical and human coherence.

Most Vosgian plants exploited lead as well as silver, as in Plancher-les-Mines and Servance, and copper, in Château-Lambert and Le Thillot. Interest in mining kept increasing during the 16th century, a time when lead and silver were greatly needed for the building sector, for the issuing of coins and for commercial exchanges with oriental countries. This century was a golden age for the Vosgian mines: the area was systematically surveyed by specialists, coming mostly from Germany. The prospecting led to the opening of new deposits, which required, for lack of a qualified local workforce, the hiring of trained miners, some of whom came from Tyrol. During that time, most of the 'grueries' (forest divisions) appeared, which seems to be directly linked to the mining activity.[5]

These mines were soon to become rather autonomous from their owners as they built their own foundries in order to meet their needs in tools (hammers, burins). This is how a real metallurgical sector developed, perfectly illustrated by the Lorraine painter Heinrich Gross. The collection of drawings concerning mining activity are not numerous but the most famous one is certainly the 'Schwarzerbergbuch' (Book of the mine of Schwaz), which was offered to Emperor Ferdinand I in 1556. Heinrich Gross's work was original since the artist chose a synthetic process. Each picture shows a scene representing the landscape, the plants or galleries and the miners with their tools in a working situation. This chronological display shows each trade in real-life situations. Carpenters or coalmen are pictured in their working place, which is of great interest to the historian.

The glass-makers who had settled in the Vosges mountains form a most interesting subject, as they represented pioneering communities in a forestry environment. The abbeys of Murbach in Alsace and Lure in Franche-Comté, united in 1554 and under the leadership of the same abbot, played a major part regarding the settlement of the Vosgian glass factories.

As early as 1699, the monks of Murbach started this policy by settling a community of glass-makers on Oderen's territory. Their aim was to take advantage of a huge forestry heritage exploited only by the nearby villages, but this action resulted in a serious conflict between the newcomers and the inhabitants of the Thur valley. The group leader was the Hug family, whose members quickly took social and political control until the factory closed down at the end of the 19th century. In order to set up their village, originally made of huts, as suggested by the toponym of the time, Glasshutte (nowadays Wildenstein), the settlers were allowed to girdle trees to increase their grazing lands. This rather common process consisted in cutting a circular notch around trunks so that the trees dried out and died.

A look at the map showing the sites of the glass factories in the forest of Saint-Antoine in the second half of the 18th century is particularly telling. This area of nearly 1515 ha was directly threatened by these three factories, situated either in its close periphery or right in its heart.

3 Wood-devouring industries

The intensive exploitation of the wood resources of the Vosgian mountains was not carried out in an anarchic way. Glass-makers and ironmasters all acted legally and used well-proved techniques.

The first efficient means manufacturers had at their disposal was undoubtedly Colbert's Great Edict of the Forests, passed in 1669 by King Louis XIV's secretary of state. This basic act in French forest history was an authoritarian and centralizing response to the fear of lack of firewood and navy wood. Moreover, it confirmed the rise of full capitalist forestry property, gradually freed from its users' rights, in order to meet the needs of factories. The enforcement of that act led to a forestry normalization, contrary to the practices in use in Franche-Comté and Lorraine.

This centralizing policy aimed at organizing the woods to make their exploitation easier. It introduced a mode of operating completely unfamiliar to locals: tree-felling exploitation which suppressed the traditional 'jardinage', a practice based on the cutting of trees one by one according to the needs. The latter was particularly well adapted to mountain forests, which required careful and limited cutting. For that reason it has been used again recently as a natural way of exploiting the forests situated in natural reserves. The implementation of the 1669 edict required a forest organization necessary to conduct large-scale fellings. It therefore involved a space and time discipline. The woods, once delimited on location by stones and on paper through topographical maps, were then divided into numbered plots (called 'plates': 'assiettes') corresponding to their rotation.

Concerned about regular supplies, many glass-makers and ironmasters strove to get specific plots. The latter, also called 'assurance cantons', were wooded areas turned into mixed woods (high-grown trees and coppice), which were to supply factories with fuel. The latter could also use the woods of plots whose organization, time of rotation and exploitation were defined by contract. The owner, the king or the abbey, promised to sell to the manufacturer the output of a given area at a fixed price, most often below that of the market. In the enclosed Vosgian region, forest owners certainly saw there a means of economic development of a heritage that had been underexploited before. The map of Saint-Antoine forest, showing the forest divisions allocated to factories and drawn upon archive documents, proves that the whole area, which is now a state biological reserve, was subjected to intensive exploitation during the 16th and 18th centuries. Some divisions were even subsequently allocated to two different factories within a 50-year period. The apparent inaccessibility of some parts of the forest is puzzling and raises the question of technology used in difficult areas.

In order to meet their needs for fuel, industrialists resorted to a local highly qualified workforce, which used techniques well-adapted to the topographical constraints and to climatic changes, which only allowed short working periods. Chief woodcutters were helped by a team of workers. Big trees, such as firs, were 'surcenes', which means a notch was made around the trunk with an axe. With the same tool, woodcutters then made a deeper cut to mark the side on which trees were to fall. Finally two men sawed them with a long double-handled blade called 'passe-partout' ('cross-cut saw'), while a third man drove a wedge to prevent the saw from being blocked. Once trees were felled, the team had to peel and prune the trunks. This last process was particularly dangerous, since the trunks were unstable and could slide at any time, taking the men with them.

Often, throwing was the sole way of bringing trees to the sawmills. This was a traditional way of conveying timbers in mountainous areas, and consisted of letting timbers slide down a slope, from the felling place to the nearest conveying track. This practice, still in use occasionally, required tracks cleared of any obstacle. Chief Carpenter Jacquemin undertook the extension of this track, called a 'glissoir' (slide-track), down to the sawmill situated in Sechenat forest (on the Lorraine side). Hurtling down at full speed, timbers were often crushed by rocks on reaching the end of the slope and were therefore unusable.

Sledging was a way of conveying timbers typical of the Vosges but also in use in some German regions (Tyrol, Bavaria). With the help of sledges, called, after the German word, 'schlitte', firewood and huge trunks slide on special lug ways. This most dangerous operation is still used by a few farmers. Sledging ways looked like horizontal ladders made of beech rungs ('ravetons') tied by pegs with a space of 60 cm in between. The slopes were 7–9% steep and the sledges used to carry about 1 or 2 steres of wood.[6]

Yet the sites of forest exploitation did not necessarily coincide with the communication routes of the time. In the Vosges, as in the whole of France, even the smallest rivers were used for floating timber. The Moselotte valley was a telling example, as it was, between the 16th and 17th centuries, a real pre-industrial centre, founding its growth on its closeness to huge beech–fir forests. Maps of the time, together with archives, show a great density of sawmills, since no fewer than 20 of them were dotted along a 20-km-long portion of river. The combination of sawing and floating was made possible thanks to the technical equipment of sawmills, with a double use: it operated the wheel of the saw and it allowed the movement of rafts of timber, locally called 'sails' ('voiles'), from their dams downwards. This equipment was sometimes built on waterways that are now mere brooks, but it once permitted the exploitation of seemingly inaccessible forest areas. Such was the case of the spruce plantation situated in the glacial cirque of Neuweiher Lake above an altitude of 800 m and which used to be considered as subnatural by some specialists. In the 18th century, Alsatian ironmasters built there a sophisticated system of dams in order to float the firewood they needed for their factories in Masevaux, 10 km away.

These factories, benefiting from a conciliating legislation and efficient means of exploitation, proved to be a serious threat to the forest ecosystems of the Vosges mountains. Fir trees were greatly needed by the mining industry, as they supplied it with props, rails and planks to fit out the shafts. Once again, Heinrich Gross's drawings (1529) are essential. They show us miners working under the direction of carpenters. The props had to be put together with extreme precision and security in the mines. These materials were conveyed underground on trolleys by carpenters who had to cut timbers and make sure the walls were strong enough. A cross-section view enables us to visualize the uses of fir trees in the mines. The shafts, which served as vertical junctions between the galleries, were covered with planks to prevent the vats from rubbing the walls while going up and down. The galleries were reinforced by strong props made of coniferous trees, where miners could also hang their lamps. Wood was largely used for outer equipment. Pestles and foundries needed large amounts of planks sawn by local sawmills. For example, at the beginning of the 18th century, the sawmill of Saint-Antoine worked only for the mines of Plancher-les-Mines. In 1733, the English manufacturer Floide had several hundred of his firs processed there.

Firs were also the subject of a wide-scale business, as proved by the floating that was operating in the Lorraine valleys of Moselle and Moselotte. There, a transport organization was established which made possible the export of Vosgian firs in the form of planks and masts to Dutch arsenals in the first half of the 17th century. Following annexation of this region by France in the years 1680–1690, the Royal Navy on the Atlantic coast took over this trade. The toll certificates of the Moselotte valley testify to the extent of this new phenomenon. Annually, 1600 m^3 of timber were floated, together with 1500 fir trees, used for the masts of war vessels.

The development of metallurgy in the Vosges mountains and their periphery (Belfort and Mulhouse regions) greatly increased the need for firewood. The best species for yielding charcoal is beech, useful both to glass-factories and foundries. Beeches were much sought after when plots were given to industrialists. Coalmen of the mines and glass-factories, as well as woodcutters, were traditional workers in forests. They were housed in sheds near the felling areas for several months and had to produce charcoal from felled trees. To do so they methodically built up a stack with split pieces of 6–20 cm in diameter, with a chimney necessary for the burning. The stacks looked like a cone, at the base of which several holes leading to the chimney were made. The last operation consisted in covering the stack with moss and earth to facilitate a slow burning of the wood. Today we can still see the marks of these sites, called 'charcoal sites', by the elderly. They are all situated on flat areas and the earth is burnt tens of centimetres deep.

By decree of the Royal Council of 28 March 1730, Miellin's glass-factory (Saint-Antoine forest) received a 500 ha beech–fir forest, exploited on a 50-year rotation.[7] Glass-makers were allowed to transform felled trees into charcoal. Yet, as early as 1765, a crisis threatened the factory because of the overexploitation by industrialists. Contrary to the 1730 decree, the wood affected had been exploited on a 32-year rotation, resulting in the disappearance of the original full-grown forest. This shortage highlights the problem of consumption in Miellin's glass-factory. The inquiry carried out by the intendancy in the 1780s speaks for itself, since it concluded that the factory consumed about 32,000 steres a year!

We can therefore assert that, at the dawn of the French Revolution of 1789, the forest ecosystem of the Vosgian mountains was globally damaged. This situation stems from the conjunction of overall industrial development and significant population growth among village communities. Clearing new stretches and multiplying industrial firms led to a fuzzy landscape, in which 'moth-eaten' edges contrasted with forest hearts, which had been preserved because of their inaccessibility.

Regular fellings, sometimes repeated every 12 or 15 years, made high-grown trees more vulnerable. Little by little this process generated a forest with a young growth stage, which exhausted soils. It also led to a form of recessive selection of trees to the benefit of less demanding species (thorny bushes and birches). Beeches were the chief victims of the silviculture of the 18th century. These shadow-loving trees could not bear and thus did not adapt to the expansion of mixed forests (high-grown trees and coppice). Moreover, it was a much sought-after firewood. White-wood trees, which produce no fruit, grew in the damaged, overexploited and over-grazed forests. Birches and thornbushes survived in the form of warped and misshapen trees since their normal regeneration was impeded by grazing cattle.[8]

Fir forests were no better off and, between the end of the 18th century and the first half of the 19th, fir trees visibly regressed and, around 1820, they were confined to a residual area much inferior to its spatial potential. Fir trees were indeed the chief victims of blunt fellings (the cutting of all trees in an area). Firs generally require moderate exploitation, whereas in the case of blunt fellings the remaining trees face strong winds in winter and sunshine in summer, which kills them. Under these conditions wind-fallen trees became more and more numerous, while young shoots came to life on a barren, arid soil, scattered with rocks. Reports indicate that, in 1750, 400 fir trees had been uprooted in Saint-Antoine forest alone. Beside this inadequate silvicultural policy, fir forests suffered from a systematic exploitation by the Royal Navy in the 17th and 18th centuries. Under those conditions it was not surprising to hear foresters complain about the scarcity of big fir trees.

Little by little and especially at the end of the 18th century, the Vosgian forest suffered deforestation and various sorts of degradations and therefore ceased to exercise a regulatory role in the water cycle. Incidentally, natural disasters, such as flooding and hurricanes, punctuated the end of that century. The Vosgian mountain villages were not spared. For example, a disastrous storm hit Gerardmer on the nights of 25–26 July 1770. The water uprooted great quantities of trees in the neighbouring woods. Theses trees were then swept away to the fields downhill, together with rocks, earth and sand.[9]

In conclusion, historical analysis, if founded on reliable sources, gives precious information to allow us to better understand the ecological impact of pre-industrial development in Europe on forest populations. It also highlights the problem of forest naturalness, a theme of controversies among specialists today. The study of the Vosgian case shows that only a few remote places have been spared from anthropic action. On these sites, the full-grown forest, exploited extensively tree by tree, contains many mature trees and follows a natural regeneration cycle, thus offering a more varied landscape.

Notes

1 Woronoff, D. (1984) Forges prédatrices, forges protectrices. *Revue Géographique des Pyrénées et du Sud-Ouest* 55, 213–218; Houzard, G. (1980) Les grosses forges ont-elles mangé la forêt? *Annales de Normandie* 4, 245–269; Badre, L. (1992) Les forêts et les industries en Lorraine à la fin du XVIIIe siècle. *Revue Forestière Française, Nancy* 4, 365–369.

2 Archives Départementales de Meurthe-et-Moselle. B 8335.

3 Garnier, E. (1998) *Un Massif Forestier et son Histoire: la Forêt de Saint-Antoine (Permanences, Mutations et Enjeux)*. Office National des Forêts, Paris, 135 pp.

4 Archives Départementales de Haute-Saône. H 600.

5 Garnier, E. (1994) L'homme et la forêt dans la gruerie de Ramonchamp (XVIe–XVIIIe siècles). *Bulletin de la Haute-Moselle* 21, 6–7.

6 Garnier, E. (1994) *L'Homme et son Milieu: le Massif du Grand Ventron à Travers les âges*. Parc Naturel Regional des Ballons des Vosges, Le Thillot, 76 pp.

7 Archives Départementales de Haute-Saône. B 9400.

8 Garnier, E. (1999) The naturalness of forests to the test of history: the example of the Vosgian uplands. *Global Ecology and Biogeography Letters, Oxford* (in press).

9 Archives Communales de Gerardmer. DD. 10.

The War of the Plants: Botanical Exchange and Agricultural Conquest of the New World during Colonial Times

3

Emanuele Amodio

Escuela de Antropología, Universidad Central de Venezuela, Caracas, Venezuela

The arrival of Spanish conquerors in America meant a change in the concept of world in the whole of Europe. The impact with a different nature, the tropical forest, imposed the creation of new representations and images. In the beginning, known categories, mainly the religious ones, were applied to the new reality, but, after a while, the widening of the geographical horizons caused substantial changes at a cultural level.

During the three centuries of the conquest, thanks to the impulse of the new epistemological paradigms, the idea of nature of the different social groups changed progressively: from a religious view, in which nature was considered as the creation and message of God to humans, to a secular view, where natural spaces are to be conquered and transformed.

There are infinite islands and abundance
Of sweet lakes, roomy fields,
Very distanced Sierras,
Sublime mountains, scary forests,
Lands to plough of great substance,
Green forests, delightful grasslands,
Of crystalline waters sweet sources,
Diversity of excellent fruits.

Juan of Castellanos[1]

1 The primordial images of the tropical forest in the mythical construction of the New World

The arrival of the Spaniards in the New World was motivated and justified by knowledge and images derived from books like Marco Polo's *Milione* and Cardinal Pierre D'Ailly's *Imago Mundi.* The trip was carried out in order to reach Katay and to be able to trace a new route for the market of spices (Taviani, 1989: 201). In Columbus's library, which is conserved in Seville, it is possible to find a copy of Pliny's *Naturalis Historia* and Seneca's *Tragedies*, annotated by Columbus, who had acquired them to confirm D'Ailly's quotes. However, it is necessary to consider other popular literary images derived from the different medieval bestiaries, in which the main characteristic of that other world was represented by an 'unsightly' and 'monstrous' nature. In fact, the first representation of the New World depends much on these mythical images and it deeply affects subsequent ones (Amodio, 1992).

The process of mythical construction of the New World also includes nature, which fluctuates between the 'recognition' of likeness (the landscape similar to that of Andalucía or Sicily) and the stupor caused by novelties, which, in order to be perceived and used, should be reduced to the literary categories of the medieval imaginary worlds (Fig. 3.1). Columbus writes: 'They saw many ways of trees, herbs and odoriferous flowers. They saw birds in many other ways than those from Spain, except for partridges and nightingales that sang and too many geese'[2] (Columbus, 1984: 53). The whole of nature is represented according to the 'deformed' and 'extraordinary' categories. Columbus writes on 16 October 1492:

> I saw many trees very dissimilar to those of our country, and many of them had branches of different sorts upon the same trunk; and such a diversity was among them that it was the greatest wonder in the world to behold. Thus, for instance, one branch of a tree bore leaves like those of a cane, another branch of the same tree, leaves similar to those of the lentisk. In this manner a single tree bears five or six different kinds. Nor is this done by grafting, for that is a work of art, whereas these trees grow wild, and the natives take no care about them.[3]
>
> (Columbus, 1984: 36–37)

The 'deformed' tree appears as sign of an extraordinary and overabundant world, immensely fertile and vital, also associated with the idea of the 'paradisaical garden', abundantly present in Columbus's writings as well as in the later Spanish authors (Rubió y Tudurí, 1981). These images are associated in Columbus's and Spaniards' perception with the Dantesque images of the earthly paradise and its corollary, the underground hell (the heat of the weather and the arid lands). In medieval cartography (the maps in a T shape), the earthly paradise was placed geographically to the east of Europe. As this cartography of religious origin was losing its religious characterization, the representation of the paradise was removed. However, by the time of the arrival of the Spaniards in America, as Sergio Buarque de Holanda writes, 'the belief in the physical and current reality of Eden seemed incontrastable then' (Buarque de Holanda, 1986: 196). So it is not surprising that Columbus, on his trip toward the 'Finis Orientae', ended up finding the earthly paradise on his fantastic route (Buarque de Holanda, 1986: 62).

> The eternal spring that Columbus imagines, and the birds whose singing he hears resounding throughout the jungle, are the terms of this latent analogy that, however, the very structure of the description seems to confirm . . . The landscape of Haiti, as Columbus saw it, was the materialization of this literary outline that Dante had extracted from a long tradition, transforming it in his imagination and in his style. In medieval literature, the variations of this motif are endless, a motif that is repeated in descriptions of paradise, gardens of love and delights, of an immobile nature between perennial trees and always melodious birds in an unalterable spring.
>
> (Olschki, 1937: 17–18)

The paradisaical models are applied to the New World, as an image of the description that Dante had elaborated in canto 20 of the *Purgatory*: 'Already anxious from searching inside and around the divine thick and living forest that tempered the new day to the eyes.'[4]

Related to the earthly paradise myth is the idea of the presence of great hidden treasures, kingdoms of gold, mainly following the Spanish arrival in Peru and Mexico (Fig. 3.1). These images, closely related to the tropical forest, are fused with myths and stories of indigenous rituals, originating the myth of El Dorado. A great number of expeditions took place to chase the illusion of the mythical kingdom – so many that it is possible to affirm that a great part of the conquest of the American continent was carried out thanks to this impulse.

For our purposes, the expeditions of interest are those directed toward the plains and the tropical forest, because it is mainly through these that the image of abundant gold became associated with the particular Amazon environment. Two of the main characters that embody this search at the end of the 16th century and the beginning of the 17th are Antonio Berrío, the great conqueror of the Orinoco River, and Walter Raleigh, the English corsair. What seems evident in the stories of the two expeditions, one by land and the other by sea and rivers, is the force of the myth that impels them to enter difficult and unknown environments. And the experience is also traumatic, because of the failure of the search and because of the great number of victims that the forest took during 50 years of trials. The plains, Guyana and the Amazon forest were associated with the search for gold, but also with the sacrifice of thousands of men to the difficulties and failures. The forest, from earthly paradise, came to be the 'damned forest', the green hell. The poet of the conquest, Juan de Castellanos, wrote (1974) in Elegy IX, part I: 'In many that looking for their Dorado, they were razed and lost.'[5] Here, the two images, one of the 'paradisaical garden' and the other of the 'dark forest,' are opposed, they superimpose and they cross, and these are the images that were to accompany the evolution of the cultural representation of the Amazon forest until the present.

Fig. 3.1. Arrival in the New World, *Lettera di Amerigo Vespucci*, Florence, 1505.

2 The war of the plants: natural landscapes vs. artificial landscapes

Although it seems evident, it is necessary to underline that what Spanish conquerors found and 'saw' was not a 'natural landscape', the forest included, but the results of interactions between that environment and the peoples that lived there. It was a 'cultured nature', the result of human action, a 'historical nature' that the Spaniards 'naturalized'. Also, to recognize the existence of human action upon it automatically implied the existence of a preceding occupation and, therefore, of a right previous to their arrival.

The opposition between the 'humanized natural world' and the 'natural wild world', in the context of the cultural images held by the conquerors, has its counterpart in a true 'war among plants' that takes place as a result of the conscious action of the European, and also, in an automatic way, on a natural level. In fact, from the beginning of the conquest, impelled by the need for alimentary self-sufficiency and because of the desire to reproduce the traditional diets of their original areas (bread, olive oil and wine), the Spaniards had to define which natural local products were edible and which of them were poisonous, and to investigate the possibilities of growing and reproducing European plants.

The possibility of growing wheat was so limited during the first phase of the conquest that casabe, bread obtained from the tuber of the yucca by the natives, was substituted for the 'daily bread' and the dry biscuits usually eaten by the sailors. Wheat bread became a rarity, developing the myth of bread as a valuable food, a sign of wealth and a status symbol. The process of acclimatization of wheat took almost two centuries, until the 1700s, when it was possible to select the species adapted to the new lands, mainly in Central America and in the south.

A contradiction emerges in the making of the image of the tropical forest. On the one hand, the forest is the place from which to extract vegetable products, wood and, naturally, gold. This implies that the conservation of the forest is indispensable to its continued use as a source of primary supplies. On the other hand, the development of the agricultural frontier implied the progressive destruction of the forest (Reboratti, 1990). The contradiction between the two positions, mostly implicit, is solved in the 18th century when, hope being lost of finding the kingdom of El Dorado, as well as the precious spices that had directed the initial expeditions, projects were expanded to achieve the progressive elimination of the forest.

The case of the importation of citrus plants is an example of these processes. From the early years of the conquest, the Spaniards, mainly the Andalusians, brought abundant quantities of citrus fruits to the Antilles. In some cases, as Gonzalo Fernández de Oviedo (1996, pp. 86–87) relates, the seeds were planted in vegetable gardens with the purpose of having a local natural product and eliminating the need to import it. The seeds of the citrus fruits spread by themselves and, having found favourable lands, they invaded the Caribbean islands and the mainland. In the popular representation, the impenetrable and unknown tropical forest opposed the forests of citrus trees, equally impenetrable, but recognized as proper landscape, with the aid of fantasy, since oranges and lemons were also cultivated in the Mediterranean. The reproductive success of the species was so great that the Spaniards had doubts about having imported it by the mid-16th century, hypothesizing that the citrus fruits were autochthonous.

The impact between the two vegetable worlds did not only refer to the implicit or explicit image made from them. The material referent of these images was constituted by a true struggle of the plants to survive:

> In some places the ecological consequences were terrible. For example, in the Caribbean, when the sugar cane was introduced, the forests practically disappeared given the continuous necessity of firewood to obtain sugar. In Nueva España, when the best lands were used for the cultivation of wheat, the Indians had to carry out their corn planting in the highlands, which were felled and suffered erosion quickly. In Peru, the horticultural farming occupied the semi-desert coastal valleys but altered the ecosystem of the plains notably, mainly because the autochthonous species were the first felled in order to obtain wood. Definitely, the cultivation of new plants enlarged the cultural shock of the

> conquest. With it a different feeding model was introduced, compelling the natives to participate in an unknown way of production, in an economy that tended to obtain earnings and to monopolize large amounts of lands to the detriment of the weakest ones.
>
> (Del Río, 1990: 34)

Along with the imported European plants came soil containing weeds and gramineous plants. Some of these plants had a natural evolution, so that the clover and some gramineous species occupied more and more land, perturbing the agricultural activities, contributing to the elimination of local species and deeply modifying the landscape.

3 The 'forest of the simple ones': searching, identifying and experimenting

Parallel to the images described above, another one developed during the 18th century. This image refers to the search for spices and medicinal substances in the forests of the New World, directed by the Crown itself and centralized in two fundamental institutions dedicated to investigating the botany of the New World: the Botica Real (Royal Pharmacy) and the Jardín Botánico (Botanical Garden), with headquarters in Madrid (see Urteaga, 1991; Tarble *et al.*, 1994: 155–169).

One of the most important purposes of Columbus's trip was the search for oriental spices. It is not odd, then, that, from the beginning of the conquest, the search for natural medicines from plants was a constant yearning of the Spaniards and of those who were associated with their undertaking. The Sicilian doctor Nicolas Scillacio, who had never been to America but, from his cathedra in Pavia at the end of the 15th century, had a fantastic geographical vision for his time. He became very excited about the American trees, 'silvestribus pomis rami curvati', and the aromatic and medicinal plants (Gerbi, 1978: 42–43). However, it was the doctor Alvarez Chanca from Seville, accompanying Columbus's second expedition, who brought the American medicinal world to Europe, with the image of the forest full of trees that were always green and of unknown varieties. Chanca in his *Letter* (written during the trip in 1494) defines the American flora according to its utilitarian type: a tree is like the laurel, not being it, another one like aloe, another like cinnamon, etc. The 'like' defines the perception of the new, with the intent of making the American flora homologous to the European one, but not accomplishing it completely and opening the way for new attributions. Truly, it is a New Vegetable World and this was immediately clear, mainly to the Spanish doctors involved in the production of knowledge from the new lands.

With Nicolás Monardes, American medical knowledge acquired its first organization and categorization. From Seville and the son of a Genoese, after having studied the Spanish medicinal herbs, he devoted himself to gathering data on the American ones, creating in Seville a garden of acclimatization of samples brought by the ships that arrived from the New World. In 1565 he finished and printed the first part of the *Historia Medicinal de las Cosas que Traen de Nuestras Indias Occidentales que Sirven en Medicina . . .*, which would be continued in 1571 with a second and a third part. The success of the book by Monardes was large, so much so that it was quickly translated into several European languages, because it was the first organized collection of this type. Monardes moves between two different ways of representing nature: on the one hand, he maintains some medieval elements inserted in the Renaissance context (he translated to Spanish the *Materia Medica* by Discorides), while, on the other hand, he refers to new ways of curing, beyond the Galenic and Hippocratic tradition. He experimented with the American medicinal plants, which he manipulated and applied to sick bodies, from the few data that he had obtained from the American peoples.

As a result, the image of the Amazon forest as a reserve of 'medical solutions', where it is necessary to know how to search, to identify and to experiment, also using indigenous knowledge, was developed. The Crown was directly interested in the topic, giving precise instructions (true geographical–medical questionnaires) on what kind of data should be gathered, how to send the mineral and vegetable products to Spain, how the living plants would be acclimatized in Madrid and, finally, bringing about the creation of experimentation centres, such as the Botica Real (Royal Pharmacy).

In the same years as the publication of Monardes's text, the interest in 'things' American had already involved a number of European specialists, from botanists to doctors. In Bologna, the naturalist Ulisse Aldrovandi, famous for his drawings of biological monsters, had created a botanical garden and an 'antiquities cabinet' of American interest. He ordered a large number of samples of the New World vegetables and botanical data, as he was convinced that the literature on this topic was lacking in method and exhaustive description (Cermentari, 1906). Finally, he decided to look personally for the American data and in 1569 requested the Duke of Tuscany to organize a scientific expedition to America directed by himself. He did not obtain an affirmative answer, but 2 years later Spain would carry out the project.

In 1570 Felipe II, King of Spain, named Francisco Hernández Protomédico (licensed medical examiner) of the Indies with the responsibility of travelling in America to gather data and samples

> of all the herbs, trees and medicinal plants that there had been in the province where you were . . . and about the use and ability and quantity that is given of these medicines and about the places where they are born and how they are cultivated, and if they are born in dry or humid places, or about other trees and plants, and if there are different species of them; and you will write the notes and signs.
>
> (Instructions to Hernández, in Álvarez Peláez, 1991: 92)

In 1571 Hernández arrived in New Spain, where he stayed for 3 years, producing an enormous quantity of naturalistic and medical data gathered in 24 books. When he returned to Spain he did not find a favourable environment for his ideas, although his texts found greater acceptance in the Italian academic world, where, in 1615, Antonio Recchi published a summary of his books.

Between Monardes and Hernández we find an entire series of chroniclers interested in describing the 'things from the Indies,' including news on the medicinal use of plants by local people. It is in this context that botanical description and ethnographic description were strongly intertwined, forcing the Spaniards to be interested not only in the American natural world but also in the autochthonous cultural worlds. This way, in the face of American nature, different images take place and they impress as much in their origin as in their form. Nevertheless, European variations on the image of the tropical forest were produced by one cultural horizon (the European world), and the same can be said about the several images produced by the American societies, particularly those that had developed in the 'lowlands' of the Amazon basin. The images produced by the Spaniards were influenced, although sometimes in a minimal and implicit way, by representations of the forests pre-existing before their arrival. For example, in the case of the use of vegetable substances to solve medical problems (either of a physical or a cultural origin), the Amazon indigenous people derived their identification from a complex system of representations, in which the forest was seen as a place inhabited by protective spirits. This 'living forest' and its associated rituals were not very different from those imagined by the Spanish peasants, who also believed in spiritual guardians of the forests and of the springs. The two images were reciprocally reinforced in the daily contact among the several ethnic groups. Indigenous elements became part of the vegetable mythology of the colonists, and elements of popular Christianity passed to the mythical world of the natives.

The situation is presented in a different way when the enlightened images of the forest produced by intellectuals and Spanish specialists are considered. In this case, from the very beginning of the conquest a secular image had been produced in which the plant that cures does so because of the chemical substances that it contains and not because of its particular association with some saint of the Christian pantheon (it was a 300-year process leading to the 'clinical pharmacy'). This secular vision hindered the creation of syncretism with indigenous elements, but it also fed from the local medical–ritual knowledge. The tropical forest comes to be thought of as a natural pharmacy (the 'forest of the simple ones'), a place to look for the solutions to the problems that Hippocratic medicine tried to understand and to solve. This image achieved its highest level of development in the 18th century, impelled by the Crown in Madrid and the renovation of botanical and medical thought.

It is during the second half of the 18th century, with Carlos III, that the development of the natural sciences and an interest in the American territories becomes a main characteristic of the scientific politics of the Enlightened Despotism. In 1762 José Quer published his monumental *Flora Española*, in which the botany of the Peninsula received its first true systematization. José Quer was the director of the Jardín Botánico and, just like his successors, some of them 'pharmacists', he looked at the Spanish world without forgetting the American world. The natural sciences acquired new perspectives, being liberated from the theocratic vision that had dominated Spanish thought of the previous century. This new perspective was imposed through a great number of scientific expeditions to the New World (La Fuente, 1988). The first, and one of the most important, was carried out on the Orinoco River (1754) to determine the limits between the Spanish American domains and the Portuguese domains, with the participation of Loefling, one of Linnaeus's pupils (Lucena Giraldo, 1991b; Amodio, 1998a).

In a larger context, the forest becomes a 'reservoir' of useful products, from beautiful wood to the inks used to colour clothing; from precious metals to a place for agricultural development, influenced by the French physiocratic ideas of Quesnay and Turgot (in Spain, authors like Juan de Arrieta and Martín González Navarrete and state ministers such as Floridablanca and Campomanes) (Puerto Sarmiento, 1988; Amodio, 1998b). All this is carried out in the shadow of an ambitious project to catalogue American nature, which exploits the relationship among metropolitan scientists, officials and missionaries – the colonial agents in America. This illustrated rediscovery of American nature (González Bueno, 1992: 79) has its centre in the medical material, as it is possible to discover in the *Instrucción sobre el Modo más Seguro y Económico de Transportar Plantas Vivas . . .* (Instruction on the surest and most economical way of transporting living plants . . .), published by Gómez Ortega in 1779, where the list of medicinal plants (47%) required by Spain surpasses those of nutritious and industrial use. Apart from containing drawings of the mechanisms for transportation of living plants and the illnesses referred to in the pamphlet by Gómez, the document mainly requested American plants that could serve as anti-fever and anti-venereal medicine and stimulants (cocoa, for example) and the utilitarian ones (e.g. fibres) (González Bueno, 1992: 80–82).

Among the most frequently requested plants was the keno (quina; cinchona sp.), which Francisco Javier (Puerto Sarmiento, 1988) defined as a true 'vegetable Dorado'. The merit of the success of the bark of the keno is largely owed to Mutis and the expedition to the New Kingdom of Granada at the end of the 1700s and, naturally, to the demonstration that the vegetable remedy worked for a series of endemic illnesses – those 'fevers' – in their tertiary and quaternary forms. The tree of the keno that was useful for the extraction of its medicinal bark grew spontaneously in hill forests (the reason why it can be included among the products that were extracted directly from the wild forest), thereby reinforcing the image of the 'useful forest'.

4 A forest of useful woods: knowing, sorting and using

Spanish population growth during the 17th and 18th centuries and its progressive urbanization accelerated the deforestation of the Spanish peninsula.

Wood had to be imported from the north of Europe, mainly to build ships. In response, the Crown passed ordinances for the conservation of the forests and their rational use, particularly starting from 1748 with the 'Reales Ordenanzas de Montes' (Royal Ordinances of woods), which were effective until 1812 (Urteaga, 1991: 17). Fundamentally, the ordinances were intended to organize and control the felling of the forests and to promote reforestation through the creation of specialized plant nurseries. The legislation of 1748 was composed of two different ordinances: 'Ordenanza para la conservación y aumento de los montes de la marina' (Ordinance for the conservation and increase of the woods of the fleet) and 'Real Ordenanza para el aumento y conservación de los montes y plantíos' (Royal Ordinance for the increase and conservation of the woods and plantations). The reference to the 'marina' (fleet) was because the coastal forests were under the jurisdiction of the Fleet Administration (Intendencia de Marina), responsible for the conservation and exploitation of these forests to supply the shipyards where the warships were built. The forests were a resource of royal interest, tutored by the institutions of the state and denied free exploitation by the private ones. It will be against these types of controls that, at the end of the 1700s, Campomanes (1763) and Jovellanos (1794), royal ministers, declared themselves in favour of the abolition of the above-mentioned ordinances and of the participation of private institutions in the exploitation of the forests (Casals Costa, 1991: 63–64).

Following the decrease of woods on the Spanish peninsula and because of the growing difficulty of carrying timber all the way from northern Europe, interest turned toward the overseas colonies for its supply. For example, the engineer Francisco Gautier stated explicitly in his *Estado de los Bosques en España* (State of the Forests in Spain), written in 1772, that 'America alone offers us the aids that are needed; the good quality of its wood is known' (Pimentel Igea, 1991: 46). Also, the conservationist laws did not seem to be applicable in the overseas lands, so that it was possible to think of the 'existence of two divided mentalities in relation to the forest issue at the level of ideas' (Casals Costa, 1991: 61). The American 'Sociedades Económicas de Amigos del País' (Economic societies of friends of the country) created in the several local capitals during the second half of the 18th century did little work on this subject. The case of Cuba is emblematic here, since the presence of a large dry dock and of an economy based on sugar cane practically razed the forests of the island. For example, between 1724 and 1794, 125 warships were built; for a single frigate nearly 14,000 m^3 of wood was necessary, and a 150-year-old oak tree can offer a maximum of 2 m^3. The continuous shipment of wood to Spain contributed heavily to the elimination of the American forests (the Escorial itself was built with Cuban wood). For example, in 1620, before the peninsular crisis, '107,688 quintals of brazil tree (*Caesalpinia echinata*), that is to say 5384 tons, 14,838,100 maravedíes worth' (Aranda y Antón, 1993), were disembarked in Seville. This happened every year and the orders were increasing.

European predatory action toward the American forests progressively advanced from the coasts toward several inland regions of the continent, having drained the insular reservations. 'To know in order to use' became the programme that motivated the scientific expeditions to America, as much in the case of the 'expedition of limits' to the Orinoco (1754) as in that of Malaspina around the world (1789). *To know* was to inventory, to register, to calculate and to elaborate an 'American natural history', in a context clearly different from the one that had moved the first chroniclers. *To use* meant that, for an indebted Spain with already scarce forest resources, America was presented once again as the anchor of economic salvation and the tropics as a 'useful forest', together with the newly illustrated ideas on the need to transform nature to make it truly usable and about the function that agriculture should carry out in the new modern state.

This perspective directly affected the Amazon forest, particularly on the Spanish side (in Guyana), although the Portuguese were not behindhand (Portuguese pressure on the Amazon was carried out from the south-east toward the north, while the Spanish was from the north to the south-east). Guyana had been a relatively isolated and impenetrable region until the first half of the 1700s, although missionaries had already advanced toward the Orinoco River, along with some military outposts. It is with the 1754 expedition that the true occupation began, continued by the governors of the second half of the century. The stabilization of the conquest is fundamentally due to Manuel Centurión, who organized the last expedition in search of El Dorado, and to Miguel Marmión, who elaborated an entire series of development plans and of exploitation of the tropical forest, considering Guyana as the 'key of America' in economic and warlike terms. In his *Descripción Corográfico-Mixta de la Provincia de GuaPyana* (Corographic-mixed description of the Guyana Province) (1788), he affirmed that the region of the Orinoco delta and Sierra de Imataca 'is all composed of grass and groves of excellent quality wood for building and any other uses anyone would want to apply them for' (Arellano Moreno, 1964: 444).[6]

Thanks to the agreement between Miguel Marmión, Governor of Guyana, and José María Chacón, Governor of the island of Trinidad, an 'Expedición del reconocimiento de maderas de construcción en los montes de Trinidad y Bajo Orinoco' (Expedition for the identification of construction wood in the mountains of Trinidad and the lower Orinoco) was organized and authorized by the Royal Order of 31 May 1786. Several experts in wood and engineers, including a North American, were hired to carry out the reconnaissance, and a ship was bought to sail the river (Lucena Giraldo, 1991a, 1992). The expedition took place in the first months of 1788, giving origin to 'Estado general de las maderas marcadas en las bocas y caños del Orinoco' (General state of the wood marked in the mouths and springs of the Orinoco) (Servicio Histórico Militar, Madrid, 5–3–12–13), with the list of the trees that had been identified being mainly those usable to build ships. Naval shipyards existed on Margarita

Island and in Cumaná. Among the identified species were cedars, carob trees, gourd trees, sassafras, etc. Beyond the relative failure of the project, what remains important to us is that the perception of the forest definitely changed. It was now an ecosystem to be rationally used in productive terms, through increased cultivation and space in which to experiment with new cultivation.

A 'real offensive orientated towards regional development', as defined by Manuel Lucena Giraldo (1992: 465), was developed during the time of Marmión in Guyana, helped by Fermín de Sansinenea, 'Comisario Especial para el Desarrollo de Guyana' (Special commissary for the development of Guyana): population censuses, new relationships with the natives, a balance in the relationships with the missionaries, trade in regional products, such as tobacco, cocoa, coffee, copaiba, leather, cotton, resins and wood, and the introduction of new cultivation in the plains and lands 'stolen' from the forest (Arellano Moreno, 1964: 399). Regarding the new cultivation, returning from a trip to Martinique, Fermín de Sansinenea took to Guyana seeds of clove, Ceylon's cinnamon, a pepper plant, another that produced resin, another that produced a red substance and 'a plant and the seeds of a special fruit from Africa' (Archivo General de Simancas, Guerra Moderna, 7090). It was not an isolated case and in fact a farmer from the region, at the time when the new seeds were delivered, communicated that he had already begun cultivation of the Ceylon cinnamon on his own (*idem*).

The introduction of new species can be considered the symptom of a new general attitude. In fact, after the import of plants from Spain in the first period of the conquest and the progressive change of the agricultural landscape, mainly on the coast and in the Andes, the efforts to obtain local productive forest species had not diminished. Numberless orders on behalf of the Botica Real (Royal Pharmacy) for the shipment of vegetable matter were largely responded to by the whole mainland territory. However, all this effort had not given many results, apart from the keno, true 'oro amargo' (bitter gold) (Puig Samper, 1991: 230).

The botanical world reacted with disappointment toward the vegetable universe, which seemed to have promised so much and had carried out so little. However, this new attitude is not surprising, especially if one considers the polemic on the nature of the American world impelled by the works of the Frenchmen Buffón and De Pauw, to whom the pluvial forest had a degenerate atmosphere because of the humidity that slowed its development of animals and humans (Buffón, 1826–1828; Gerbi, 1985). Certainly, there were quite violent reactions to this kind of image from Spanish (Feijoó, before Buffon himself) and American (Robertson and Franklin) parts. But, at the officials' and the colonists' level, a doubt remained about American nature, particularly in the tropics. From this image, the consequent concepts of the illustrated posture: to rationally modify that natural panorama, to turn it into something recognizable and manageable, into something productive, including the 'import' of European white peasants, that the republic, worthy heiress of the illustrated thought, would try to put in practice.

5 Conclusions

The image of the tropical forest was originally linked to medieval culture, which tried to produce an adherence with reality and coherence between perception and imagination. It was not only an earthly paradise but also a 'green hell,' a 'useful forest' and a barrier to the development of agricultural exploitation. With transformations in European culture, more at the educated level than at the popular level, the attitude toward the tropical forest changed and so did the images held. The ecological and economic danger that the indiscriminate felling of the trees implied was noticed during the 1700s, but there was little it was possible to do, and so the young republics of the 19th century inherited the problems and the contradictions of the old régime.

These images are not exclusive. Very often they overlap and interact, according to the social group that participates in the action, as well to the imperial or local changes in politics. The same utilitarian vision of the 1700s does not exclude the re-realizing of religious images, just as the optimistic image of progress based on the rational exploitation of nature does not eliminate a certain pessimism about the capacity of humans to take it to a good conclusion. As Malaspina wrote, 'in big monarchies the order of nature is transformed' (Pimentel Igea, 1991: 56), referring to the indiscriminate exploitation of American resources. But this conscience had little influence on the people in Cuba, Margarita or Cumaná, where they kept building 'floating forests', as the big ships had been called.

Two exemplary images of the Amazon forest illustrate different forms of thinking, one almost at the beginning of the Spanish conquest and the other at its end, shortly before the beginning of the wars of American independence. The first is that of Walter Raleigh, an Englishman who attempted the conquest of Guyana at the end of the 16th century on behalf of the English Crown. The book that he wrote after the failure of his project, *The Discoverie of the Big, Rich and Bewtiful Empyre of Guiana . . .*, was translated into several European languages and excited the spirit, impelling thousands of people to go to the Amazon in search of El Dorado and influencing political and cultural attitudes. Among the many descriptions, it is worthwhile quoting the one which is perhaps the most metaphorical of them, the one that feminizes the Guyana region:

> To conclude, Guiana is a country that hath yet her maidenhead, never sacked, turned, nor wrought; the face of the earth hath not been torn, nor the virtue and salt of the soil spent by manurance. The graves have not been opened for gold; the mines not broken with sledges, nor their images pulled down out of their temples.
>
> (Raleigh, 1596: 96)

Guyana as a woman/nature (opposed to the man/culture) that offers herself to the European man is a metaphor deeply rooted in western thought and, more than once, had been implicitly used to justify the conquest and the looting of American wealth.

The other image, which closes the colonial period, refers to nature as an 'obstacle' and it belongs to Jovellanos, minister of Carlos III, in his 'Report about the Agrarian law' of 1795:

> Without a doubt, to her [the agriculture] nature owes big improvements. Anywhere the eyes turn to, it is embellished and perfected by man's hand. Everywhere forests are plucked up, the wild animals are driven away, the lakes are dried up, the rivers are grooved, the seas are restrained, the whole surface of the earth is cultivated and full of farms and villages and beautiful, and you magnificent populations are offered as an admirable spectacle the monuments of human industry and the efforts of the common interest to protect and to make easy the individual interest.
>
> (Jovellanos, 1968: 171)[7]

Only wild nature represents an obstacle to human progress. Hence, it is necessary to control it and, if necessary, to eliminate it. These kinds of enlightened postures were as common in Spain as in America and they marked in a definitive way the later times. It is true that Alexander of Humboldt, the German scholar who travelled down to the Amazon regions at the end of the 18th century, tried to produce a different image: that of a harmonic nature with which one can cohabit and which it is necessary to admire and to respect. It is also true that romanticism would recover these images of the beauty and the mystery of the tropical forest (consider the German painter Bellerman in Venezuela in the middle of the 19th century), but Jovellanos's image, produced at the beginning of our modern times, survived and caused violent results, leaving us with a beaten nature that we will scarcely be able to cure.

Notes

1 'Hay infinitas islas y abundancia/De lagos dulces, campos espaciosos,/Sierras de prolijísima distancia,/Montes excelsos, bosques tenebrosos./Tierras para labrar de gran sustancia,/Verdes florestas, prados deleitosos,/De cristalinas aguas dulces fuentes, /Diversidad de frutos excelentes.'

2 'Vieron muchas maneras de árboles, yervas y flores odoríferas. Vieron aves de muchas maneras diversas de las d'España, salvo perdizes y ruiseñores que cantavan y ánsares, que d'estos ay allí hartos.'

3 'Y vide muchos árboles muy diformes de los nuestros, d'ellos muchos que tenían los ramos de muchas maneras y todo en un pie, y un ramito es de una manera y otro de otra; y tan disforme, que es la mayor maravilla del mundo cuánta es la diversidad de la una manera a la otra. Verbigracia: un ramo tenía las fojas de manera de cañas, y otro de manera de lantisco y así en un solo árbol de çinco o seis d'estas maneras, y todos tan diversos; ni estos son enxeridos porque se pueda dezir que el enxerto lo haze, antes son por los montes, ni cura d'ellos esta gente.'

4 'Vago già de cercar dentro y dintorno la divina foresta spessa y viva, ch'alli occhi temperava il novo giorno.'

5 'En muchos que buscando su Dorado, Quedaron asolados y perdidos.'

6 'Toda ella se compone de montes y arboledas de excelentes cualidades de maderas de construcción y demás usos a que se quieren aplicar.'
7 'Sin duda que a ella [la agricultura] debe la naturaleza grandes mejoras. Á doquiera que se vuelva la vista, se ve hermoseada y perfeccionada por la mano del hombre. Por todas partes descuajados los bosques, ahuyentadas las fieras, secos los lagos, acanalados los ríos, refrenados los mares, cultivada toda la superficie de la tierra, y llena de alquerías y aldeas, y de bellas y magníficas poblaciones, se ofrecen en admirable espectáculo los monumentos de la industria humana, y los esfuerzos del interés común, para proteger y facilitar el interés individual.'

Acknowledgements

Special thanks go to Alejandra Martínez for translating this article into English, and to Tania Granadillo and Kay Tarble for their revisions and comments.

References

Álvarez Peláez, R. (1991) La historia natural en los siglos XVI y XVII. In: Diez, A.R., Mallo, T., Pacheco Fernández, D. and Alonso Flecha, A. (eds) *La Ciencia Española en Ultramar*. Doce Calles, Madrid, pp. 89–95.

Amodio, E. (1992) *Formas de la Alteridad. Construcción y Difusión de la Imagen del Indio Americano en Europa Durante el Primer Siglo de la Conquista de América*. Abya Yala, Quito, Ecuador.

Amodio, E. (1998a) El paraíso vegetal. Las fronteras étnicas de un botánico ilustrado. In: Pilar de San Pío Aladrén, M. (ed.) *La Comisión Naturalista de Löefling en la Expedición de Límites al Orinoco*. Real Jardín Botánico, Lunwerg Editores y Caja Madrid, Madrid, pp. 61–84.

Amodio, E. (1998b) La agricultura ilustrada en el Oriente de Venezuela durante la segunda mitad del siglo XVIII. In: Molina, L. and Amodio, E. (eds) *Técnicas y Tecnología en Venezuela durante la Época Colonial*. Número monográfico (6) del Boletín del Museo Arqueológico de Quibor, Venezuela, pp. 67–98.

Aranda y Antón, G. (1993) Las maderas de Indias. *Asclepio*, 45(1), 243. Madrid.

Arellano Moreno, A. (1964) *Relaciones Geográficas de Venezuela*. Academia Nacional de la Historia, Caracas.

Buarque de Holanda, S. (1986) *Visión del Paraíso*. Biblioteca Ayacucho, Caracas.

Buffon, G.-L. (1826–1828) *Oeuvres Complètes de Buffon*, vol. 15. Delangle frères, París.

Casals Costa, V. (1991) Del cultivo de los árboles a las leyes de la espesura. In: Lucena Giraldo, M. (ed.) *El Bosque Ilustrado. Estudios sobre Política Forestal Española en América*. Instituto para la conservación de la naturaleza, Madrid, pp. 63–89.

Cermentari, M. (1906) Ulisse Aldrovanti e l'America. *Annali de Botanica, Rome*, 4, 313–366.

Columbus, C. (Colón, C.) (1984) *Textos y Documentos Completos*. Prólogos y notas de Consuelo Varela. Alianza Universidad, Madrid.

de Castellanos, J. (1974) *Elegías de Varones Ilustres de Indias*. Academia Nacional de la Historia, Caracas.

Del Río, J. (1990) La transformación ecológica indiana. In: Fernández Pérez, J. and González Tascón, I. (eds) *La Agricultura Viajera. Cultivos y Manufacturas de Plantas Industriales y Alimentarias en España y en la América Virreinal.* Consejo Superior de Investigaciones Científicas, Madrid.

Fernández de Oviedo, G. (1996) *Sumario de la Natural Historia de las Indias.* Fondo de Cultura Económica, México City.

Gerbi, A. (1978) *La Naturaleza de las Indias Nuevas.* Fondo de Cultura Económica, Mexico City.

Gerbi, A. (1985) *La Disputa del Nuevo Mundo.* Fondo de Cultura Económica, Mexico City.

González Bueno, A. (1992) Virtudes y usos de la flora americana: una aproximación al carácter utilitario de las expediciones botánicas en la España ilustrada. In: La Fuente, A. and Salas Catalá, J. (eds) *Ciencias Colonial en América.* Alianza Universidad, Madrid.

Jovellanos, G.M. (1968) *Informe sobre la Ley Agraria (1795).* Edime, Madrid.

La Fuente, A. (1988) Las expediciones científicas del setecientos y la nueva relación del científico con el estado. In: del Pino, F. (ed.) *Ciencia y Contexto Histórico Nacional en las Expediciones Ilustradas a América.* Consejo Superior de Investigaciones Científicas, Madrid.

Lucena Giraldo, M. (1991a) Defensa del territorio y conservación forestal en la Guayana (1758–1793). In: Lucena Giraldo, M. (ed.) *El Bosque Ilustrado. Estudios sobre Política Forestal Española en América.* Instituto para la Conservación de la Naturaleza, Madrid, pp. 137–147.

Lucena Giraldo, M. (1991b) *Laboratorio Tropical. La Expedición de Límite al Orinoco (1750–1767).* Monte Ávila/CSIC, Caracas.

Lucena Giraldo, M. (1992) Exploraciones científicas y políticas de fomento en Guayana (1784–1791). In: *IX Congreso Internacional de Historia de América.* AHILA, Seville, pp. 461–479.

Olschki, L. (1937) *Storia Letteraria delle Scoperte Geografiche.* Olschki Editore, Florence.

Pimentel Igea, J. (1991) La riqueza forestal de las costas del Pacífico. Noticias e informes sobre maderas en la expedición Malaspina (1789–1794). In: Lucena Giraldo, M. (ed.) *El Bosque Ilustrado. Estudios sobre Política Forestal Española en América.* Instituto para la Conservación de la Naturaleza, Madrid, pp. 45–63.

Puerto Sarmiento, M. (1988) *La Ilusión Quebrada: Botánica, Sanidad y Política Científica en la España Ilustrada.* Serbal, Barcelona.

Puig Samper, M.A. (1991) El oro amargo. La protección de los quinares americanos y los proyectos de estanco de la quina. In: Lucena Giraldo, M. (ed.) *El Bosque Ilustrado. Estudios sobre Política Forestal Española en América.* Instituto para la Conservación de la Naturaleza, Madrid, pp. 219–241.

Raleigh, W. (1596) *The Discoverie of the Large, Rich, and Bewtiful Empyre of Guiana.* Robert Robinson, London (facsimile edition: World Publishing Company, Cleveland, Ohio, 1966).

Reboratti, C. (1990) *Fronteras Agrarias en América Latina.* Geo Crítica, Barcelona.

Rubió y Tudurí, N.M. (1981) *Del Paraíso al Jardín Latino.* Tusquets editores, Barcelona.

Taviani, P.E. (1989) *Cristóbal Colón: Génesis del Descubrimiento.* Editorial Bip, Caracas.

Urteaga, L. (1991) La política forestal del reformismo Borbónico. In: Lucena Giraldo, M. (ed.) *El Bosque Ilustrado. Estudios sobre Política Forestal Española en América.* Instituto para la Conservación de la Naturaleza, Madrid, pp. 17–44.

Urteaga, L. (1997) *La Tierra Esquilmada. Las Ideas sobre la Conservación de la Naturaleza en la Cultura Española del Siglo XVIII.* Serbal/CSIC, Madrid.

The Historic Evolution of the Haitian Forest

4

Françoise Hatzenberger

Université Pierre et Marie Curie (Paris VI), 10 rue Laugier, 75017 Paris, France

During their settlement, the Indians moderately modified the repartition of the vegetal ecosystems on Espaniola island. They only worked with and, above all, introduced and multiplied some domestic and edible species. At the time of its discovery by Europeans, the island was covered by mesophytic and hygrophytic timber trees, on the slopes of the mountains. On the plains and in the lowlands of the high hills, a more or less dry forest, made up of lower and generally evergreen trees, developed. This littoral forest, constituted of hardwood trees or species with medicinal and tinctorial qualities, was the first to be felled. So, before the end of the 16th century, the brazilleto of the southern peninsula and some lignum-vitae in the north country disappeared. But almost all the species, like the *Arecaceae*, were grubbed up only to enable the settlement of the new populations without considering the commercial price of the cut-down trees. Little by little, the littoral was re-covered by secondary woodland, where numerous introduced species increased, such as lemon and the orange trees in the 16th century and the logwood in the 18th century. During this period, the grubbing up reached the mountains because of the development of the coffee tree. At times, the mahogany trees abounding on the high hills were dispatched to Europe. Other timber woods were used in Espaniola to build the sugar estates and the mills. After the period of the Revolution, the clearing of the mountains' summits increased and the forest resources, mahogany trees and logwoods, were foolishly pulled down, particularly from 1880 to 1890. In the 20th century, export has decreased, but cutting for firewood continues because of domestic and artisan needs. Today, the forest no longer exists. Of all the species which formerly constituted this formation only a few subsist, but they rarely reach 5 m high.

1 Introduction

When the Spaniards sailed along the north coast of Espaniola for the first time in December 1492, they discovered a very varied ground cover, different from that of the Bahamas or Cuba. The forests of the north-west coast were lower and their architecture was reminiscent of the Mediterranean forest but, as the winds pushed the Europeans eastward, cultivated fields and high dark forests mingled. The floristic richness of the island could have resulted from the native vegetation planting from Amazonian, Gondwanian affinity stock and from many Laurasian genera, such as *Pinus* from the North American continent. The drier period of the Quaternary could explain why the xeric biocenosis persisted on the plains as well as the high rate of endemism. The Amerindians favoured species of economic interest, thus diversifying the flora of Espaniola. It would be subjected for 6 months to an unbearable heat without any western wind and to frequent rains, which created a very uncomfortable moisture. The climatic and edaphic factors enabled the development of the morphologically and floristically different forests in rather close environments. With the European conquest, humans became an important factor in the environment. The Spanish occupation and an agricultural economy until the 18th century that was based on revenue and the post-revolutionary economic and political upheavals repeatedly disrupted the native floristic composition of the island, as well as the geographical repartition of the forest community.

2 The Spaniards and the land clearing of the littoral dry forest in the 15th and 16th centuries

In the 15th century, the Spaniards discovered a vast forest environment. Because the Europeans were interested in dye-woods and medicinal plants and species, the conquistadores exploited the coastal areas, which were easy to reach from the sea, and multiplied trade plants from the Old World.

Before the European settlement, the plains close to the sea were covered with different arborescent communities. When the annual rainfall rate was below 1200 mm and the hydric reserves low, twisted shrubs lying more or less by the sea rose progressively as wind strength diminished inland. This planting of mostly evergreen species with small leathery leaves developed into a xerophytic forest or even xeromesophytic in the moister areas. That forest, made of a stratum of lower trees, which were often spinate, could develop on the low slopes south of the high hills. On riparious ground and on the northern plains, which were subjected to moist trade winds, the dry forest scattered into a sub-evergreen mesophytic facies, or even an evergreen one.

The xerophytic forest comes in the form of 'raques', that is to say, a combination of 'stunted woods, brambles, white woods, shrubs, spongious vegetals', small trees and various *Cactaceae*. The lands covered with this vegetation were called 'fredoches'. This vegetation included leguminous species, such as the 'acacia sauvage', *Acacia scleroxyla* Tuss., and heliophilous species with deciduous leaves, such as the West Indian birch, *Bursera simaruba* (L.) Sarg., but also evergreen species, such as the small lignum-vitae, *Guaiacum sanctum* L. The pitch-apple, *Clusia rosea* Jacq., and the yellow sandalwood, *Zanthoxylum flavum* Vahl., towered above shrubs, such as the torchwood, *Amyris elemifera* L. But palm trees, belonging to the *Thrinax* genus, and endemic species, such as *Zombia antillarum* Bailey or *Pseusophoenix vinifera* (Mart.) Becc., were numerous. Despite the thinness of the soil, trees like the ironwood, *Krugiodendron ferreum* (Vahl.) Urb., and the mastic tree, *Sideroxylon foetidissimum* Jacq., survived with the aid of roots spreading over a large diameter.[1] Most of these hardwoods were described as 'incorruptible' and were used for building.

From the 15th century onwards, some of these woods were sent to Europe. Fleets carried 'brazil-wood', 'rouge de braise', used for dyeing. From 1569 to 1608, 10.7 tons of non-tinctorial woods were sent to the Spanish harbours, including 'palo-brasil', probably of the *Caesalpinia* genus, and 3 tons of various tinctorial woods: the yellow sandalwood for its much sought-after yellow tincture, the guaiacum, which gave a black colouring, and the logwood. But the brazilletto or logwood, which could be found at Yaquimo (Aquin) in the south of the southern peninsula, quickly disappeared. Discovered in 1494, it was exploited as early as 1499.[2] The town of Port of Brazil, situated in the Cayes plain, was already abandoned in 1606.[3] But guaiacum, *Guaiacum* sp., represented the most important tonnage sent to Europe. Lignum-vitae, *Guaiacum sanctum*, and guaiacum, *G. officinale* L., also called 'wood of Indies', 'likely to cure bubons' and syphilis, were developed in such huge quantities that '20,000 or 100,000' quintals could be taken 'if trade for so much wood were to be'. Of the 800 tons navigating toward the Iberian peninsula from 1568 to 1608, nearly 40% came from Santo Domingo and was sent between 1581 and 1596.[4] In those same Spanish harbours, 1.6 tons of sarsaparilla, medicinal roots recommended for venereal diseases, arrived from the islands. It was certainly *Smilax havanensis* Jacq., a littoral plant with long stems and spinate leaves. That tonnage gives an idea of the important clearing in littoral areas. In 1589, 0.2 tons of ebony, *Brya buxifolia* (Murr.) Urb., left for Seville. West Indian ebony or 'gratte-gale' is a shrub from 2 to 2.5 m in height, whose ramous, often tortuous, stem is slightly bigger than sugar cane. Its hard wood, black-purple sprinkled with yellow, was much sought after for inlay work. The ebony shrub, with a few small leaves and yellow flowers looking like peas, settled in rocky, dry and arid places as well as in woods. It was a dominant species in the shrub communities.

The introduced species multiplied abundantly, especially on the Leogane plain, probably in the gardens that were abandoned by the Amerindians. Fifty tons of cassia-stick tree, *Cassia fistula* L. from Asia, were sent to Seville from 1581 to 1596. That vegetation was still proliferating in 1665.[3] Another introduced species occupied that area 150 years later: the coconut, *Cocos nucifera* L. Those figures only represented a part of the transatlantic trade. All the European harbours trading with the West Indies received cargoes from the East Indies. The Spanish settlement modified the vegetation in the plains quickly and in depth. If, at the end of the 16th century, the higher-altitude forests remained untouched for the large part, cuts along the littoral were important near Cap Français and around Les Cayes in the south and Leogane in the west. The technical means at the time did not allow intensive cutting of straight timber or the transportation of the felled trees over long distances. None of the species sent to Europe were over 10 m high; they were closer to shrubs than to trees.

The cleared areas were rapidly invaded by introduced species, which occupied an important place in the coastal plains, while the native shrub stratum progressively disappeared. In cleared areas, the Spaniards multiplied species from the Old and the New World. Since the beginning of the 16th century, the citrus species, used to make up hedges, had formed sweet and sour orange woods in a few decades. American species, such as the guava, developed very quickly. According to de Acosta, 'that tree did not exist before the Spaniards arrived'.[5] The guava multiplied to infinity because animals did not eat the seeds. The cattle from Europe, growing in large numbers and grazing freely on these islands, which had lacked quadrupeds, were responsible for the multiplication of certain plant species.

3 The development of cultivations and disappearance of the mesophytic forests

If some ridges were covered with xerophilous species, the Limbé mountains 'situated at more than one league (almost 4 km) from the sea were only covered with soft wood'.[6] These trees developed in clay and were not decay-resistant. They were considered as proof of a rich soil.[7] These forests could also be found in plains where sugar estates multiplied rapidly. North-east winds, the trade winds, carrying rain, enabled xeromesophilous and mesophilous forests to settle in the moist low areas and on the north side of the high hills. On the southern sides, which were less subjected to rain, mesophilous communities developed at a higher altitude. The towering forest grew when the rainfall ranged from 1500 mm to 2000 mm annually.

In the mesophytic forest, the *Arecaceae* was the best represented family. Palm trees belonging to the *Roystonea*, *Acrocomia*, *Coccothrinax* or *Thrinax* genera could be frequently found. With other species, they made a rather thick forest which, however, remains quite open despite the persisting canopy leaves. The highest trees in the mesophytic Haitian forest could rise over 20 m. This stratum included many species: the 'chêne d'Amérique', *Catalpa longissima* (Jacq.) Sims, the West Indian cedar, *Cedrela odorata* L., the candlewood tree, *Cupania americana* L., and the mahogany tree, *Swietenia mahagoni* (L.) Jacq. Below, the shrub layer could be made of roucou, *Bixa orellana* L., and bwa glo savann, *Ixora ferrea* (Jacq.) Benth, among others. In the drier areas, the xeromesophytic vegetation developed mostly on calcareous soils. Its floristic composition was close enough to that of the dry forest. But these higher forests also countained many *Arecaceae*. Palm trees, such as *Sabal domingensis* Beccari., mixed with species like the wild cinnamon, *Canella winterana* (L.) Gaertn., the sweetwood, *Dipholis salicifolia* (L.) A.DC., the 'bois de lance', *Oxandra lanceolata* (Sw.) Baillon, the 'bois d'orme', *Guazuma tomentosa* HBK., the cedar, *Tabebuia heterophylla* (DC) Britt., the kitwood, *Gymnanthes lucida* Sw., or the torchwood, *Amyris balsamifera* L.

During the first years of settlement, the trees were cut without proper judgement in order to clear the land for cultivation. But the plots of land were not entirely cleared of trees, a part of them was kept as 'bois-debout' and was used as a firewood reserve. Timber from the western and northern littoral forests was used to build houses in the estates and in towns. The houses in Cap Français and in Petite-Rivière-de-Nippes were built of palm wood, particularly thatch palms, coming from the nearby plain.[3] Progressively, 'incorruptible wood' disappeared to make buildings. An extremely tall thatch palm, the endemic species *Copernicia berteroana* Becc., became sparse. Until the middle of the 20th century, it would subsist on Mount Bienac.[3] The West Indian cedar, *Cedrela odorata*, was used to make 'essente', shingles that covered almost all the houses because of the lack of tiles or slates. Very hard woods, such as the gris-gris, *Vitex divaricata* Sw., or the locust-tree, *Hymenaea courbaril* L., were very much appreciated, although they were difficult to work when making tables and sugar-cane mill rollers. Because of the growing shortage of incorruptible species, wood was bought in Canada or, from 1750, cut in the southern peninsula – in the Baradères plain – which was less settled.[8]

The military protection of the colony also modified the vegetal environment. Along the sand coasts, the towns were protected by the mangrove swamps and elsewhere by forts armed with guns. To make carriages and cap-squares, West Indian cedar and chêne-pays, which were common in the Artibonite plain, but also in Gonaïves, at Jacmel and at Les Anglais near Cap Tiburon, were used. The acacia sauvage was used for the big parts. Being difficult to penetrate, the mangrove swamps played a protecting role, which, however, did not stop exploitation. But there were various ways of using the mangrove trees: the mangle, *Rhizophora mangle* L., and the 'jambe-de-chien' mangle, were barked to produce tan, which killed them. The bark from the branches of the 'mahot d'herbe' was spun and also used to make rigging. So was that of the seaside mahoe, to make all the bonds needed on the estates. The heart of the old woods – which was beautiful – was used to make pistol handles. Near towns, the mangrove swamp was usually eradicated to clean up the district, as in Leogane. But, to protect the country from enemy attacks, only a few clearings were made, like that around Petite-Rivière-de-Nippes village.[9] In the middle of the 18th century, when there were too many clearings, protection of the coastal tidal area became compulsory. The use of the wood in Cavaillon Bay, which was exploited at the beginning of the settlement, was forbidden after 1745. Even if the mangrove swamps still covered leagues or kilometres of coastal areas, they progressively lost their richness and the purity of their population. The main reason was the lack of firewood, used to heat and to cook. The sugar estates needed more and more combustibles to make sugar, so bagasse – cane trash – was used to feed the boilers. So a part of the firewood was supplied by the xeromesophytic forest on the hills, the calcareous formations north of Saint-Marc, such as those at Minguet, at Pipe, on the ridge at Pierrot and in the Tranquillité. Multiplied by the Spaniards, guava, though 'worth nothing', was also used in the making of charcoal. In the estates, where oil and candle-wood were needed, trumpet-wood, *Cecropia peltata* L., was burnt during the night on an iron plate placed in the middle of the sugar-house for light.

The xeromesophytic forest on the plains and on the hills also produced the lignum-vitae sought by the navy of the *ancien régime*. In the middle of the 18th century, the Port-au-Prince district sent hundreds of blocks of that wood to Rochefort to manufacture sheaves.[10] To build ships, galba trunks, white or red fig and sometimes silk cotton-tree or pin-tree, which only lasts 2 or 3 years, were used. If the woods near the harbours were lacking at the end of the century, trees situated on remote heights still remained, because they were more difficult to exploit.

A part of the most precious woods was sent to Europe. Before the Revolution, the largest exports seemed to take place in the middle of the 18th century. At Le Havre and Rouen, the imports of exotic woods from the French West Indies went from 12,000 'livres-poids' in 1730 to 905,000 in 1767 to 643,000 in 1769, before going down steadily to 7000 livres in 1776.[11] The wood then became sparse or more difficult to exploit. The woods sent to France were from various trees, mainly red cedar or mahogany, logwood and guaiacum. In 1791, Bordeaux received from the French colonies 215 mahogany beams, 73,161 pounds of logwood, 53,568 pounds of guaiacum and 4229 pounds of Sainte-Marthe (*Caesalpinia* sp.). 'Quinquina' bark from the heights of Port-au-Prince was also sent to the home country.[12]

From the middle of the 18th century on, coffee plantations invaded the mountains[13] and whole forests were cut down. But the soil gave up quickly. At the end of the 18th century, many coffee plantations on the mountains of Port-au-Prince and Haut-Moustique collapsed after 30 years or so of development. On those same mountains, some 'hattes' – cattle-pens – could be found, as well many indigo plantations and a few cotton plantations on the drier lands. The hills of Saint-Marc were cleared for the same reason but also for the purpose of market gardening, at least in the last part of the century. Coffee trees were also cultivated on the eastern heights. In the Abricot country, at the end of the south peninsula, the cutting of mammee-apples, *Mammea americana* L., began around 1770.[8] Almost all the trees of the mesophytic forest would have disappeared in 20 years or so.

Faced with the progressive disappearance of the primitive forest, statutes and rules were published at the beginning of the 18th century. Every inhabitant 'must leave a quantity equal to one hundred American square feet in woods aimed for building or other uses' or they must:

> plant seeds and take care of the woods on an area equal to 100 square paces and to replant it again whenever it would be cut to meet their needs, otherwise a hundred pounds fine will have to be paid to seed and plant again in order to replace the missing trees.

If they did not want to pay similar fines the inhabitants had to plant or seed a minimum of 100 square paces in various species of dye-wood, brazil, fustic tree, cacao tree, cotton tree, guaiacum, 'bois-marbré', 'bois-de-fer', 'bois-de-chêne-violet', cedar, 'acajou-bâtard' or 'grenadille' in the hedges of their lands.[3] But, at the end of the 18th century, the problem still remained. Other projects suggested reforesting with the bay-rum tree.[14] The king even wished that some tree plantations should be aimed at trading. Those species could be 'acajou, acomat, balata, le courbois (courbaril)' and the 'tamarin', as well as 'amandier-pays' and 'poirier-pays', species which already existed in the island. But cultivating species which would only be worth money in 100 years did not suit the views of various owners who were looking for short-term revenues. To burn it all for an immediate profit was the policy of the planters at the time. However, a real concern for the natural equilibrium and interactions between vegetal and animal species had already been expressed before the revolutionary period.

> The bird species named 'bout-de-tabac' is strongly diminished in all the island, especially in the north part, where almost none can be found; it is believed that the cutting of the woods covering the reefs helps in striking them off by destroying their favourite sheltering grounds.[6]

In these dry areas, where the old xerophytic and xeromesophytic forests stood, many introduced self-propagating species proliferated. These were mainly leguminous, such as the logwood, *Haematoxylon campechianum* L. In the Plaine-du-Nord parish, south of Cap-Haitien, the first logwood was naturalized in Santo-Domingo. The seedling taken in the Campêche bay was brought to the Cap around 1730, to a sugar-house in the Morne Rouge, where lemon-tree hedges grew with difficulty. From then on, in various parts of the settlement, the logwood took over the lemon-tree, which it perfectly replaced as a means of enclosure.[3] Additionally, the wood of this evergreen spinate tree of rapid growth could be used for dyeing. It was only a small part of the trade with Europe. But for use in hedges the tree had to be pruned three or four times a year, otherwise the branches grew too high and the seeds poisoned the cultivated lands around. The problem with these hedges was that nothing could grow around them and frequent care was necessary to prevent the logwood from invading the nearby grounds. Once it settled, it was very hard to destroy, as it grew easily and quickly. It poisoned many plains, because, trimmed like the hedges of yoke-elm trees, it lined many paths, as in the flat part of the Croix-des-Bouquets or the Cul-de-Sac plain.[3] In the 18th century, the logwood was mostly exploited in the southern peninsula, in the Acquin and Saint-Louis areas, where the wood became scarce along the coasts as in Leogane. Large logwood also developed in incredible quantities on the coast between Miragoâne and Grande-Rivière-de-Nippes and the Petit-Trou parish.[3] Progressively, the logwood invaded the surviving dry forests.

Guaiacum, logwood, acacias and tamarinds formed a mixed forest of introduced and native species. But another close species, the bayahonde, *Prosopis juliflora* (Sw.) DC., from South America, appeared particularly in the plain separating Saint-Marc from Artibonite. In 30 years or so, the *Mimosaceae*, very common in savannahs or in hattes, multiplied profusely in dry areas, at the end of the 18th century.[15] A new xerophytic facies appeared; *Leguminosae* and *Cactaceae* multiplied. The disappearance of the ground cover profoundly modified the environmental factors. During May 1754, the effect of the drought was such that a fire started in the woods and the 'savannahs' and persisted for months. The districts of Jacmel and Acquin in the south of the island and that of Grande-Anse in the west were among those which suffered the most.[10] This catastrophe was repeated several times during the following decades. In 1786, after many months of an 'extraordinary and ravaging drought', fire continued for more than a month in the mountains and hills around Port-au-Prince. Since the beginning of European settlement, the island only suffered periodically from drought. The conversions of forests increased both the evaporation of the hydric soil reserve and the solar radiation on the cleared land. The intensive grubbing up associated with the length of important dry periods could explain aridization.

At the end of the 18th century, the rain forest still covered the tops of the hills. On the slopes of the hills, forest areas alternated with cultivated areas. Except for the secondary woodland made of introduced species, forests started being situated far from the coast. A few species were protected in the 'hattes' on small stations. The sugar estates stretched over the plains, while the cultivation of coffee trees developed on the slope of the high hills, mainly in the areas occupied by mesophytic forests. In the 19th century, the future of the population would be in higher uncleared areas: the rain forest. Where the soil was deep enough, the rain forest covered the slopes on the northern hills, the ridges of the central massifs and, in the south peninsula, a belt situated between 1000 and 1800 m. That environment was difficult to penetrate.

4 Deforestation of the summits and trade of the woods in the 19th and 20th centuries

After the revolutionary period, slavery disappeared in Santo-Domingo. Profound mutations transformed the economic system, thus affecting agricultural and forest activities. The properties confiscated from the French settlers were redistributed. The rural code tried to maintain labour in the important estates but could not stop their decline. Instead, small properties where coffee and food crops were produced developed on the mountains. Although the mountains where the coffee plant liked to grow were as rich as the plains where the sugar cane was cultivated, 'high lands and hilltops, where it was exclusively developed until now, were promptly degraded and stripped from their vegetal cover by excessive rains falling within the tropics'. A coffee estate could not be maintained for more than 20 years or so. The only way to maintain one's income was to abandon the ageing estates and to harvest the virgin forests.[16] In the 19th century, the coffee plant was not cultivated 'à découvert' – on an open area – any more and it was not pruned. Left to themselves the seedlings reproduced at random where the seeds fell on the ground. The landscape turned into a coffee-tree maquis, which progressively deteriorated. The new peasants, who had escaped the former plantations in the plains, cleared the lower part of the surviving native forest until 1840–1850. In the face of the lack of new lands, the peasants settled on the highlands in the following decades, first on the sunny slopes and then on the north-facing ones. In 1880, the forest ecosystem was destroyed by 10–20 ha coffee and food-crop properties, where coffee trees and food crops were cultivated in long rotations. It was then the limit of cultivable lands.

In 1825, France recognized Haiti, providing that the French government would be financially compensated. In the 1880s, while the peasants cultivated coffee, the Haitian forests were exploited for their precious woods and their dyeing wood as well. Haiti exported logwood and mahogany or West Indian cedar. After the coffee-planters cleared the forest, exports of woods for cabinet-work quickly devastated all the mountainous districts that the coffee plantations had not already largely damaged. In 1904, Saint-Marc harbour exported 15,149,000 pounds of logwood, 11,762,000 of roots transported by rafts along the river and 2,279,000 pounds of guaiacum. The logwood was introduced 150 years ago. It developed in plains and multiplied in thick clumps in the Artibonite grassland. Even if the red tincture had lost its value long ago, those woods enabled Haiti to reimburse its debts. The steep sides of the gullies, which were protected by a perennial vegetation of bushes and shrubs, had been unexploited because they were difficult to access. But in the 1930s they were being cultivated. The persisting forest maquis disappeared with the cutting down of the arborescent elements. The clearing gave way to vast 'savannahs' where cattle could graze. But, as the grazing areas diminished, the cattle were replaced by goats, which aggravated the erosion effects. The lack of fossil energy forced the peasants to use charcoal as the sole domestic fuel.

Secondary forests were cleared and the exposed soil ran to the sea with the first rains. At first the woody and then the agricultural landscape was transformed into a semi-desert plain, because of the degradation of the hydrographical network and the progressive salinization of the lands. The deforestation and bare soil at high altitudes produced innumerable 'ravines'. This gully erosion swept away all the vegetation in a wide river bed. Long scars cut the sides of the mountains, calcareous portions fell, rocks piled at the bottom of the gullies, blocking them. Water could run inside the alluvial cones, producing springs or swamps spreading out the plains. During torrential rains, it destroyed everything in its way, giving birth to real rivers of mud.

The removal of a forest turned xerophytic increased dryness. The only remaining forest cover consisted mainly of introduced species, where coffee plants made up the extremely thick undergrowth. If many species disappeared, others, like the sucrin, *Inga vera* (Sw.) Willd., a shade tree, were very common. The quickly growing mambe, *Colubrina arborescens* (Mill.) Sarg., the avocado-pear, *Persea americana* Mill., the guava, *Psidium guajava* L., the palm *Roystonea regia* (HBK.) Cook., and at times the mango, *Mangifera indica* L., were used for shading and as ornamentals. The extreme clearing favoured the growth of heliophilous species, which were more xeric, like the cabrita, *Bunchosia glandulosa* (Cav.) L.C.Rich., the blackthorn, *Pisonia aculeata* L., and the leaf-of-life, *Bryophyllum pinnatum* (Lam.) Oken. Conveyed by humans, their seeds developed on turned skeletal soils, which prevented the native species from regenerating, despite temperature conditions and rain rates still favourable today. The surviving forests were mostly situated in areas difficult to access and subjected to moist winds. Plant families which used to be represented in large numbers still subsist. The dominance of *Sapotaceae* could be explained by their latex production, which protected them from cutting. The *Annonaceae*, *Anacardiaceae*, *Sapindaceae* and *Moraceae* have many edible species and seeds which enable their proliferation without humans intervening or destroying them. Some isolated timbers, like the silk-cotton tree or trees belonging to the *Areceae* family, were protected because they were considered as 'street-altar trees', or cult objects. As for the 'bois d'orme', *Guazuma ulmifolia*, its leaves were used for washing because of their emulsifying agent. All these facts explain why these families still exist today in that environment.

In arid areas, intensive clearing can lead to a tree herbaceous layer where *Acacia* spp. and *Cactaceae* develop. The evolution towards spinate bush or spinate shrub 'savannah' in coastal zones was favoured by the introduction of the bayahonde, *Prosopis juliflora*, in the 18th century. Because the ecological conditions were favourable, the newly introduced species strongly competed with the native plants and quickly occupied the biotopes that suited them. These biotopes were free because the native vegetation had been destroyed. Without the introduced species, the native vegetation would have conquered the environment again. The adaptability of these foreigners was often so important that they formed almost pure stands and determined vegetal landscapes very distinct from the indigenous vegetation.

Where the soil had disappeared because of erosion, species remain despite their 'raques' facies, that is to say, a vegetation which was continuously cut but not rooted out. However, different canopy strata appear with some balance, but, apart from a few exceptions, those trees do not have their normal height. Many decades would be necessary to restore those pre-existing plant communities. If cutting trees is forbidden for 10 years, growth and the beginning of vertical stratification can be observed. The term forest only corresponds to the existence of the vegetal species constituting the formation. The superior stratum starts at 5 m, while the trees which make it up should rise over 15 m. The middle stratum is situated between 3 and 4 m and the shrubs are only 1 m high. The herbaceous stratum is very developed and often comes from the neighbouring cultivated or barren lands. It progressively invades opened spaces.

5 The proliferation of softwoods

When the European settlement began, pines were covering a few areas at low altitude, maybe even along the sea. They would have been used to make masts of ships. But, in the 18th century, *Pinus occidentalis* Sw. occupied mainly the middle-height mountains[17] and the massifs of the northern peninsula (Port-de-Paix and Port-à-Piment heights) and of the north-east on volcanic or metamorphic soils. They were not pure formations, but the pine contributed to the floristic composition of the mesophytic forest. In the centre, the calcareous ridges of Montagnes Noires were generally covered with thick woods, but the tall and perfect trees were confined to small areas. In the north-west of Haiti, the main forest was situated between 600 and 1200 m on the summits of the Massif de Vallières. However, some drier mountains with a calcareous soil – such as those situated in the north-west of the Etang-Saumâtre – were particularly rich in pines. On the south massifs, the thick forests stretched out under 1600 m to the summits, on sandstone or basaltic soil. Towards the east, at Fond-Verrette, they made up a tropical forest between 1000 and 1400 m.

From the 19th century on, *Pinus occidentalis* multiplied when the broad-leaved trees disappeared and the soil was poor. On isolated plots of lands, it developed in woodlands or even in 'savannahs'. The thick groups on the ridges were cut across by xerophytic bushes. The forests, like those of Chaîne des Matheux, confined at an altitude of 1000 m or higher above sea level, were cut on account of coffee cultivation at the end of the 18th century.[3] The population became used to setting fire to the foot of the trees to cut them more easily and then they would divide the wood into tiny pieces used to make 'bois-chandelle' for cooking and lighting. Progressively, the drier and lower areas where the soil was poor were colonized by scattered pines. They settled sparsely, probably owing to anemophilous seeds from the altitude forest but also to human action.[18] By the mid-20th century, at Santo-Domingo, the pine was the most common arborescent species and, in some places, the only constituting actual forests. Today some residual forests still survive but they are more and more attacked by the development of migrant cultivation, in areas where the vegetation is evolving towards a savannah facies.

Since the Tertiary period, the presence and persistence of *Pinus occidentalis* on high mountains can be explained by a climatic factor. At very low temperatures, the tree cannot compete with broad-leaved tropical trees, and so the vegetation at about 1800 m drastically changes. In many low areas, *Pinus occidentalis* cannot compete with the calcifugous *Pinus caribaea* Morelet, which did not exist at first in Espaniola (it was introduced as a reforestation tree). Today, mixed pine–hardwood woodlands are common. It has taken over the degraded native broad-leaved forest invaded by pines. It colonized poor, sterile soils which became even poorer.

6 Conclusion

With European settlement, a mesophilous facies has evolved, 'enriched by' many introduced fruit-trees. Because of the clearing, a large part, even the whole, of that formation disappeared progressively. When the plantations were abandoned, the pre-introduced species reclaimed the area, but the open environment was less moist and a xeromesophilous series, largely deciduous, appeared. The slow-growing tree species were replaced by a more colonizing shrub stratum. Repeated burnings to clean the land have destroyed the actual potential forest. A regressing shrub stratum took over the soil. Today, wide 'savannahs' occupy a large part of Haiti. The evolution favours a vegetation richer in spinescent species. This degraded shrub vegetation made of local and foreign species, such as the logwood, has disappeared too and only subsists on the coast as a *Prosopis juliflora* open-spinate plant formation. Shrubs with a branch diameter greater than 2.5 cm are still exploited to make posts and to produce charcoal. This vegetation is a spinate maquis, where *Cactaceae* and herbaceous species tower over a low thicket 1–3 m high. The shrub stratum covers almost 30% of the surface, the remaining 70% being occupied by herbaceous species and *Poaceae*. It is dominated by a few trees, which only survive because of voodoo. The current xeric vegetation spreads on high slopes, where the South American species *Prosopis juliflora* proliferates. It was made up of real 'forests' alternating with cultivated fields before the settlement of the 'savannah' facies. Remnants of a more or less native vegetation survive on slopes which are difficult to reach or in small basins where the soil is not eroded.

Today, erosion is caused by deforestation and overgrazing. In littoral areas, these practices provoke salinization in layers, due to intense evaporation, and sterilize the soil. The haline and hyperxerophytic vegetation cannot survive. The vegetal cover is scattered and not one broad-leaved tree survives, as in the Grand-Piquant plain between Cabaret and Port-à-l'Ecu in the north-west peninsula.

The anthropization is marked by the proliferation of fruit-trees introduced from the hedges of neighbouring houses. When abandoned, a shrub association reappeared made of species from the secondary and sometimes climactic forests. The progressive disappearance of the forest can be explained partly by the cutting which took place during the previous centuries and partly by the diminished humidity due to the increase of cleared land. Like the evergreen mesophytic forest, the rain forest disappeared when the clearing, which enabled the settlement of coffee estates at the tops of the high hills, took place. Those practices still continued in the massifs in the south of the islands. In some valleys in the Montagnes de la Selle, all the lands are cultivated, even the river beds during the dry season.[19]

Notes

1 Hilliard d'Auberteuil, M.-R. (1776–1777) *Considérations sur la Colonie Française de Saint-Domingue*, 2 vols. Grangé, Paris, I: 327 pp., II: 368 pp.

2 Heers, J. (1981) *Christophe Colomb*. Hachette, Paris, 666 pp.

3 Moreau de Saint-Mery, M.L.E. (1797) *Description Topographique, Physique, Civile, Politique et Historique de la Partie Française de l'Isle de Saint-Domingue*, Paris, Maurel, B. and republished by Taillemite, E. (1984) Société Française d'Histoire d'Outre-Mer, 3 vols, 1565 pp., 1 map.

4 Chaunu, H. and Chaunu, P. (1956) *Seville et l'Atlantique (1504–1650)*, SEVPEN, Paris, 1097 pp.

5 de Acosta, J. (1598) *Histoire Naturelle et Moralle des Indes tant Orientalles qu'Occidentalles*, Paris, republished by Orry, M. trans. R. Régnault, 405 pp.

6 The Limbé district was cleared around 1712 and 1714 (Arch. Nat., Colonies, F/3/132).

7 Pouppé-Desportes, J.-B. (1770) *Traité ou Abrégé des Plantes Usuelles de Saint-Domingue*, vol. III. Lejay, Paris, 453 pp.

8 Nicholson, R.P. (1776) *Essai sur l'Histoire Naturelle de l'Isle de Saint-Domingue*. Gobreau, Paris, 375 pp., 10 plates.

9 Labat, J. (1722, 1742) *Nouveau Voyage aux Isles de l'Amérique, Contenant l'Histoire Naturelle de ces Pays, l'Origine, les Moeurs, la Religion et le Gouvernement des Habitans Anciens et Modernes, les Guerres et les Évènements Singuliers qui y sont Arrivez . . . le Commerce et les Manufactures qui y Sont Établies*, 6 vols, 1st edn. 4 vols, 2nd edn. G. Cavelier, Paris; J.-B. Delespine, Paris.

10 Arch. Nat., Colonies, C/9A/95, De La Lalanne au Ministre, Port-au-Prince, 8 May 1754.

11 Dardel, P. (1963) *Navires et Marchandises dans les Ports de Rouen et du Havre au XVIIIème Siècle*, SEVPEN, Paris, 787 pp.
12 Arch. Nat., Colonies, C/9A/163, Port-au-Prince, juillet 1789, envoi de quinquina de Saint-Domingue (Lettre de Marbois à M. de Vaivre reçue le 20 Septembre 1789).
13 Pluchon, P. (1997) L'économie d'habitation à Saint-Domingue. *Revue d'Histoire Maritime*, Paris, 1, 197–241.
14 Malouet, V.-P. (1801–1802) *Collection de mémoires sur les colonies, et particulièrement sur Saint-Domingue*, Vol. IV. Baudoin, 378 pp.
15 Descourtilz, M.-E. (1821–1829) *Flore Médicinale des Antilles*, 8 vols. Casimir, Paris.
16 Arch. Nat., Colonies, F/3/152 f°15, Histoire du café.
17 de Humboldt, A. (1826) *Essai Politique sur l'Île de Cuba*. Gide fils, Paris, republished by Erasme, Paris-Nanterre (1989), 211 pp., 1 map.
18 Robart, G. (1984) Végétation de la République d'Haiti. Thèse Doctorat es Sciences, Université scientifique et médicale de Grenoble, 2 vols.
19 Hatzenberger, F. (2000) *Paysages et Végétations des Antilles*. Karthala, Paris, 350 pp.

Forests and Forestry Culture in Tuscany in the 18th and 19th Centuries

5

Paolo Nanni

Rivista di Storia dell'Agricoltura, Accademia dei Georgofili, Logge degli Uffizi Corti, 50122 Florence, Italy

Extended deforestations occurred all over Europe at the end of the 18th century. The whole Italian peninsula experienced this phenomenon. In this context, Tuscany brought its original contribution, characterized by a double reform. The first was introduced at the end of the 18th century by Grand Duke Pietro Leopoldo. The other was promoted by the Accademia dei Georgofili and addressed the problems related to the management of the forestal surfaces (deforestations and afforestations), promoting a modern 'forestal culture', until the creation in 1869 of Vallombrosa Forestal Institute. The availability of documents of different kinds (printed periodicals, *catasto* and censuses) stored in Tuscany archives and libraries allows us to reconstruct, with a good level of reliability, the forestal surface, the property distribution and the people employed in the forests in the 19th century.

1 Introduction

In recent years, forestry history has attracted a growing interest from scholars belonging to different fields of study. The sources and the specific methods of research offer the possibility of an integration of these many different points of view. This synergy is indispensable in order to get a deeper knowledge of the different dimensions of the wood (typology, dynamics) and of the relation of humans with the forest in its double aspect as both a resource (ownership,

(eds M. Agnoletti and S. Anderson)

control, management of the forestry resources and forestry economy) and a cultural and social value.[1]

The pages that follow are intended firstly to give brief hints about the changes that affected the forestry areas of Tuscany in the 18th and 19th centuries and then to specify the fundamental lines in the formation of a modern forestry culture. Though these issues have already been the object of various studies, we do not deem it devoid of interest to attempt to outline a synthetic picture of the Tuscan forests starting from the documentary sources.

We shall focus our attention on two elements: the first concerns the importance of integrating the various documentary sources that permit the reconstruction of the state of the forestry areas in relation to the agrarian ones and the jobs and employees related to the management of the forestry resources, with the analysis of the registers of landed properties and of the census taken in Tuscany during the 19th century; the second is the specification of the original contribution given by the Georgofili to the creation of the Italian forestry school, which found in the Forestry Institute of Vallombrosa, established in 1869, a first expression of great importance. The role played by the Institute of Vallombrosa in relation to the principal European schools, the French and the German, was of great importance in order to outline a methodology and a forestry technique suitable for the forestry reality of Italy. Generoso Patrone himself, on the occasion of the 100th anniversary of the school, gave prominence to the elements that gave Italian forestry very particular conditions: the density of population, the medium to small dimension of the forestry areas, the prevalence of private property and the presence of civic uses, the strict relations between forestry and stock raising, the location in mountain areas and on poor lands of the majority of the Italian woods, the important protective (apart from the productive) function of the forests and the variety of species. The teaching at Vallombrosa was centred much more 'on some aspects of the forestry management such as the forestry statics, the ground arrangement, the forestry assessment and the hydraulics arrangements of the forests than on what is properly called silviculture'.[2]

The echo of the academic discussions and of the Tuscan experience in its complexity reached the parliamentary discussion of 1877, after the unification of Italy, when both in the Chamber and in the Senate the Tuscan liberal model was contested in favour of the approbation of a law that extended the forestry regulations all over the lands of the new unitary country. The Georgofili did not stop their interventions and the long debate regarding the forestry law saw the Accademia committed to intense activity. In 1877 the Accademia proposed a series of questions about the bill presented in the Chamber and the Senate. Between May and June of the same year the Georgofili convoked five public conferences on the forestry law, where these subjects were discussed; a voluminous correspondence on these conferences is stored in the historic archives of the Georgofili.[3] The long tradition of debates and battles on 'forestry culture' was being renewed in the new national context and its new perspectives.

Before going into the treatment of the proposed subject, we need to hint at some important phenomena regarding the relation between humans and forests during the period that we are treating, phenomena that concerned the whole European continent, from which Italy was not excluded. The whole of Europe, starting from the end of the 18th century, was affected by large deforestation in a measure and for reasons that were still unknown. Demographic pressure, with the consequent need for land to be put in cultivation, in addition to the use of the forestry resources, were, during the course of the centuries, the main factors that had determined a swing between the agrarian and the forestry areas. At the end of the 18th century and the beginning of the 19th century, the new needs dictated by the rise of industry and the opening of new markets provoked by the liberalization of commerce were the causes of the deforestation of a vast part of the continent's forestry area; this brought quite serious consequences for environmental conditions. In order to face these events, the already consolidated laws regarding forests of every single nation went through many adaptations, on the one hand, to favour an economically profitable management of the forestry resources and, on the other, to find a remedy for the great damage provoked by this large deforestation. These remedies mirrored the peculiar cultural, social and political contexts of every single state, as well as their particular geomorphological characteristics.[4]

Furthermore, the 19th century was the culminating period of a change in the mentality of humans, which had its roots in the far past; this change was recognizable in a progressive detachment of humans from the natural environment, so that the unity of humans with nature, characteristic of the past century, was mediated by a new 'scientific' culture. From the Middle Ages to the modern age, humans have had different attitudes towards the forestry spaces, but, while in the Middle Ages, as Fumagalli rightly observed:

> people of all ranks shared the same resistance to cutting down trees, woods and forests, and to draining swamps in order to replace them with the more profitable farmed lands, during the 18th century man started to face the environment without any impediment of any kind, to the detriment of its natural components; finally, starting from last century, this latest attitude became common, radical and dissolving'.[5]

2 Forests and forestry jobs

Tuscany, situated between the Tyrrhenian Sea and the Apennines in the centre of Italy, can be divided into three zones: the plain (along the coasts and the main rivers), the hills (the largest part of its territory) and the mountains (on the north-east side of the region).

Forests and woods, in the Tuscan agrarian countryside of the 19th century, took in those areas where cereal cultivation, together with the vines and olive trees, could not grow. The forestry areas mainly stretched over the mountains and hillsides, where the chestnut woods were largely distributed; crossing and crowning the hills, they gave the agrarian landscape a typical chromatic aspect, which contrasted with the different tonalities of the green of the olive trees.

The cultivation of chestnuts was the principal resource for the population living on the mountains. Their importance was due to the powdered chestnuts, which were used partly as a food and partly for the markets. Also, chestnut wood was very much used both in agriculture (e.g. poles for the vineyards) and for the construction of casks and vats. The other resource was stock-raising; above all, sheep in winter were led from the pastures on the Apennines to the pastures in Maremma, called 'transhumance'. Pigs too were left grazing in the woods, especially in the oak groves, where they found their food. This cohabitation of woods and pastures, even if diffused over a large area, was not exempt from criticism.

Forestry areas and wastelands covered more than half of the surface of the Grand Duchy of Tuscany.[5] The data coming from the censuses and the statistical studies of famous agronomists in the first part of the 19th century, thoroughly considered, offer the possibility to specify the trends in the distribution of the agrarian and the forestry areas. It is much more difficult to specify the main characteristics and the conditions of the many wooded areas. Under the definitions of woods, waste woods and 'grazing woods', one could find very different realities, and it is not easy, starting from the written sources, to describe the individual forestry species that were really present and the age of every single wood. An exception is represented by the chestnuts woods, which, because of the importance they had for the economy and for the diet of the population living on the mountains, were more studied.

A source offering synthetic data on the agrarian and forestry areas is a document of 1847[6] held in the archives of the Finance Ministry; this document was based on data collected by the Tuscan 'Catasto'[7] of the first half of the 19th century (Table 5.1). In order to get an approximate value concerning the survey of woods, one needs to subtract the areas corresponding to the wastelands from the reported data. It is likely, then, that the forestry area, chestnut woods included, was about 850,000 ha. These data confirm what Del Noce affirmed in his *Trattato Istorico Scientifico ed Economico delle Macchie e Foreste del Granducato Toscano*[8] when he described the deforestation during the first part of the 19th century as the third part of the total forestry area surveyed by him at the half-century (Table 5.2).

A substantial part, resulting in a great increase in comparing the two documents, was that occupied by the forests of chestnuts (from little more than 123,000 ha to more than double that). From the data coming from the above-quoted *Dimostrazione dell'Annua Media Raccolta*, the powdered

Table 5.1. Agrarian and forestry areas (1817–1834).

	Hectares	Percentage
Lands sown with vines, olive trees, fruit trees	422,061.1	19.8
Lands fit to be sown	458,296.4	21.5
Chestnut woods	123,949.5	5.7
Forests, pastures with stables	1,123,876.1	52.8

Source: *Dimostrazione approssimativa dell'annua media raccolta che suole ottenersi in generi frumentari, legumi, biade, e tuberi nelle terre seminative di ciascuna comunità ed in farina di castagne dalle selve esistenti nelle medesime comunità comprese le due parti domenicale e colonica.* Archivio di Stato di Firenze, Ministero delle Finanze, Miscellanea di piante dei Ministeri dell'Interno e Finanze, no. 6, 17 July 1847.

Table 5.2. Forestry land in Tuscany (1842).

	Hectares	Percentage
Beech woods	87,246.5	15.2
Fir woods	22,974.5	4.0
Oak groves, groves of turkey oaks, holm oak woods	78,221.0	13.6
Woods in sharecropping farms	28,717.5	5.0
Chestnut woods	263,380.5	46.0
Pine woods	91,622.5	16.0
Total	572,162.0	100.0

Source: G. Del Noce (1851) *Trattato Istorico Scientifico ed Economico delle Macchie e Foreste del Granducato Toscano,* Florence.

chestnuts amounted to 1,933,201.91 bushels, as compared with a harvest of 19,783,287.58 bushels of different species of 'grains of wheat' coming from the land put to cultivation. The chestnut woods were present in almost two-thirds of the 248 communities of the Granducato. These data confirm the importance of the cultivation of chestnuts in the agrarian economy of Tuscany.

The other phenomenon happening in the course of the 19th century was the increase of the lands put to cultivation to the detriment of the wastelands and in part also of the wooded areas. These data are confirmed by the documents of the Agrarian Inquiry, put forward by C.M. Mazzini for the provinces of Tuscany (Table 5.3), and in the 'catasti' of the 20th century that followed.[9] During this time the forestry area was lowered to half of the total land surface. From a document attached to the Parliamentary Acts of 1877, we can see a comparative view of the wooded areas in the regions of Italy. From such a prospect, Tuscany emerged as the region with the largest wooded area (594.852 ha) and also in comparison with its land surface (26.7%), after Sardinia (43.11%) and Liguria (29.77%).

Table 5.3. Distribution of cultivation (1880).

Province	Area (ha)	Towns, roads, water (ha)	Cultivated area (ha)	Wooded area (ha)	Wasteland (ha)
Florence	587,174	27,074	275,200	156,200	128,700
Arezzo	329,746	14,546	161,400	79,500	74,300
Siena	379,448	13,848	191,100	109,300	65,200
Pisa	309,063	14,263	196,100	61,800	36,900
Livorno	31,658	1,758	16,900	6,900	6,100
Lucca	143,042	6,142	73,400	54,300	9,200
Grosseto*	456,037	15,526	109,003	149,173	176,996
Total	2,236,168	93,157	1,023,133	617,173	497,396

Source: C.M. Mazzini (1882) *La Toscana Agricola,* Rome.
*The data concerning Grosseto come from Jacini's Inquiry.

These data describe a process set into action during the 19th century, consisting of tillage for the cultivation of the wastelands, deforestation for the new requests by the market for wood and, finally, the inner transformation of the forestry areas, characterized, for example, by an increase in chestnut woods.

Forestry property was characterized by a net predominance of private property. From the same document attached to the Parliamentary Acts, there can be seen a high percentage of private property in woods (88.27% versus 5.8% state and municipal property), the most consistent in the whole nation. Even today this phenomenon characterizes the distribution of forestry property in Tuscany.[10]

The forests gave work to the majority of the population living in the mountains: full-time and, above all, seasonal forestry workers (carpenters, wood-cutters, sawyers, tanners of beams, bearers of beams and hatchets), craftsmen, wood merchants and coalmen.

The census of 1841 permits us to specify significant data for every single person, family, parish and community. From this source we can obtain information on forestry jobs. Comparing two communities, Pratovecchio and Bagno di Romagna, as samples of the two sides of the forest of the grand duke situated on the Apennines between Tuscany and Romagna (Foresta Casentinese-Romagnola), we can note the following factors: both were covered for more than two-thirds of their surface by woods, but, while at Pratovecchio chestnut woods represented an important presence (20.6%), with a consequent production of powdered chestnuts of the order of 29,481.5 bushels versus 36,398.4 bushels of wheat, at Bagno the chestnut woods represented only 3.4% of the total forestry area. Furthermore, for both of them, the area fit to be 'naked-sown' prevailed over that to be sown with trees, even if this prevalence was more marked on the side of Bagno di Romagna. The larger extension of woods, unbroken prairies and 'naked-sown' lands in the community of Bagno can be explained by the consistent presence of stock-raising.

The data coming from the census obviously 'hide' the seasonal jobs, being concerned instead only with permanent or major jobs. In addition, we must note that the presence of forestry areas interested every farm worked by smallholders or conducted as *mezzadria*; so the woods, above all the chestnut wood, occupied the majority of the activities of the farmers. The operations to be carried on in the forests, both that of the grand duke and that of the monks of Camaldoli, were performed by hired seasonal workers. The census of 1841 presents families in which some members declared that they carried on forestry jobs: 40 families of 676 at Pratovecchio and 21 of 1200 at Bagno. Mostly, these were men's jobs, covering all ages, which were handed down from father to son. The jobs for the women belonging to these families were no different from those of the farm families; they worked full-time at home or in manufacturing (spinners, weavers, etc.). In Poppi, for example, one could find a great number of spinners married to coalmen, pruners, wood-cutters, etc. This confirms the importance of the work of women belonging to the rural families to the manufacturing sector.

3 The 'agrarian reform' from Pietro Leopoldo to the Georgofili

At the end of the 18th century and in the first half of the 19th, Tuscany was the site of a double reform regarding the agrarian–economic field. The first was introduced by Grand Duke Pietro Leopoldo[11] at the end of the 18th century; the second was promoted by the Accademia dei Georgofili[12], where the new principles of a modern rational agriculture found their foundation. In the Accademia dei Georgofili the problems regarding the correct economic and silvicultural management of the forests were discussed also; the Accademia then became the centre promoting the new trends in the field of forestry.

The new grand duke, inspired by physiocratic ideas, dedicated himself with determination to facing the problems that affected Florence and Tuscany. Commercial freedom and the elimination of ecclesiastical privileges were among the first provisions, while special attention was given to the processes that had blocked agriculture, the main resource of Tuscany. The lack of good roads, transportation, canals and drainage of territories, the working conditions of the farmers, the economic fiscal system, which impeded the investment of money in agriculture and took the owners away from their lands and the necessity for agricultural instruction and education to help introduce new techniques in cultivation were only a few of the many problems that had to be dealt with. However, the reforms of the grand duke were strongly criticized; the criticism became even stronger when the new policy of the grand duke turned towards the forests.[13] In spite of the immediate and serious consequences that the first measures of 1768 had caused, after a brief second thought with the law of 1776, permission to cut the forests was finally approved in 1780.[14] In the few years in which Tuscany was part of the French Empire, the forests were put under the rule of an administration to which the owners had to refer before cutting. Once the reign of the grand duke was re-established, his measures were restored in 1815.

The innovative aspects of the grand duke's policies in the forestry sector had to do with various issues. Above all, as already mentioned, permission to cut the forests and the creation of new roads that facilitated transportation gave a new impulse to the exploitation of the forestry resources. Such events naturally had, as a first consequence, that of a profound attack on the forests. Secondly, a progressive tendency in the privatization of agrarian and forestry ownership occurred; such privatization was obtained through transfers and contracts of renting lands of the grand duke, of common ownership and above all of the ecclesiastic institutes and orders. This determined the progressive disappearance of common goods and civil uses, changing an economic and social equilibrium that had existed for centuries.[15]

Besides the legislative measures, the Lorenese grand dukes, being the owners of the forests, were the promoters of some significant actions regarding forestry, according to the indications of the new 'forestry culture'. In particular, one must remember the work of Pietro Municchi, superintendent of the Royal Possessions, and Carlo Siemoni, inspector of the Casentinese-Romagnola forest.

The Leopoldian reforms caused the jolting of a static situation. In this context, the answer of the Tuscan managerial class that formed itself around the Accademia dei Georgofili was to correct and straighten out the choices carried out by the owners, using arguments of an economic and social character and demonstrating how correct management of the forest area would bring about a usefulness in both the public and the private sectors.

Furthermore, the tendency that seems to be confirmed by more than one party is that the transfers and the contracts of land that were made in the 18th and 19th centuries, rather than creating a class of small landowners, due to the lack of capital did nothing but favour the large landowners.

4 The 'forestry culture' in Tuscany before the unification of Italy

In the first decades of the 19th century, at least in the academic field, one could notice the clear awareness of the economic and environmental value of correct forestry management and 'forestry culture', no longer only as a complementary sector of agricultural activity but as an important sector of the correct and complete management of territory in order to optimize the cultivation of barren land.[16] The territory specialists of the era considered the 'art of forestry' as a specific sector of the more vast 'agrarian art'. In any case, this cultivation, the cultivation of forests, did not get the same type of attention as that reserved for other types.[17] Many denounced the lack of 'forestry art'.

As we have already mentioned, the changes in the economy and in marketing changed the management of forestry resources. Some significant events that punctuated an irrational deforestation of Tuscany were quoted in the Accademia dei Georgofili at the end of the 18th and the beginning of the 19th century. Giovanni Bettoni in his letter of 7 March 1824 recorded that, according to some news that he had gathered, in the previous decade the number of oaks that had been cut in Mugello[18] amounted to 20,000 units, all in the correct dimension for the construction of 'vessels and boats'. The price of oak was doubled. Once again, the landowners' lack of thinking in the long term was denounced; in fact, they were much more interested in instant earnings than in carefully managing the forestry resources. The request for potash[19] and the high prices offered for maritime wood were indicated as possible causes of the intense deforestation. Furthermore, the phenomenon provoked significant consequences on the wood market in all of Europe. The availability of construction wood was lacking in the whole continent, including those countries that traditionally were the main suppliers. Again, De Bardi underlined the fact that Tuscany itself spent a lot on the importation of wood from America, India and Russia. Furthermore, because of the lack of charcoal the consumption of wood was increased.

Because of these events, the attempt to put in effect the foundations of new forest techniques and of a rational economic management of the forestry resources was developed.[20] The terms of the debate articulated in Tuscany in the Accademia dei Georgofili were published by authors such as Savi, Meguscher, Siemoni, Lambruschini and De'Ricci in the *Atti dell' Accademia* and in the *Giornale Agrario Toscano*. The necessity of constant afforestation, the improvement of cutting and rotation and the creation of forestry nurseries were only some of the main themes.[21] Many articles had the goal of demonstrating, with instructions from the silvicultural and economic point of view, the convenience of this cultivation. More and more, the importance of the forests as a form of prevention of hydrogeological ruin and as climatic regulators was emphasized. The conflict between public and private interests and the conflict on the need for legislative action regarding the limitation of deforestation were the principal arguments in this debate. In the face of the request to limit the damage provoked by the widespread deforestation through the adoption of strict laws, the Georgofili affirmed that it was necessary for the landowners to understand that correct management of the territory would bring about greater long-term benefits. But the Georgofili's interest did not stop merely at theory. Particular attention was given to achievements in the grand duke's forests, obviously permitted by favourable conditions, such as the dimensions of the forests and the ownership by the grand duke. Pietro Municchi, superintendent of the Royal Possessions, presented his speech to the Georgofili in 1845 exactly on this argument.[22] He also considered that the recent afforestations should not be performed by private owners but rather by the state, by the community and by monastic orders. Then the author illustrated to the public the interventions that had been carried out under his management in four areas. Orazio Ciancio, commenting on the speech, underlined the will of Municchi to found what is now defined as 'forestry culture', emphasizing a profound respect towards the forests, which he considered to be a precise 'civil responsibility'.[23]

The forests that were administered by Municchi amounted to approximately 10,220 ha. He showed the Georgofili the works carried out in four of the largest forests in terms of extension, position and quality: the Casentinese-Romagnola forest of the province of Pratovecchio (3910 ha), the forests of Boscolungo and Abetone (1972 ha), San Rossore (2431 ha), and the maritime zones of Vada and Cecina (153 ha); they covered a total of 8466 ha.

In the forest of Boscolungo and Abetone, from 1835 to 1845, 131 acres had been sown and 56 had been planted with different species of fir trees, which would become in a period of 40 years 235,000 trunks. This forest, having been transferred from the administration of Magona, which used it as a source of charcoal needed by the ironworks, to the administration of the Royal Possessions, was covered with 520,000 fir trees of an average age of 10–50 years.

In the San Rossore forest, the predecessors had already planted 43 acres at Maddalena and San Bartolomeo, leaving a heritage of 19,000 oak, ash and elm approximately 30 years old. There one could also find young pine trunks (570,000), domestic pines (435,000), wild pines (135,000), oak (117,000), helm oak (8,000), elm (45,000), ash (22,000), alder (24,000) and white and black poplar (36,000), in addition to all the others trees already planted on the edge of the Arno and Serchio rivers. In the maritime zones of Cecina and Vada, 102 acres had been planted, mostly with pines, with the expectation of obtaining 7500 plants once they had matured.

It was Carlo Siemoni, a corresponding member of the Georgofili since 1842, who contributed a vast and articulate work on the structure of the Casentinese forest.[24] The large forest of beech and fir trees was in a bad condition, even though the Casentinese wood belonging to Santa Maria del Fiore had been, at the beginning of the 19th century, granted to the Camaldolesi Monks.[25] Some years after the lease contract was drawn up in 1818, the situation of the forest had not improved at all and the grand duke decided to call in two experts, Antonio Seeland and Carlo Siemoni, so that they could carry out an inspection. The contract was revised in a short time and the forests of the Opera del Duomo passed to the ownership of the Royal Possessions; in 1839 the project of Siemoni was started.

Municchi, administrator of the Royal Possessions, said that the cost of opening roads in order to solve the serious problem of transportation was £124,000. Then he calculated that already, in the first 5 years of carrying out the project, various species of fir trees were sown (common fir, red fir, larch) on

Table 5.4. Industries of the administration of the Royal Casentinese forest.

	Profits (Italian lire)	Expenses for cutting, processing and transportation (Italian lire)
Beams	49,000	27,222
Trees for building	19,000	11,400
Circular-section wooden poles and arches	10,000	7,143
Logs	48,000	30,000
Staves, poles, oars, bars	15,000	12,000
Total	141,000	87,765

Source: F. Mariotti, Intorno alle coltivazioni e industrie introdotte dal 1839 in poi nella Foresta Casentinese, Notizie statistiche. In: AAG, *Letture, Memorie . . .*, b.81, ins.1414, 2 April 1854.

a total of 1112 acres, and 79 acres were planted with a total of 3,600,000 young fir trees. In this situation the work of Siemone becomes evident in the agricultural and forestry fields; for example, various industries and manufactures were introduced in Casentino. The industry of beams and of wood for construction was able, in a few years, not only to satisfy the local need but also to supply exports to Genoa, France, Spain and the East. Filippo Mariotti[26] wrote a speech, which is conserved in the archives of the Accademia, that gives a detailed description of these industries, of the earnings that were brought to the forestry administration, and of the earnings of the forest workers (Table 5.4). To the earnings and expenses of the forestry administration one should add those of the private industries: refining, small woodworks, and transportations, according to Mariotti, earned £62,000.

The other news given by Mariotti concerns the number and the qualifications of the forest workers, a total of about 2000 units, which were divided thus: 850 seasonal workers (from March to October) were appointed to the opening of new roads, the maintenance of those already existing and the shovelling of snow in spring; 12 masons and stone-cutters; 19 grooms; 36 wood-cutters; 50 coalmen; about 120 sawyers; 45 transporters of building wood; about 180 bearers of staves; 64 hoop-makers with 40 lads; and 200 people who, after having paid a small licence fee, worked on their own account picking strawberries, raspberries, mushrooms and medicinal herbs.

Notes

1 For a wide treatment of this subject for the whole European continent, see Cavaciocchi, S. (1996) *L'Uomo e la Foresta. Secoli XIII-XVIII*, Le Monnier, Florence.

2 Patrone, G. (1970) Il contributo tecnico scientifico della scuola forestale italiana dal 1869 al 1924. In: *L'Italia Forestale nel Centenario della Fondazione della Scuola di Vallombrosa*, Accademia Italiana di Scienze Forestali, Florence, p. 13.
3 *Conferenze Intorno alla Legge Forestale (13, 20, 27, 31 maggio, 10 giugno 1877)*, Atti dei Georgofili, Series IV, 1877, pp. 1–350; Carteggio Relativo alla Conferenza intorno alla Legge Forestale Tenuta dall'Accademia. In: Archivio Accademia dei Georgofili (= AAG), *Pareri richiesti all'Accademia*, 99.54, May–September 1877, cc.19. See also Lambruschini, R. (1866) Sopra i danni di una legge forestale, *Atti dei Georgofili, Nuova Serie* 13, 205; Balestrieri, D. (1871) Sopra la cultura dei boschi e sua relazione con la legge forestale, *Atti dei Georgofili*, IV(1), 125.
4 Romualdo Trifone, in his treatise on the history of forestry law, specified the criteria of every single piece of legislation:

> On one side, you found the usual prohibitions or negative measures accompanied by positive obligations dictated, not so much by the law but by local authorities appointed to the forestry services and to the public works; on the other, they tried to defend the status quo of the same lands, in order to better guarantee their consistency without troubling the life conditions of the mountain and, above all, without interrupting the existing agrarian cultures.

In: Trifone, R. (1957) *Storia del Diritto Forestale in Italia*, Accademia Italiana di Scienze Forestali, Florence, p. 155.
5 In Fumagalli, V. (1992) *L'Uomo e l'Ambiente nel Medioevo*, Laterza, Rome, pp. IX–X. The quotation goes on:

> There is now one who writes that the defence of nature and the love from which such defence has its origin result from a sort of remorse, a sense of fault, deriving from the detachment of man from nature, from his leaving that closed relationship with her, that unity typical of the past. So, for one who proposes this theory, in the past man did not 'love' nature almost obliging himself to do that, he could not even think about that love as a duty, man was so bound to the natural world that he had no necessity to realize this fact, much less to love what he considered part of himself; indeed, man in the Middle Ages, and even if in different ways in the following centuries, loved nature although he was sure he was not detached from her, he was not other from her, indeed just for this fact.

6 The document has been studied and published by Pazzagli and has been used here, even if with caution, for the possibility of synthesis that it offers. See Pazzagli, C. (1973) *L'Agricoltura Toscana nella Prima Metà dell'Ottocento. Tecniche di Produzione e Rapporti Mezzadrili*, Olschki, Florence.
7 Register of landed property.
8 Del Noce, G. (1851) *Trattato Istorico Scientifico ed Economico delle Macchie e Foreste del Granducato Toscano*, Florence.
9 Pazzagli, C. (1979) *Per la Storia dell'Agricoltura Toscana nei Secoli XIX e XX. Dal Catasto Particellare Lorenese al Catasto Agrario del 1929*, Einaudi, Turin.
10 The forestry private property amounts to the 83%. The farms with forestry areas (chestnut woods and poplar groves included) are 85,338 and cover 767,857 of the 888,507 ha of Tuscany, with a mean area of 9 ha per farm. See *Il Sistema Foresta-Legno in Toscana*, ETSAF, Florence, 1993.
11 François Fejto writes:

> The work of modernization [started by Francesco di Lorena] was kept on and intensified by the second son of Maria Teresa, Pietro Leopoldo. With the help of the *Accademia dei Georgofili*, founded in 1753, he realized in Tuscany the most perfect example of a reformism imposed from above. His effort clashed with two events: his leaving for Wien where he succeeded to Giuseppe II and the French Revolution.

Fejto, F. (1991) *Requiem per un Impero Defunto*, Mondadori, Milan, pp. 85–86.

12 Born in the second half of the 18th century, the Accademia dei Georgofili distinguished itself in its 'practical attitude'; such an attitude represented its main characteristic, as well as the main distinctive character of the so-called '*moderati*' in the Tuscany of the 19th century. See Ciuffoletti, Z. (1977) *L'Accademia Economico-agraria dei Georgofili*, 'Quaderni storici', Florence, sett., pp. 865–875; Coppini, R.P. and Volpi, A. (1991) *Cosimo Ridolfi. La Formazione e le Intraprese Economiche*. In: *La facoltà di agraria di Pisa. Dall'Istituto agrario di Cosimo Ridolfi ai Nostri Giorni*, Pacini, Pisa, pp. 9–42; Coppini, R.P. (1993) *Il Granducato di Toscana dagli 'Anni Francesi' all'Unità*, Turin; UTET; Conti, F. (1994) *Introduzione*. In: *Carteggio Ridolfi–Viesseux (1821–1838)*, Florence, Le Monnier, pp. 1–32.

13 For a detailed description of the measures undertaken by the grand duke towards the forests, see Pietro Leopoldo D'Asburgo Lorena (1977) *Relazioni sul Governo della Toscana*, Vol. I, Olschki, Florence, pp. 354–355.

14 See Trifone, R. (1986) *Storia del diritto forestale in Italia*, quoted in Sulli, A.Z. and Sulli, M. (1986) La legislazione del settore forestale in Toscana nel secolo XVIII, *Rivista di Storia dell'Agricoltura*, 26(1), 117–153.

15 Rombai, L. and Sorelli, M. (1997) *La Romagna toscana e il Casentino nei tempi granducali. Assetto paesistico-agrario, viabilità e contrabbando*. In: Corradi, G.L. and Graziani, N. (eds) *Il Bosco e lo Schioppo. Vicende di una Terra di Confine tra Romagna e Toscana*, Le Lettere, Florence, pp. 13–106.

16 See Saltini, A. (1989) *Storia delle Scienze Agrarie*, Vol. 2, *I Secoli della Rivoluzione Agraria*, Edagricole, Bologna; Vecchio, B. (1974) *Il Bosco negli Scrittori Italiani del Settecento e dell'Età Napoleonica*, Einaudi, Turin; Sansa, R. (1997) La trattatistica selvicolturale del XIX secolo: indicazioni e polemiche sull'uso ideale del bosco, *Rivista di Storia dell'Agricoltura*, 37(1), 97–144.

17 See Molinelli, G. (1779) *Sopra la necessità dei boschi*. In: AAG, *Letture, Memorie e Discussioni Pubbliche*, Florence, 57.57.

18 An area in the north-east of Tuscany.

19 De Bardi said that, especially in Maremma, in a few years, large forests were destroyed in order to get potash. De Bardi, G. (1825) Dell'utilità che si riceverebbe dal vestire le cime e il dorso degli Appennini non meno che le coste marittime di piante boschive. In: AAG, *Letture, Memorie, Comunicazioni . . .*, 68.727.

20 Among the most significant contributions to the Tuscan forestry: Savi, G. (1811) *Trattato degli Alberi della Toscana*, Florence; Megushcher, F. (1833) *Sul Miglior Sistema per la Cultura dei Boschi in Toscana*, Florence; Siemoni, C. (1864) *Manuale Storico Pratico d'Arte Forestale*, Florence.

21 See Nanni, P. (1997) Disboscamenti e reimpianti: ricorsi storici. Il contributo dell'Accademia dei Georgofili fra Sette e Ottocento, in *Agricoltura e Rimboschimento*, Atti della Giornata di Studio (Latronico, 25 October 1996), Accademia dei Georgofili, Florence.

22 Municchi, P. (1845) Cenni sulla necessità della conservazione dei boschi, sull'opportunità di coltivarne dei nuovi e sui mezzi per giungere a questo duplice scopo, *Continuazione degli Atti dei Georgofili* 23, 100–116.
23 Ciancio, O. (1995) Commento a 'Cenni sulla necessità della conservazione dei boschi, sull'opportunità di coltivarne dei nuovi e sui mezzi per giungere a questo duplice scopo' di Pietro Municchi. In: *Memorie dei Georgofili (1753–1853) rilette oggi*, Accademia dei Georgofili, Florence, p. 602.
24 See Gabbrielli, A. (1978) L'opera rinnovatrice di Carlo Siemoni selvicoltore granducale, *Annali dell'Accademia Italiana di Scienze Forestali*, Florence, pp. 173–194.
25 See Siemoni, M.C. (1975) *Carlo Siemoni (Karl Simon, 1805–1878)*. Una figura da ricordare nella riorganizzazione della foresta dell'Opera di S. Maria del Fiore durante il dominio dei Lorena, *Rivista di Storia dell'Agricoltura* 15(2), 67–78.
26 Mariotti, F. (1854) Intorno alle coltivazioni e industrie introdotte dal 1839 in poi nella Foresta Casentinese. Notizie statistiche. In: AAG, *Letture, Memorie*..., 81.1414.

Earlier Images of the Amazon in the USA 6

Lise Fernanda Sedrez

History Department, Stanford University, Stanford, CA 94305-2024, USA

The text presents selected representations of the Amazon in the American mentality. These representations resulted from the reports of naturalists, adventurers, travellers and businessmen. The text also explores the experience of the former US president Roosevelt in the charting of an unknown river in the Amazon region and the Ford Company's frustrated attempt at producing rubber on a large scale in the Amazon. By analysing primary sources and commentators, the chapter underlines how these different agents projected their own idea of wilderness on the Amazon reality. Thus their experiences and projects for the region were mediated by their own projections.

1 Introduction

The image of the Amazon region in the American mentality has been shaped by naturalists, travellers, wilderness lovers and capitalists. This chapter explores earlier and different representations of the Amazon that had a major impact on the American mentality. These representations do not exclude or complement each other: they rather compose, even in their contradictions, a complex image. Some elements of these representations – such as the universality of the scientific Amazon – are still present in the American mentality; others have been replaced.

I have avoided representations from the literature and singled out personal and historical experiences. However, in these cases, the experience,

(eds M. Agnoletti and S. Anderson)

or the accuracy of its report, is less important than the representation it created.

2 The naturalists

The first tales of the Amazon arrived in the USA in the form of reports by naturalists in the 18th century. Before that, information regarding the Amazon was jealously filtered by the Spanish and Portuguese rulers. For Americans who took the time to forget their own problems with the British Crown and looked south, the Amazon region seemed a mysterious forest in the maps, with incredible legends[1] and little more information.

The first true scientific exploration in the Amazon was launched in 1736 by the French Académie des Sciences, intended to resolve some of Newton's theories regarding the size and shape of the earth. The party included Charles Marie de la Condamine and ten other 'natural philosophers'. La Condamine's journey differed from earlier ones in that it was sponsored by a scientific institution and in principle 'concerned the accumulation of pure knowledge' (Hecht and Cockburn, 1990). Also in the 18th century, the Amazon valley would be visited by the naturalists Alexander von Humboldt and Aimé Bonpland (1799).

It was in the next century that such enterprises really expanded. In the 19th century, a feverish interest in Brazil, or in Brazil's tropical nature, brought to the Amazon several scientific expeditions. A. Reis (1960), a Brazilian historian, calls this century 'the revolutionary century' for the knowledge of Brazil and the Amazon. In Humboldt's footsteps came Karl von Martius and Johann von Spix from Bavaria, Hercule Florence, Jean Louis and Elizabeth Agassiz from France, Langsdorff, Gaetano Osculati, Richard Spruce from England, and then Henry Walter Bates, Wallace, Spix and many others. Joseph Ewan, historian from the Missouri Botanical Garden, lists over 600 naturalists of all nationalities who visited and published on the South American continent. Among those, North Americans were a minority until the 20th century. Ewan identifies as significant 19th-century American naturalists in the Amazon only James Orton (1830–1877), William Henry Edwards (1822–1909) and Frank Michler Chapman (1864–1945) (Ewan, 1989, 1992).

Why were there so few American naturalists in the Amazon forest? Perhaps because, before the end of the 19th century, the wilderness in America was as exciting, mysterious and unexplored in the civilized mind as any other place. Americans were busy exploring, discovering, praising and collecting their own wilderness. By 1850 British noblemen adventurously vacationed in the wilderness of the upper Missouri River; about 40 years later, the wilderness frontier in America was declared dissolved by the historian Frederick Jackson Turner. By the beginning of the 20th century, rich

Americans, such as the former president Theodore Roosevelt, had begun to explore the Amazon rain forest. Between the British noblemen in 1850 and Roosevelt in 1910, 'the United States changed from an exporter to an importer of wild nature' (Nash, 1982).

If not naturalists, Americans were present in Navy-sponsored surveys in the Brazilian forest. In 1849, Lieutenant Lewis Herndon and Midshipman Lardner Gibbon were requested by the US Navy to explore the entire watershed of the Amazon, with regard to navigability and also to the terrain's possibilities in 'the field, the forest, the river or of the mine'. They were also required to bring back any specimens or seeds anticipated to do well on American shores. Herndon's report was published in 1854, arousing great excitement for the description of the riches of the Amazon and the beauty of its women. Samuel Clemens (the American writer Mark Twain) was one of several young men who, after reading Herndon's report, rode south to New Orleans, hoping to find a ship that would take them to Pará, in the mouth of the Amazon. Fortunately for American literature, Twain could not find any ship going to Brazil then and decided to become a river pilot (Hecht and Cockburn, 1990).

Most naturalists were also true heirs of Rousseau. Together with their descriptions of plants and seeds, the naturalists reported that there were plenty of Amazon tribes that 'frolic in their innocent beauty, have the leisure to appreciate life, and are attuned to the deeper verities of human existence. Tropical exuberance honors moral perfection,' but does not ensure civilization (Hecht and Cockburn, 1990).

3 The traveller

The Revd James C. Fletcher and his co-writer Revd D.P. Kidder wrote in 1857 the first book about Brazil that received a wide reception by the American public. In fact, the author wrote in the sixth edition in 1866 that 'the favourable reception which five editions of this work have had in the United States, England, and Brazil, indicates a growing interest in the largest and most stable country of South America' (Fletcher and Kidder, 1866). The book, called *Brazil and the Brazilians*, challenges any easy classification. It includes some useful information on the Portuguese idiom, currency, prices and Brazilian etiquette. It is at once historical narrative, religious analysis, travel tale and tourist guide.

Fletcher's book is especially interesting because of his attempts to make Brazil attractive to his American readers. Brazil's geographical and natural characteristics were his major enticement. After all, there were many similarities between the temperate and young USA and the tropical and young empire of Brazil.[2] Both were, for instance, continentally large and unexplored:

> Brazil has neither been explored nor surveyed, and its full extent cannot be accurately ascertained; but, according to the best calculations made in 1845 for

> the Diccionario Geographico Brasileiro, the Empire contains within its borders 3,004,460 square miles. The United States, according to the latest computations of the Topographical Bureau in Washington, DC, has an area of 3,002,013 square miles.

For the farm-driven Americans, Fletcher has an extra bait: 'It is not', he continues, '[Brazil's] extent which should attract our attention so much as the fact that no portion of the globe is so available for cultivation and for the sustaining of man' (Fletcher and Kidder, 1866). By the time of his sixth edition, Fletcher had visited Brazil six times. Although most of the time he stayed in the capital of the empire, Rio de Janeiro, he also went to the temperate south, to the north along the coast and to the explored part of the Amazon and central Brazil. It was no mean feat, considering the precarious transport of the middle of the 19th century.

In his descriptions, Fletcher paid special attention to forest products, by then an important item in the trade relations between Brazil and USA, together with coffee. It was, however, the tropical Atlantic rain forest that was the major source of such products, because of its easy access from and to the sea. For instance, the rosewood-tree, today an endangered species, was 'cut down, deprived of its branches, and conveyed to market generally by floating it to some seaport-town, whence it is shipped to North America and Europe . . . The United States annually purchase of Brazil eighty thousand dollars' worth of rosewood' (Fletcher and Kidder, 1866). In his chapter on the Amazon region, however, Fletcher focused not on the actual limited trade or relationship with the USA, but on the potential for future trade or relationship. Therefore, Fletcher overcame the prejudice of a common belief that tropical climates were unhealthy for Europeans' descendants. By doing so, he also foresaw a future for the Amazon, a vision that would be shared by many Americans and Brazilians after him:

> Some people who have given much attention to this subject [tropical climate] argue from the nature of the case that the provinces of Pará and Amazonas can never become flourishing rendezvous for Northerners. But, as Brazil differs from all other tropical countries, it may be that the 'howling wilderness'[3] of the Amazon will yet smile with Industry and civilization. This was my conviction when in the valley in 1862.
>
> (Fletcher and Kidder, 1866)

The Amazon valley, its potential and the wonder of its creatures, greatly impressed Fletcher. Nature in the Amazon has always been in the realm of the wonderful, be it annoying, such as vampire bats, or magnificent, such as the Victoria Regia (Fletcher and Kidder, 1866). Notwithstanding the wonderful characteristics, however, for Fletcher this nature and its original inhabitants were doomed to disappear and to be replaced by the Amazon's real destiny: civilization. This process could already be seen as settlers advanced into the territory and entered into conflict with the former inhabitants of the land. In the civilized territory of Goyas, named after a nearly extinct tribe, Fletcher noted that:

> various other tribes still exist within its borders, several of which cherish a deadly hatred to the people who have invaded their domains and disturbed them in their native haunts. Settlements are often laid waste by the hostile incursions of these Indians . . . I do not know that it holds true in Brazil, as in North America, that the bee precedes a few miles the onward march of civilization – advances as the Indian and the wild beast prepare to take their departure – and thus is the pioneer of a better state of things.

The USA, as well as other nations, could play a part in developing the future potential of the Amazon:

> It would be an interesting expedition, and great good would be accomplished, if the Government of Brazil would consent to send out, with England, France, and the United States, a joint scientific commission, to explore thoroughly the whole district of central Brazil, from Bolivia to Bahia, with particular reference to the navigability of the waters, that here interlace, of those vast rivers which irrigate such a wide extent of territory.
>
> (Fletcher and Kidder, 1866)

Although Fletcher was careful enough to underline the need for approval from the Brazilian Government, he was in troubled waters. The control of natural resources was a prerogative about which the Brazilian Government was extremely sensitive. In the 1850s, the letters of the American Lieutenant Maury, who advocated the opening of the Amazon for international navigation, had dangerously affected relations between Brazil and the USA. As Fletcher witnessed, Americans felt the repercussion in Brazil of the so-called Maury letters throughout Brazilian territory. For instance:

> outrages committed upon citizens of the United States in the distant portions of the Empire in 1858 very tardily met with redress from the interior magistrates, whose feelings toward 'Norte Americanos' were embittered by the conclusion arrived at after reading the letters of 'Tenente Maury'.
>
> (Fletcher and Kidder, 1866)

The letters of Lieutenant Matthew Fontaine Maury deserve a special parenthesis for their impact on American public opinion, as well as on the Brazilian officers. Maury was a Navy officer, head of the National Observatory in Washington. His letters were motivated by the reports from two different American government-sponsored expeditions: Lieutenant Thomas J. Page, USN, in 1853 on the Paraná, Paraguay, and the 1851/52 expedition when Lieutenants (USN) Herndon and Gibbon descended the Amazon – one by its Peruvian and the other by its Bolivian tributaries. Maury was Herndon's brother-in-law. According to A. Reis (1960), Maury depicted the Amazon as a generous Eden of natural resources, waiting for 'strong races' to carry on its scientific and economic conquest. Such a paradise could not be closed to humanity by a 'Chinese policy'. Reis, Fletcher and Kidder (1866) and Hecht and Cockburn (1990) 100 years later agreed on the impact that Maury's letters had on the American image of the Amazon: all that great potential for wealth in the Amazon valley and the lack of the right people to develop it.[4] After letters to newspapers and conferences with businessmen in several cities, Maury put all his ideas, advice and suggestions in a book named *The Amazon River and the Atlantic Slopes of South America*, published in 1853 and with a Brazilian edition the same year. The Brazilian government was understandably cautious and suspicious of such American enthusiasm for the Amazon. The reaction was no different from when Portugal or Spain ruled the Amazon states: their resources should be protected from international greed. This nationalist defensive argument would be an integral part of any policy for the Amazon in the future.

The nationalist and somewhat xenophobic argument can be found in Teixeira de Macedo, then Brazilian Minister in Washington, when he warned his superior in Brazil on the American intentions toward the Amazon:

> The Anglo-American is totally convinced that he should regenerate the whole world, give a new form of government to all human society, and rule by his influence all parts of the world, of which he holds today the centre, because his position, which dominates the two big oceans, the Gulf of Mexico and the Antilles Sea . . . In his work, Lieutenant Maury claims and proves that communications between Pará and New York are easier and shorter than between Pará and Rio de Janeiro, and consequently it is easier to rule the regions served by the Amazonas River from Washington than from the capital of the Empire of Brazil.
>
> (Macedo, 1960)

It is easy to infer how such ideas affected the diplomatic relationship between Brazil and the USA. Maury's letters, however, correctly exemplify how Amazon fitted in with the US 'manifest destiny' doctrine: Americans' manifest destiny was to bring development to all the regions that, in their turn, had the manifest destiny of being developed. The non-development of the Amazon was a waste that the Americans had the right and duty to try to reverse. As we can see in Fletcher's conclusions about Brazil, although more polite (and diplomatic!) than Lt Maury, he and the other US Navy officers shared the same opinion: the capabilities of the Amazon valley were immense and navigation should be opened to other nations. Explicitly, Fletcher wrote that 'the Amazon should, with proper restrictions, be opened to the commerce of the world. The grandest valley of our globe would, under such conditions, reap, in the course of the next ten years, incalculable benefits' (Fletcher and Kidder, 1866).

4 The wilderness lover

In 1913, the conservation champion of the Progressive era, the former American president Theodore Roosevelt, had his own experience in the Amazon forest. Invited by the Brazilian government and under the scientific patronage of the American Museum of Natural History, he and his party joined an expedition to chart the course of an unknown river, a tributary of the Amazon River basin. The 2-month expedition, named the Roosevelt–Rondon Scientific Expedition, resulted in a magnificent collection of some 2500 birds, 500 mammals and several scores of reptiles. Many of these were previously unknown to science, and most were new to the museum's collection. Results also included a wound from which Roosevelt never completely recovered, a frustrated hunting experience and an exciting report by Roosevelt in his *Through the Brazilian Wilderness*, published in 1914.

Roosevelt's party included himself, George K. Cherrie, Leo E. Miller and Anthony Fiala (three expert naturalists from the American Museum of Natural History), and Roosevelt's son Kermit Roosevelt. Their travel to the Brazilian wilderness would be initially a pure adventure to collect new specimens for the Museum, and to provide some hunting opportunities for a depressed T. Roosevelt, after the failure of his 1912 presidential bid. Instead, it turned into an exploratory expedition when the Brazilian Government requested Colonel Cândido Mariano da Silva Rondon to escort the former president. Rondon defined a new goal for the expedition: to chart the course of the 629-mile-long river unknown to cartographers, from the River Paraná, in Mato Grosso, to the River Amazon. The river was called Rio da Dúvida (River of Doubt), as all rivers were before their charting, and it was renamed Roosevelt River after the expedition.[5] Roosevelt described Rondon as 'an officer and a gentleman, also a particularly hardy and competent explorer, a good field naturalist and scientific man, a student and a philosopher'. Rondon was also probably the man who best knew the extended Brazilian territory by that time. For 20 years before he met Roosevelt, he had explored the Brazilian interior, recorded its natural history, built telegraph lines and peacefully pacified primitive tribes (Williams, 1986).[6]

The expedition was not like any exciting hunting vacation or a well-organized safari. Roosevelt faced:

> piranhas, jaguars, crocodile and a catalog of nasty insect life. He emerged from the jungle wracked by fever, weakened by severe malnutrition, hollow-eyed from exhaustion and with a tube draining pus from an abscessed leg. Consummate naturalist and compulsive adventurer, T.R. had loved every minute of it.
>
> (Williams, 1986)

Roosevelt may have loved it, but the expedition was different from his expectations. For several reasons, hunting was not an easy activity during the trip. First, nobody knew how long the river was, and how long it would take them to reach any civilized area to get new provisions. Therefore, time was a critical issue and could not be wasted in leisure hunting. Secondly, the expedition was carried out during the rainy season. On the one hand, it provided a better speed down the river; on the other hand, animals did not need to stay close to the river. After almost 1 month on the river, Roosevelt noted:

> So far the game, fish, and fruit had been too scarce to be an element of weight in our food-supply. In an exploring trip like ours, through a difficult and utterly unknown country, especially if densely forested, there is little time to halt, and game cannot be counted on. It is only in lands like our own West 30 years ago, like South Africa in the middle of the last century, like East Africa today that game can be made the chief food-supply.
>
> (Roosevelt, 1914/1955)

Roosevelt's comparisons showed the Amazon as a wilderness different from others. There was no big game, little hunting. The real foes were the insects, the small and far from noble carregadoras ants and borrachudos mosquitoes.

> The mammals were a great contrast to what I had seen in Africa. Africa is the country for great game. There is nothing like that in South America. The animals in South America are of interest to the naturalist more than to the person who is traveling through the country and takes the ordinary layman's point of view.
>
> (Roosevelt, 1914/1955)

Thirdly, the obstacles during the trip – not at all uncommon for similar expeditions – stripped Roosevelt little by little of most of his material assurances from civilization. His clothes (including underwear) were mostly eaten by ants; his books – 'the last two volumes of Gibbon, the plays of Sophocles, Moore's Utopia, Marcus Aurelius, and Epictetus' (Roosevelt, 1914/1955) – were abandoned in the forest when the party had to redistribute the cargo among the remaining canoes.

However, it was exactly this direct and non-mediated experience with wilderness that Roosevelt sought. Theodore Roosevelt was part of the American romantic tradition, which praises the US wilderness, almost as in a funerary eulogy. The frontier was gone and it was mourned by its admirers; at the same time, civilization and progress were considered the biggest achievements of humanity.[7] In the vision of James Fenimore Cooper, the wilderness writer of the prairie and another member of this tradition, 'the elimination of wilderness was tragic, but it was a necessary tragedy; civilization was the greater good' (Nash, 1982). Roosevelt went further. For him, the most valuable of the goods was not civilization, or wilderness, but the virile qualities, the strength of character that the continuous contact with the wilderness – and its conquest – offered to the American people. The greatness of the American civilization was a direct result of the greatness of this frontiersmen's virtu. 'Under the hard conditions of life in the wilderness,' Roosevelt wrote, those who migrated to the New World 'lost all remembrance of Europe' and became new men 'in dress, in costumes, and in mode of life'. He also realized that, by the 1890s, 'the frontier had come to an end; it had vanished'. This alarmed Roosevelt, chiefly because of its anticipated effect on 'national virility and greatness' (Nash, 1982). This virtu was lost when the civilization concluded its task, and its children mourned the loss. The American historian Roderick Nash quotes the 19th century historian Francis Parkman, who claims that:

> [c]ivilization had a destroying as well as a creating power. Among its casualties were the Indian, the buffalo, and the frontiersman, 'a class of men . . . so remarkable both in their virtues and their faults that few will see their extinction without regret.
>
> (Nash, 1982)

Roosevelt shared this mourning and feared the effect of the loss of the frontiersman in the American character. The continuous contact with the wilderness was an important way of keeping alive in the complex American civilization 'the fundamental frontier virtues' (Nash, 1982).

Despite the difference in the game and glory, Roosevelt saw the same virtues in his comrades of the expedition. He praised 'the way the work was done, at the good-will, the endurance, and the bull-like strength of the camaradas, and at the intelligence and the unwearied efforts of their commanders'. How pitiful was the 'ignorance of those who do not realize the energy and the power that are so often possessed by, and that may be so readily developed in, the men of the tropics'. His pity (and disdain) was also extended to those who experienced the 'false wilderness' and therefore did not develop the virile characteristics that only the wild could offer.

> Another subject of perpetual wonder is the attitude of certain men who stay at home, and still more the attitude of certain men who travel under easy conditions, and who belittle the achievements of the real explorers of, the real adventurers in, the great wilderness . . . [T]he work of the genuine explorer and wilderness wanderer is fraught with fatigue, hardship and danger.
>
> (Roosevelt, 1914/1955)

Like Fletcher, Roosevelt foresaw a glorious civilization in the South Amazon: 'Here the soil was fertile; it will be a fine site for a coffee plantation when this region is open to settlement.'[8] In fact, to do otherwise would be a waste, a sin against land-hungry humanity. 'Such a rich and fertile land cannot be permitted to remain idle, to lie as a tenantless wilderness, while there are such teeming swarms of human beings in the overcrowded, overpeopled countries of the Old World.' Technology and progress would turn what were then obstacles into clear benefits for a healthy and wealthy society.

> The very rapids and waterfalls which now make the navigation of the river so difficult and dangerous would drive electric trolleys up and down its whole length and far out on either side, and run mills and factories, and lighten the labor on farms. With the incoming of settlement and with the steady growth of knowledge how to fight and control tropical diseases, fear of danger to health would vanish.
>
> (Roosevelt, 1914/1955)

Roosevelt, however, did not envy those who would live on the tamed land, and it is not their experience that he coveted. 'A land like this is a hard land for the first explorers, and perhaps for their immediate followers, but not for the people who come after them.' It is not the conquered land as much as the conquest of the land that he praised. The virtu belongs to the 'first explorers' (Roosevelt, 1914/1955).

5 The capitalist

Henry Ford never went in person to the Amazon. His presence was, however, almost as real as the Roosevelt river, in the abandoned skeleton of the cities of Fordlandia and Belterra. From 1927 to 1945, the Ford company carried on a social and economic experiment by the margins of the rivers Amazon and Tapajós, unsuccessfully trying to turn the forest into an organized and efficient centre of rubber production.

Rubber gathering is an old activity of the Amazon people and a traditional item in Brazil–US commercial relations. In 1866 Fletcher described the process of production of the rubber by Indians. 'The use of the caotchouc or gum-elastic', Fletcher wrote, 'was learned from the Omaguas – a tribe of Brazilian Indians. These savages used it in the form of bottles and syringes (hence the name syringe-tree).' *Seringueira* is the Portuguese name for the rubber tree, or *Hevea brasiliensis*. By 1800 Belém was exporting rubber shoes to New England, with this trade reaching 450.00 by 1839 (Fletcher and Kidder, 1866).[9]

The Ford adventure began when world rubber prices sky-rocketed in 1922, when the world shortage of rubber was seriously afflicting even the South-east Asian plantations. The US Department of Agriculture carried out research in the Amazon countries, trying to find alternatives to the Asian rubber. Equally, Henry Ford looked for his own solution, which would give him autonomy for his tyre factories and a social laboratory in the middle of the Amazon.

In 1927, Ford bought from the Government of Pará 2.5 million acres, stretched for 75 miles along the east bank of the Tapajós, south of Santarém. In contrast with Roosevelt, Ford did not revel in the process of conquest of the wilderness. His idea of virtu was the ordered, union-free and almost aseptic production facility he created in Brazil. The technological city of Fordlandia was not born in the Amazon: it was transported directly from Dearborn to the Tapajós. It was a modernity cluster in the wilderness, as described by the environmental historian, Warren Dean:

> In December 1928, a company-owned freighter deposited at Fordlândia the components of an entire plantation nucleus, from diesel engines to nails and bolts. The American staff and their Brazilian workers immediately set to work to construct a small city, soon to be the third largest in the Amazon, complete with Hospital, schools, cinema, water supply, electricity, docks, machine shops and warehouses. Visitors never failed to marvel at this superb infrastructure, unmatchable for a thousand miles in any direction. The neatly aligned wooden bungalows, bunkhouses, general stores, and mess halls elicited praise from junketing Brazilian officials and went far to extract Ford from his early political difficulties.
>
> (Dean, 1987)

It was to be an enterprise in the Ford mould. Ford's industries were characterized by tight linkages with suppliers and rigorous control over labour.

Fordlandia was to be a southern extension of the Rouge Plant in Detroit, with its integrated system of production. 'In this perspective, the requirements of success were organization, capital and work disciplines of the advanced industrial world' (Hecht and Cockburn, 1990).

Ford's idea was to start everything from scratch, out of nothing, out of the ahistorical forest; and his failure is due in a great degree to the fact that the forest was none of these. Instead, the forest was located in a political context, and Ford had to deal with jealous and petty local disputes. The political context also included the workforce in the Amazon, with its own history and culture. They did not fit in with Ford's ordered scheme. Ford tried to give the worker an orderly and modern environment: health and dental care, wages twofold higher than in any other place in the Amazon, new and clean houses, schools and even soccer fields. The workers, however, were used to a more personal (even if highly exploiting) relationship with the *patrão* (patron), who was at once bossman, godfather and protector. Capitalist work relations were a bigger burden to them than all the debt servitude that was so common in the Amazon. Riots and grudges against the management became common, some caused by professional provocateurs, others by the sheer cultural gap.[10] Facing these problems, the management in Fordlandia wondered whether it would be possible to import people, as the city itself was imported. The forest was not the problem, they thought, but the forest people, or the lack of forest people available to work on capitalist terms. Immigration was a wise option: 'the Government of Pará and the federal government [were ready] to import Asian labor if it should happen that enough workers could not be found in the Amazon and the Brazilian Northeast'. LaRue, one of the first advocates of the large-scale plantation of rubber in the Amazon, wrote to the Firestone company (which was also considering starting its own operation in Brazil) in 1924:

> I know you will be permitted to import Chinese. A million Chinese in the rubber sections of Brazil would be a godsend to that country. And after Brazil saw what wonderful development came from the Chinese labor, that nation wouldn't ask for expatriation of the coolies.
>
> (Dean, 1987)

Not only did the workers not fit Ford's blueprint, but the forest also had its own environmental requirements. In its natural ecosystem, the *Hevea* (rubber tree) has a biota that lives off the tree. This means that, when planted closely in a line of production typical of industrial exploitation, the rubber tree is highly vulnerable to a number of blights and diseases, as happened in Fordlandia. These diseases do not exist in Asia, where the *Hevea* is an exotic species, and this is the major cause of the success of the Malaysian plantations. Ford also planned to process the lumber from the cleared forest as a second source of funds for Fordlandia, with no better results than in the rubber plantation. To the dismay of Ford's managers, tropical hardwood did not occur in clusters of the same species, making a systematic exploitation of the resource commercially impossible (Galey, 1989).

Ecological problems succeeded each other as in a curse. In Fordlandia, the feared caterpillars that attacked the plantations were temporarily dominated by their natural enemy, the sauva ant. The ants then turned their attention to the trees. White root disease and canker attacked the tree roots.

> Jungle underbrush also hindered tree growth and maintenance, so cover crops were planted to retard the undergrowth. These cover crops soon dried up in the Amazonian dry season and created a fire hazard. In the rainy season the cover crops absorbed water and plant foods that might ordinarily have promoted the growth of the rubber trees. By 1933 Ford officials realized the Detroit engineering approach to rubber planting was hopeless.
>
> (Galey, 1979)

These problems had not been foreseen by Ford staff, who lacked expert knowledge of the region. A factory is a factory anywhere; why would a rubber tree plantation not be the same everywhere? Even better, in the natural area of occurrence of the rubber tree? The high technology of the city contrasted with the total ignorance of the staff on the basics of the business. 'For the first five years of its existence, Fordlândia had not one resident on its staff, or even available as a consultant, with scientific training in tropical agriculture or practical experience in rubber planting' (Dean, 1987).

A radical solution was tried in 1934, when the plant pathologist James Weir was in charge of the plantation: the complete transference of the experiment. 'Modernity works,' he seemed to say, 'we are just in the wrong spot.' On 4 May 1934, 'a 281,500 ha tract of the Fordlândia concession was traded for another of equal size at Belterra. The sawmill and the machine shop began to replicate the warehouses, dock, housing, schools, hospital and the like of Fordlândia' (Dean, 1987). New seedlings were imported from Asia, in the expectation that they would resist the blight better, and then cloned. A new area was cleared, and the problems were the same. 'After thirteen years, an investment of almost $10.5 million, and the planting of 3,650,000 rubber trees, there was still almost nothing ready to tap!' (Dean, 1987). By the 1940s, the Ford company had lost all its interest in Fordlandia, after a brief revival caused by the demand for rubber because of the Second World War. In 1945, Belterra was sold to the Brazilian government for $500,000, less than 5% of the total investment. In 18 years, the plantations had failed to produce any rubber for large-scale commercial use (Galey, 1979).

6 Conclusion

The Amazon experiences of the naturalists, as well as Fletcher, Maury, Roosevelt and Ford, coexist with many others that helped to form the earlier representations of the Amazon in the American mentality. No less important is the idea of the 'green hell' that is associated with the tropical rain forest in general, be it Asian, African or American. The green hell, the dark and betraying jungle, has been better explored in literature, for instance, by Joseph Conrad's *Heart of Darkness* or the Latin-American writers of the first half of the 20th century, who wrote the 'jungle novels' ('la novela de la selva') (Marcone, 1997). However, these experiences point to some elements, sometimes contradictorily, that would characterize the relationship of the North Americans with the Amazon. These elements are, for instance, the 'perception of divine nature as virgin, unmodified by human action, and thus meet for virtuous contemplation'; the role of science in the future of the Amazon and/or the role of the Amazon in the future of science, 'inspired by a view of the Amazon as an enormous, unsullied laboratory for the scientific contemplation and classification of nature'; the plan and/or role of countries other than Brazil in the future of the Amazon; the nationalistic reaction to these plans; and the control of the resources of the Amazon. Each of these representations has also implied a project for the Amazon.

At the same time, these representations have in common the fact that they projected upon the Amazon their own ideas of what the region is or should be. These impositions 'have exacted a heavy price: a refusal to permit the Amazon to tell its own story' (Hecht and Cockburn, 1990). The American mentality thus has representations of the Amazon that largely disregarded the plans and/or conceptions that the Amazonians may have had about themselves. It is a legacy that would last in the American public opinion at least until the age of the international environmental movement in the 1980s – moreover, a legacy that could challenge the legitimacy of this movement in the Amazon.

Notes

1 The name Amazon is due to a colourful report by the expedition of Orellana (the second-in-command of the Spanish conquistador Pizarro), in approximately 1540. Orellana affirmed that his expedition was attacked by fierce warrior women, like the mythological Greek Amazons, while they descended the river.

2 Brazil became independent from Portugal in 1822.

3 The Amazon region, in the mercantilist logic of naming the regions after what they produced, was known as 'Região das Drogas do Sertão' (Region of the Drugs of the Wilderness). Fletcher translated *sertão* as 'wilderness', but this translation should be taken carefully, especially because of the change in the meanings of both words, *sertão* in Brazil and wilderness in America, in the last century. An original meaning for *sertão* is 'desert place', the opposite of civilization, and it is probably the Portuguese word with the closest meaning to wilderness. It could be applied to the Amazon region or to the north-east dry countryside. However, by the turn of the century, *sertão* had become more associated with the specific ecosystems of *caatinga* and *cerrado*. *Sertanejo* is the man of the *sertão*. The man-desert wilderness is not the best translation for *sertão* anymore, but the more accurate 'backlands', used by Warren Dean and Hecht and Cockburn (Dean, 1987; Hecht and Cockburn, 1990). The term implies the populated, uncivilized region.

4 Another of the plans of Lt Maury for the Amazon was to sell southern slaves to the Brazilian planters there:

> The slaves of the South are worth 15 hundred million. Their value is increasing at the rate of 30 or 40 million a year. It is the industrial capital of the South. Did ever a people consent to sink so much industrial capital by emancipation or any other voluntary act?
>
> (Hecht and Cockburn, 1990)

5 A small tributary of this river was named after Kermit Roosevelt.
6 Rondon was also one of the true heroes of Brazilian history – and there is not a single shade of irony in the use of the word 'hero'. His mother was from the tribes of Terena and Bororo; his father had mixed origins, Portuguese and Guará. An ardent disciple of the French philosopher August Comte, Rondon was best known for his activities as protector of the Brazilian Indian peoples. He actually succeeded in reversing the Brazilian Government's informal support for the fights against the Indians. Instead, Rondon persuaded the government to establish the Indian Protective Service, naming the colonel its director. His one iron-clad order to his subordinates – 'Die if you must, but never kill' – became the official motto of the service, and its men lived and literally died by it. He received the Livingstone Award in 1914 from the New York Geography Society. In 1913 the International Council of the Races, meeting in London, considered Rondon's accomplishment an example to be followed for the honour of universal civilization (Ribeiro, 1958/1998).
7 Frederick Jackson Turner proposed the 'frontier thesis', a landmark in American historiography, in 1893. The frontier thesis explained American history as a series of westward waves that formed the American national character. When a report in the 1890 census stated that the western land of opportunity had been filled up and the frontier had come to an end, Turner feared for the future of the USA and its national character (Opie, 1998).
8 Most of the soil in the Amazon is poor and dependent upon the ecosystems. Deforestation and agriculture have almost completely eroded the topsoil of some areas. However, other areas in Mato Grosso, close to Roosevelt's itinerary, are rich. I do not know whether this observation referred to one of these fertile areas or if it was an optimistic mistake by Roosevelt.
9 Rubber was also the Brazilian nationalists' symbol for the world's greedy eyes on the Amazon. In the 19th century Brazil had the monopoly of rubber until a British adventurer secretly (and illegally) took seedlings of *Hevea* to Kew Botanical Garden, in London. These seedlings were the origin of the British rubber tree plantations in Malaysia and the end of the Brazilian monopoly of rubber (Dean, 1987).
10 The American historian Galey describes a riot caused by the cafeteria system, where the workers were required to help themselves. 'I am a worker, not a waiter,' one of the workers reportedly complained, and what followed destroyed the whole cafeteria (Galey, 1979).

References

Ewan, J. (1989) Five American naturalists on the Andes and Amazon: a study in personality. *Archives of Natural History* 18, 1–14.

Ewan, J. (1992) Through the jungle of Amazon: travel narratives of naturalists. *Archives of Natural History* 19, 185–207.

Galey, J. (1979) Industrialist in the wilderness: Henry Ford's Amazon venture. *Journal of Interamerican Studies and World Affairs* 21, 261–289.

Hecht, S. and Cockburn, A. (1990) *The Fate of the Forest: Developers, Destroyers and Defenders of the Amazon.* Harper Perennial, New York.

Marcone, J. (1997) De retorno de lo natural: La serpiente de oro, la 'novela de la selva' y la critica ecológica [Back to nature: the golden snake, the 'novel of the jungle' and the environmental criticism]. Unpublished manuscript, Rutgers University at New Brunswick, New Jersey.

Nash, R. (1982) *Wilderness and the American Mind,* 3rd rev. edn. Yale University Press, New Haven, Connecticut.

Opie, J. (1998) *Nature's Nation: an Environmental History of the United States.* Harcourt Brace, Orlando, Florida.

Reis, A.C.F. (1968) *A Amazônia e a Cobiça International [The Amazon and international greed].* Record, Rio de Janeiro, Brazil.

Ribeiro, D. (1998) Cândido Mariano Rondon. Rondon, o marechal da paz [Rondon, the marshall of peace]. [WWW document]. URL http://www.cr-df.rnp.br/~dia/RONDON1.htm. (Original work published in 1958, *Uirá sai a procura de Deus-Ensaios de etnologia e indigenismo* [Uirá seeks for god–Essays of ethnology and indigenism]. Paz e Terra, Rio de Janeiro, Brazil.

Roosevelt, T. (1955) *Theodore Roosevelt's America: Selections from the Writings of the Oyster Bay Naturalist,* ed. Farida A. Wiley. Devin-Adair, New York. (Original work published in 1914.)

William, J.H. (1986) A bully Brazilian adventure. *Americas* 38, 36–39.

7

The Tree and the Machine: Manufacture, Communities and Institutions in the Woods of Southern Italy

Marco Armiero

Department of History, University of Naples, Via Giuseppe Bonito 29, 80129 Naples, Italy

The tree and the machine: at first glance the two seem to belong to two completely different symbolic worlds. But they have not always been as far apart as we tend to think. We are all well aware of the many contributions the woods have brought to manufacture: the tree makes the machine move, providing energy and raw materials.

However, it is not my intention to reconstruct how much fuel and especially wood was used by southern manufacturers at the beginning of the 19th century; besides, it appears it would not even be possible to carry out this type of research based on the sources available. Rather, I shall try to analyse the relationship between manufacture and forest resources, not in terms of quantity but in terms of quality. In short, my hypothesis is that manufacture did not necessarily destroy the woods, at least as far as production was limited, but it did harm the traditional balance in the way the forest was used, generating a conflict with communities and reshuffling the balance of power and interests between those communities and government agencies whose purpose was to enforce forest legislation.

Manufacture imposed its haste, its methods and interests in its exploitation of the forest; and even then it did not always employ destructive methods, or at least more destructive than others. Of course, it is closely tied to how large the production was. Within the limits of modest manufacturing the forest was able to sustain production, with the loss, of course, of the traditional environmental and social stability. There are even some scholars, such as Jean-François Belhoste,[1] who have stressed the positive impulse that manufacturing and industry gave to silviculture as a science that aided production.

What was missing was the plurality of ways in which the forest had been used: at first it had been a resource used to produce many consumer goods and now it was being gathered only in order to fuel production and increase the manufacturers' profit.

We have to ask in what way, then, the traditional users of the woods – the communities – reacted when the new way of treating the forest took over: how did they react to this fairly industrial attitude toward using forest products in an exclusive way? One of the possible outcomes of the competition between manufacturers and communities was conflict, even if it was not the only possible result, as I shall attempt to demonstrate.

In the events that I have examined, local and state institutions played an important role. These government agencies protected the forests and kept exploitation in check. Their actions certainly had an effect on the economic and social dynamics that fuelled the communities and the manufacturers. The institutions had several options by which they could intervene: they could mediate between the opposing parties, they could defend the dominant social group or they could enforce the laws protecting the territory. But of course they never acted the same in every situation; in fact, their attitude was rather fickle and contradictory.

The object of my research is the region of Abruzzo before the unification of Italy. It is a region that quite clearly represents the intricate web between environment and economy and, because of its orography, was the perfect setting for the creation of a proto-industrial fabric.

1 Discussion

There are two cases of interaction between woods and manufacture which give us insight as to how the forest resources were exploited: the ironworks located in Marsica and the majolica factory in Castelli.

There were three parties involved in the case of the ironworks: (i) a private factory (owned by a company in Marseilles); (ii) the communities around the ironworks; and (iii) the government Administration of Water and Forests.

When the Marsica ironworks, which was located in the district of Avezzano, began to exploit the wood resources of Mount Argatone – and the estimated quantity of charcoal needed daily to feed a perfectly working oven amounted to the enormous mass of 20–40 *cantaia*[2] (about 1780–3560 kg) – the problem did not seem to be so much that the factory was using astronomical amounts of charcoal as the fact that it was trying to impose laws that limited the local population's use of wood.

The felling of trees used by the ironworks was supposed to be done according to specific rules meant to protect the reproducibility of the woods on Mt Argatone. These rules limited the locals' use of the woods. According to the forest administration's calculations, those woods could provide up to 90,000 *cantaia* of charcoal[3] (about 8 Mg). Therefore, the felling rules called for the preservation of 15 saplings for every *moggio* (0.3 ha) cut, a perfect cut and the fencing of the felled area to keep out the locals' pasture animals; it was also forbidden to transport wood and charcoal across areas that had already been cut and were reproducing; and it was in any case obligatory to ask permission to carbonize. The presence of a forester paid by the ironworks sanctioned the manufacturer's control over the territory.[4]

This was not a destructive approach towards the forest resources, nor was it more destructive than the local population's treatment of the forest. It was simply an alternative economic approach to the traditional one. Indeed, the ironworks seemed to be rather conscious of maintaining a sustainable exploitation of the forests of Argatone; this at least is what the director of the ironworks wrote in a letter to the Administration of Bridges and Roads. Naturally he needed to work for the preservation of the forest if he was to have a continuous supply of charcoal for the ironworks; so they followed the rules of forest economy.[5]

On the contrary, there were constant reports of damages done to the woods, which the population from the surrounding towns was accused of. For example, there were the forest fires of 1844[6] and illegal felling reported in 1851.[7] However, the local authorities (*decurioni*) sided with the local population in defence of tradition against the ironworks' new techniques.[8]

There were two parties involved in the case of the majolica production of Castelli, the town of Castelli, which identified itself with majolica manufacture, and the Administration of Water and Forests.

While the ironworks in Marsica had been considered an alien entity in the social and environmental context, the manufacture of majolica was firmly grounded in the local fabric. So in this case the conflict was not between manufacturers and locals but rather between the forest administration and the locals. In fact, the interests of the town administration corresponded perfectly with the interests of the majolica producers: together they fought against the protectionism of the forest administration.[9]

In the majolica industry, wood was needed in order to fire the clay as well as for glazing. Between 1812 and 1824 the Castelli factory consumed 1000–2500 *canne* of wood per year (4692–11,734 m^3) (by the end of the century they were using an estimated 4000 m^3).[10] Generally, the factory was able to obtain the amount of wood it needed.[11] However, there was certainly some conflict between the government institution, the factory owners and the local administration concerning how the supply of wood was to be obtained. In 1830 the mayor and the *decurioni* (local authorities) of Castelli complained of the restrictions imposed on their traditional rights in using the surrounding forest. Those woods had always belonged to the town and the people had always had an unlimited right to fell without charge. But now, the local authorities wrote in a petition sent to the Ministry of the Interior, they were forced by forest agents to pay for their wood and fell trees for firewood only during the felling season. For them, indeed, the tragic and inevitable outcome of the new forest regime would be the closing of the factories and poverty for the entire population.[12]

Free use of the surrounding woods had obviously caused some problems: in 1829 some areas of the forest for sale could not find a buyer because the majolica manufacturers had felled the trees randomly. The District Attorney representative of the Gran Corte dei Conti (the court that regulated finances) expressed his thoughts on the event: he wrote that it was necessary to defend the preservation of the woods as much as to defend the development of manufacturing; therefore, the only thing he could do was delegate this difficult task of mediating between the different demands to the *intendente* (the governor of the province), who was like a 'good father'.

There are two events which go to show how unsuccessful this type of management of the woods was: the landslide of 1845 due to the deforestation around Castelli,[13] and in 1858 the forester's report that the woods were deteriorating.[14]

The case of the ironworks presents the following characteristics: a private manufacturer tries to organize the exploitation of the territory; it does so mainly by dealing directly with the state administration; the communities react badly because they fear firewood will be scarce.

This case makes us re-examine some of our common ideas about the woods, such as the idea that the woods are destroyed because humans use them:[15] on the contrary, the ironworks in question had all the interest in the world in ensuring that the Argatone woods were not depleted quickly. Of course, it was a matter of keeping manufacture production within limits: to go beyond those limits of production and thus beyond the limits of the reproducibility of the resource would mean compromising the resource itself (in Abruzzo the Morino ironworks closed during the middle of the 19th century probably due more to the depletion of charcoal than to the depletion of iron).

The rules that the Marsica ironworks had to follow were established, as I said earlier, by the forest administration in agreement with the owners; so the territory was managed directly by the government and private owners. This led, of course, to antagonism from and conflict with the local communities.

Was the population's fear of running out of fuel well-grounded? From the documents it appears that it was not. However, as A. Brosselin, A. Corvol and F. Vion-Delphin have written, the idea that the resource was running out is not an objective judgement, but rather a subjective feeling – perhaps experienced by a large group of people – that is quite hard to measure.[16] Furthermore, the fear of running out of firewood during the freezing Abruzzo winter in reality hid other interests that were upset by the ironworks and its organization of the territory: in the sections of forest subjected to regular felling the communities of Avezzano would bring their grazing animals, infringing at once both the prohibitions set by the forest administration and the exclusive rights of the factory owners.[17]

In both cases, the heart of the matter appears to be conflict; the management of the territory and its resources took shape within that very conflict of interests that arose. It was not that communities were provident defenders of the woods fighting against tree-devouring capitalists, but neither was it true that the ironworks' exploitation was rational just because it was modern and that they were fighting against the communities' exploitation, which was irrational because it was archaic.

In this sense, the fact that the factory reported infringements – the forest fires and illegal fellings of 1841 and 1851 – should not be interpreted as proof of an irrational use of the woods compared with a rational one. This interpretation is incorrect, first because it is impossible to define one use of the forest as rational and another as irrational. Secondly, those infringements, those crimes, were not proof that the locals misused the forest as much as they were phenomena stemming from conflict; they should be read as an expression of the community's resistance to new technologies and new ways of using the resource that worked for new economic interests. Even though we are all well aware of the distance, not just geographical, that separates colonial India from my native Abruzzo, I believe it is still worthwhile to mention the cases of community resistance studied by R. Guha and M. Gadgil:[18] it is surprising how similar phenomena repeat themselves and at the same time how very similar the rhetoric adopted by the government agencies to stigmatize the resistance was.

However, the case of the Castelli majolica manufacturers is thoroughly different from the case of the ironworks. Here the number of parties involved is greatly reduced – here, in fact, the community and the manufacturers were in agreement; the community's livelihood depended on the factory supplying itself with wood. Hence, the dynamics involved are simplified: communities/manufacturers vs. forest administration.

The problem in this case was that the resource became the focus of only one local economic interest; the former plurality of interests had disappeared. Thus the government agencies found themselves having to represent the superior interests of the resource: they were not able to represent or involve any other party in this social drama.

The local community did not seem to be able to find the right way to take advantage of the forest's resources without compromising its reproduction; nor did it want to accept rules imposed on it from outside – which was quite common, as can be seen in what occurred in France in a case studied by Y. Rinaudo.[19] The hydrogeological hazards, as well as the poor state of the woods, as described in the 1850s, could prove that the following hypothesis is correct:

that local conflict of interest did not necessarily cause the dissolution of the collective resources nor was it just an expression of a crisis; on the contrary, it stemmed from the clash between the different uses of the woods – I would say, a positive effect. This positive effect could occur because, when conflicting interests surfaced, it became absolutely necessary to mediate between parties and to discuss different exploitation strategies. This did not happen elsewhere, for example, in Castelli, where the resource remained prisoner to a unidirectional exploitation, which, especially over a long period of time, eventually depleted it.

2 Conclusion

The plurality of contributions that the woods had brought to the local economy generally meant that there was also a plurality of interests and parties involved in the management and control of the resource. It was a situation that inevitably led to conflict between groups, individuals, classes and communities. The forest was ceasing to be a practically free good, belonging to the people by right: complementary strategies of exploiting the woods were being replaced by single exploitation techniques. These new techniques, solely by affirming themselves, excluded all the other different ways of using the forest. The affirmation of a sole technique and a sole way of making use of the resource impoverished not only the social groups that had come out as the losers in the contest, but also the resource itself.

Alongside all the different players, each with their own economic interest, the government also concerned itself with the woods of Abruzzo; this meant both the offices directly dealing with forest management and the local administration, such as the *decurionati* and *intendenze* (provincial governor's office). The two levels of government administration fulfilled different tasks: sometimes they mediated between the disagreeing parties or between the state and community, and at other times they intervened directly, usually in favour of the most influential party, even if it was not unusual for them to use their authority to defend customary rights. What matters most, however, is the presence itself of the government in the management of the natural resources. Thus, it does not appear that a free rein for economic interests, that is a free market, could have ensured a continuous exploitation of the resource, and thus its survival. Of course, the government's intervention – through its agencies in charge of the protection of the territory and its peripheral administrations – was not impartial and disinterested. The generalized corruption of the forest bureaucrats, the way in which these bureaucrats obtained their position in the first place and the social structure of the kingdom all go to show how much the government was wrapped up in the matter and how the government itself introduced the idea of property and progress. However, despite these structural limitations, precious know-how in management and control of territorial balance was acquired within the French and Bourbon bureaucracy.[20] In fact, using the woods meant dealing with a complex network of relations – water management, stability of slopes, climate and, of course, the floristic and faunal composition of entire tree-covered areas.

Nevertheless, the need to regulate the use of the resource on the government level was only one of the possible ways of managing the resource and certainly not the most commended one. On the contrary, many[21] throughout the 19th century, condemned the government's excessive interference and decried the negligence inherent in community-owned property, confirming their unshakeable faith in the liberalist creed: the well-being of the individual automatically meant the well-being of the collectivity. And, in the end, the legislation of 1826 and that proposed in the 1850s sanctioned the privatization of the forest resources.

Here one must avoid falling into the romantic idea of bad guys and good guys, placing the blame entirely on capitalism and praising the community's evolved environmental knowledge.[22] The problem lies in the transformation of a society and its economy. The community's economy was a moral economy, rooted in a limited traditional vision of obligations and social norms and based on harmony between the population and resources. The community made collective use of its goods and could not be fitted into a privatistic structure.[23] These traditional forms of property and exploitation of resources were completely overturned when the community found itself face to face with a different economy – a market economy.

The affirmation of just one form of ownership – private ownership – and just one economic system – a market economy – inevitably led to the disappearance of nature as a producer of commodities. Furthermore, it prevented a true evaluation of what is called its functional value[24] – that is, the forest's contribution to the livelihood of the communities making use of it, for example, the relationship between woods and pastures.

The resistance of the society and its customs emerges in the countless heterodox ways of using the forest's resources, in the persistence of collective forms of property and usage, in the everyday exception to the rule and in the many conflicts that sprang up around the use of environmental resources, which deserve the same attention in our country that they have received elsewhere.

Notes

Abbreviations: Archivio di Statio di Napoli: ASN; Ministero dell'Interno: MI; Ministero di Agricoltura, Industria e Commercio: MAIC.

1 Belhoste, J.-B. (1990) Une sylviculture pour les forges, XVIe–XIX siècles. In: Woronoff, D. (ed.) *Forges et Forêts*. Ecole des haute Etudes en Sciences Sociales, Paris.

2 ASN, MI (1845) *Commissione Geologica per l'Esplorazione delle Miniere degli Abruzzi, Celano 28.7*, II inv., f. 670 bis.

3 ASN, MAIC (1855) *Verbale Relativo all'Apprezzamento e Valutazione del Legname Esistente nel Perimetro della Superficie Boscosa di 1000 Moggia Napoletani, pari a Moggia Legali 4840 [. . .] di Esclusiva Pertinenza del Comune di Bisegna*, 10.8, f. 616.
4 ASN, Min. Int. (1851) *Manifesto*, Aquila 25.1, II inventario, f. 670 bis.
5 ASN, MAIC (1843) *Al sig. Direttore Generale di Ponti e Strade, Acque, Foreste e Caccia da Charles Aunè*, 21.4, f. 616.
6 ASN, MAIC (1844) *Il Sottointendente*, Avezzano 3.10, f. 616.
7 ASN, MAIC (1851) *Verbale di Verificazione dei Danni Commessi nelle Proprietà Boscose del Comune di Bisegna e Riunito Villaggio di San Sebastiano*, 18.11, f. 616.
8 ASN, MAIC (1843) *Il Segretario Generale al Direttore Generale di Ponti e Strade*, Aquila 22.8, f. 616.
9 See Vion-Delphin, F. (1987) Forêts et cahiers de doléances: l'exemple de la Franche-Comté. In: Woronoff, D. (ed.) *Révolution et espaces forestiers*, Paris.
10 Tino, P. (1989) La montagna meridionale. In: *Storia dell'Agricoltura Italiana*, vol. I, Marsilio, Venice, p. 736 fn.
11 ASN (1812) Amministrazione Forestale, *Corrispondenza coll'Ispezione della Divisione degli Abruzzi*, f. 534.
12 ASN, MI, *Doglianza dei Cittadini di Castelli sulla Inibizione di Poter Legnare nei Boschi del Comune, Nota per lo Consiglio*, II inventario, f. 580.
13 ASN, MI (1845) *Atti del Consiglio Provinciale del I Abruzzo Ulteriore*, a., II inventario, f. 4083.
14

> The main and perhaps only cause of such deterioration was the continuous illegal felling committed by many majolica makers in that town who, needing special material to feed their ovens, chose the best plants over a thirty-year period, especially during the fatal emergencies of 1848 and 1849.
>
> (ASN (1858) Amministrazione Forestale, *Il Guardia Generale all'Amministrazione Generale, Teramo 6.7*, f. 431)

15 Moreno, D. (1990) *Dal Documento al Terreno*, Il Mulino, Bologna, p. 11.
16 Brosselin, A., Corvol, A. and Vion-Delphin, F. (1990) Les Doléances contro l'Industrie. In: *Forges et Forêts*, Ecole des Hautes Etudes en Sciences Sociales, Paris, pp. 13–15.
17 ASN, MAIC (1845) *L'ispettore G. Labollita all'Amministrazione Generale di Ponti e Strade, Acque, Foreste e Caccia*, Aquila 29.9, f. 616.
18 Guha, R. and Gadgil, M. (1989) State forestry and social conflict in British India. *Past and Present*, n. 123.
19 On rejecting exogenous regulation of the exploitation of forest resources, see the French case studied by Y. Rinaudo (1987) La révolution conservatrice: les bois communaux Varois au XIXe siècle. In: Woronoff, D. (ed.) *Révolution et Espaces Forestiers*, Paris.
20 See Bevilacqua, P. (1989) Acque e terre nel Regno di Napoli. In: *L'Ambiente nella Storia dell'Italia*, Fondazione Lelio e Lisli Basso, Marsilio, Rome, Venice, p. 88.
21 'We admire with envy many towns in this province . . . that, having split up their small woods among the citizens, see to the woods' safety.' In: De Nobili, S., Fini, C. and De Vincentiis, C. (1821) Statistica agraria del circondario di Casoli in Apruzzo citeriore. *Annali di Agricoltura Italiana*, 9; 49; and also 'Since most of the woods belong to the D'Avalos, they look after them jealously.' In: Durini, N. (1820) Statistica agronomica

dei circondari di Vasto e Paglieta in Apruzzo citeriore. *Annali di Agricoltura Italiana*, 7, 244; the *intendente*'s secretary stated that he had personally verified the 'thriving state of the divided section' and the deterioration of the part which had not been split up among citizens. In: ASN, MI (1838) *Il Segretario Generale di Intendenza al Ministro*, Chieti 8.12, II inventario, f. 72.

22 Viazzo, P.P. (1990) *Comunità Alpine*, Il Mulino, Bologna, pp. 52–53.

23 Thompson, E.P. (1981) L'economia morale delle classi popolari inglesi del secolo XVIII. In: *Società Patrizia, Cultura Plebea*, Einaudi, Turin.

24 See Bevilacqua, P. (1994) Natura e lavoro. Analisi e riflessioni intorno a un libro. *Meridiana* 20; and Immler, H. (1993) *Economia della Natura*, Donzelli, Rome.

The Muniellos Forests (Asturias, Spain) in History 8

Juan Pablo Torrente

Instituto Feijoo de Estudios del Siglo XVIII, Universidad de Oviedo, Campus del Milán, s/n E-33011 Oviedo, Spain

This chapter presents a local-scale study of the forests that constitute the Muniellos Integral Reserve (Asturias, Spain, 5542 ha), which can be considered as one of the best forest areas in Europe. Its historical evolution as a rural estate is described from several points of view: (i) property and possession – the dissociation between wood and underwood in the 18th and 19th centuries and reunion in the 20th century in the hands of forestry/mining companies; (ii) management – the forests' increase in value in the 18th century, due to the demand for wood for military naval construction, and the absence of forestry or dasocratic works until the 20th century; (iii) setting up of a major forest industry in the 20th century; and (iv) preservation – productivity versus preservation in the 20th century.

1 Introduction

The Muniellos forests form the Reserva Natural Integral de Muniellos[1] (Asturias, Spain). The reserve includes the three adjoining rural estates of Valdebueyes (within the council of Ibias), Muniellos and La Viliella (situated in the council of Cangas del Narcea), which contain forests populated by dense native trees. Until recently, each one of these three forests maintained a separate historical personality, delineated by the division of their respective basins: the rivers Muniellos, Aviouga and Ibias. Both the action of the state and

(eds M. Agnoletti and S. Anderson)

that of the Asturian Regional Government have joined their histories in the last few decades.

2 Notes on method

The study of forest areas must take into account both their natural and their social histories. The flora and vegetation in Muniellos are the topic of a recently published book and a map,[2] which explore in detail the climate, soil and flora of the Muniellos Reserve. However, there is no basic research on a fourth dimension, the historical one, that is, the one that analyses human intervention in history. This would be essential in order to undertake a retrospective study of the vegetation of these famous forests.

The current forest landscape that the Muniellos forests present is the result of the action of several elements. Some have to do with forest dynamics and others are the result of the historical activity of humans. The former lays the foundations of its vegetation's natural evolution; the latter conditions the first in several ways and with different intensities.

The historical approach to these forests can take two related points of view. The first centres on the evolution of its vegetation, that is, on the retrospective study of its vegetal population, stating the stages of progression and regression, the changes in species and their causes, and the transformations of its woodland with the passing of time. The second approach would consider the Muniellos, Valdebueyes and La Viliella forests as rural estates, dedicated to livestock, agricultural or forestry uses in the past.

3 Property and possession

The legal past of the Muniellos Forest is rather obscure (as is that of Valdebueyes and La Viliella, and of many other forests in western Asturias). Historical research on the matter of its ownership moves between two extremes. One theory considers it as part of the entailed estate of the Queipo family (House of Toreno) since its foundation in the 16th century, and not having any other legitimate title than age-old possession. The other is a more likely and plausible theory, which considers the forest as belonging to the Moal village council. Its inhabitants would have enjoyed free common usage and exploitation, although the House of Toreno would have held its ownership by usurpation until the end of the modern age.

The disputes between commoners and 'the mighty' in relation to the ownership of the forest and the *brañas* (wood-pasture) resulted in never-ending lawsuits in the ancient régime. In the specific case of the litigation over Muniellos, it seems clear that, until the middle of the 18th century and coinciding with an increase in the value of the forest because of the exploitation of its wood for naval construction in the arsenal of Ferrol (Galice), the inhabitants of Moal agreed to a judicial transaction in favour of the earl of Toreno.

During the 19th century, the forest's woodland was transferred to several individuals and trading companies. Forest management under each regional industrial (forestry and mining, even railway) project had different outcomes.

Nevertheless, the underwood in Muniellos remained the property of the Queipo family until 1901, when they sold it in instalments, together with their share in the woodland of other forests in the area. The buyers were Juan Sansinenea, a Basque banker, and Luciano Borié, a French landowner. The following year, these two businessmen, together with three other partners, founded the Sociedad General de Explotaciones Forestales y Mineras Bosna Asturiana SA.[3] Muniellos Forest and the above-mentioned shares are Sansinenea's and Borié's contribution to the company, which was obliged now to pay off the debt to the Queipos. The purpose of the company was the management of its forest and mining properties in the council of Cangas del Narcea, and the manufacture of tanning extracts, dyes and other derivatives. When the Bosna company went bankrupt – because of mismanagement – Muniellos was left in the hands of the Herrero family, who had taken part in the company. In 1952 the company Muniellos SA was created, with the purpose of managing the woodland in the Muniellos Forest. In October 1964, ownership of that woodland went into partnership with Patrimonio Forestal del Estado[4] for an intended period of 40 years, after which the partnership would disappear and both wood and underwood would be owned by Muniellos SA. However, this only lasted for a decade, since, at the beginning of 1973, the Instituto Nacional para la Conservación de la Naturaleza (ICONA)[5] bought the forest from the company.

The historical evolution of the ownership of the Valdebueyes Forest follows similar steps to those of Muniellos. This was an indivisible property owned by the stately houses of Ron, Tormaleo and Mon, with no title other than age-old possession. The inhabitants of the village of Valdebueyes brought a lawsuit against them, which was settled in the 18th century.

During the 19th century, the undivided parts which formed Valdebueyes were bequeathed or bought by successive generations. The owners managed the forest either by taking part in trading companies, with the aim of exploiting its woodland, or by allotting it for a determined period of time with the same purpose. In 1969, the Compañía Anónima Española Explotación de Robles del Valledor SA,[6] shortened to Robledor SA, bought the majority of the forest. At the beginning of 1983, ICONA bought it from this company.[7]

The La Viliella Forest is also an indivisible property, probably arising from a former redeemed leasehold. However, the fact that it consists of both the forest and the municipal area of the village of La Viliella may induce one to think of the usurpation of common goods. Its last owners were Santiago Parras

Iglesias (from Plasencia) and Otilio Lucio Vicente Gómez Hernández (from Madrid), who had gradually bought different parts of the forest since the end of the 1960s. ICONA bought it at the beginning of 1980.[8]

4 Management

The resources of Muniellos, Valdebueyes and La Viliella have always been exploited. Some working mines are still present in some parts of the reserve.

Both the cultivation and the management of the forests (silviculture and dasocracy) started during the middle of the 20th century. Before this century, historical sources reveal that human intervention consisted primarily of the mere extraction of timber. They also show a preference for oak and the use of culling as a felling method. Selected trunks were sawn to make pieces for naval construction or to fit the use for which they were destined. To a smaller extent, other secondary uses and exploitations also appear, such as wood-pasture or grazing land. In general, human presence during these years can only be spotted in easy-access areas, that is, it is limited by the existence of penetration routes into its woodland or high pastures.

The 18th century and the first half of the 19th century are characterized, first, by the use of timber for naval construction and, secondly, by the absence of silvicultural or dasocratic works. In the period in which the special jurisdiction of the Navy was in effect (1748–1821), the woods in all coastal regions in Spain were directly linked to naval construction. The criteria to follow when culling trees can be defined as dendromorphic, that is, trees were selected for their shape and were cut and sawn in the forest, their final destination being some of the wooden pieces of warships. At the time, Muniellos and Valdebueyes enjoyed outstanding prominence in the history of the Navy's exploitation of Asturian forests.

From the middle of the 19th century, the timber extracted from Valdebueyes and Muniellos was mainly used for railway sleepers and mine buttresses. Gradually, their woods began to receive the attention of capitalists and industrialists, both at home and from abroad. The local press at the time talks about the 'eager look' of bankers in France, England and Switzerland upon the forest.[9] The regeneration of the forest was mainly left to the action of nature. The management of the forest during this period was merely extractive, being closely related to the dissociation of the ownership of its underwood, owned during these two centuries by the House of Toreno, and its wood, allotted to capitalists and trading companies for long periods of time.

At the beginning of the 20th century this capitalist interest in the forests led to the reunion of wood and underwood in the hands of forestry/mining industries. In this century, an important forest industry settled in Muniellos, with its origin in the Bosna Asturiana. The amount of timber extracted by this company is unknown. The final forest products obtained were railway sleepers, planks and dressed timber to make staves. The Bosna company built two dwellings in Tablizas, one for the manager and the other for the forest-keeper. They also built a mechanized sawmill, activated by hydraulic power by means of a weir canal from the river Muniellos with a length of more than 1 km, achieving a 30 m fall.

After 1952, Muniellos SA managed the forest to a larger extent than Bosna, with plans for an unlimited period of time. They invested in large-scale, stable improvements and infrastructure, building an industrial plant with a reinforced concrete base, a new sawmill, a hydroelectric plant with a canal, an overhead electric line, and a transformer plant with a turbine and an electric generator to activate the factory. The plans were to create a major forest industry in Muniellos. Once the quota of railway sleepers for the Red Nacional de los Ferrocarriles Españoles (RENFE)[10] was achieved, the company stopped making sleepers in order to manufacture dressed timber for the staves that wine companies from southern Spain demanded. They also made parquetry for flooring from saw scraps, and planks and reels to make pieces of board and props for gallery mines, demanded by the coal-bearing area of Cangas del Narcea. Beech and birch were used to make planks, and their larger parts were transformed into board reels. When the exploitation stopped, at the beginning of the 1960s, the sawmill was dismantled and the generating plant was abandoned.

5 Preservation

During the 1960s, several forces converged upon Muniellos: its appeal as a tourist area, its low economic return at the time, the existence of several conservationist claims – with some attention from the media – together with disagreements between the forest's regional and national administrations. All this led to a confrontation between conservationist and productive forces, which was settled in 1973 with the acquisition of the forest by ICONA and the dissolution of the partnership, in order to develop reforestation with fast-growing species.

From the beginning of the 1960s, the forest managers themselves saw the need to transfer the ownership of Muniellos to the state, for public-utility reasons, since it is essentially the protector of a basin. The idea was to prevent a 'weakening of woodland preservation, even if temporary, which could give way to erosive phenomena with irreparable consequences', as a 1962 official report explains. The same report claims that, if it remains as the property of a trading company:

> its evolution will go from bad to worse, and the picture of bare basins, rocky places and flooding will expand, with Muniellos very likely to follow this sad evolution because of its topography, its soil and the noble species that populate it.[11]

Therefore, the economic (private) and social (public) interests confronted each other.

It seems that the forest was not very profitable for the owning company at the time, because very few types of exploitation were allowed and because timber prices had decreased. Towards 1960, Muniellos SA proposed, as a first measure to make the exploitation productive, the intensification of tree felling. Although forest managers admitted the legitimate right of the company to make its business profitable, they were also aware that the geographical features of Muniellos and its protective function had to limit its exploitation and therefore its short- and medium-term profits. The forest demanded more care and improvements than those observed under the trading strategy of Muniellos SA.

In 1962, forest managers offered its purchase as a solution. They also thought of a temporary solution which consisted of a partnership with Patrimonio Forestal del Estado in order to care for the forest's needs – reforestation, preservation and improvement. Internal differences among shareholders excluded both solutions, and they only allowed a 'Project of Technical Aid'[12] from Patrimonio Forestal, which would regulate the management, preservation and reforestation of the forest. Under this project, the Patrimonio's deputy director drew up a joint plan of reforestation in October 1962, which ended up becoming a 'Partnership for Reforestation' in October 1964. This had been common procedure in the area since the beginning of the decade (that is what had happened in the case of other forests, such as Oballo, Larna and San Pedro de las Montañas). In this way, the forest enjoyed some degree of protection of its woodland, even if the 'Terms of the Partnership' stated that:

> taking into account the possible existence of mining areas in their estate and the surveys on prospecting undertaken by Muniellos S.A., Patrimonio Forestal del Estado explicitly allows them, should they see the convenience of it, to open pit heads, exploit quarries and use land as dumps, open trenches or any other type of resort, provided that these works will not damage the reforestation of the area.

Its integrity as a woodland full of leafy native trees was not protected, since the partnership explicitly admits a 'change to fast-growing species' due to the 'urgent need to renew its old and decrepit woodland'. This change in species was neither drastic nor absolute, since it also expressed the purpose of working for the 'preservation, improvement and exploitation of the already existing woodland'.[13] In any case, the content of the terms of the partnership regarding the 5-year felling plans were interpreted in different ways: Muniellos SA and some forest managers supported a complete felling of the forest, followed by its reforestation with pine trees, while other forest managers did not agree with a change in species or the above-mentioned felling method. This is an additional aspect of the history of these forests that has yet to be clarified.

In these decisive years for the forest, new forces emerged, with the regional and national press as their means of expression. Some voices claimed that the estate should be converted into a national park, for two main reasons: first, the beauty of the forest, which had been observed by the Asamblea Provincial de Turismo[14] in 1964, and, secondly, the existence of several conservationist and scientific lobbies that were asking for an integral preservation of the forest. During the middle of the decade, the International Biological Programme and the World Wildlife Fund showed some interest in buying the forest for its preservation for scientific purposes. At the same time, biologists, such as Ivar Ottosson from the Växtbiologiska Institutionen in Upsala and José A. Valverde from the Estación de Doñana, visited Muniellos and worked on its flora and fauna. At the end of the decade, Miguel Ángel García Dory warned about the dangers that threatened the integrity of the forest, and he saw its salvation in the tourist potential of a nationalized Muniellos.[15] In the same way, Félix Rodríguez de la Fuente asked for the preservation of the forest, because of both its natural values and its tourist potential.

In 1964, Franco's government put forth the 9 July 2395/1964 Decree, declaring the 'wonderful virgin Muniellos forest' as a picturesque spot, 'with a surface of fifty square kilometres'. The Cangas del Narcea city council and the owning company were therefore compelled to 'strictly observe the regulations of the Leyes del Tesoro Artístico, Municipal y Ensanche de Poblaciones'.[16] Muniellos was therefore under the protection of the state, and its tutelage was exercised by the Ministerio de Educación Nacional[17] from that moment onwards. Although little is known about the effectiveness of that tutelage, it is at least known that in February 1970 the civil governor transmitted an order coming from the Dirección General de Bellas Artes[18] to the Mayor of Cangas de Narcea, which commanded the stoppage of all felling in the forest and the submission of a project of study following the current legislation on historical and artistic heritage.[19]

In such a situation, Muniellos SA had tried to sell the forest to the Ministerio de Información y Turismo,[20] to the Patrimonio Forestal del Estado and to the Diputación Provincial de Asturias several times in the 1960s, but no agreement on its price was reached. At the beginning of the 1970s, the Sociedad Protectora de Animales y Plantas[21] in Oviedo, with the aid of ADENA, also prepared a project for the creation of a national park in Muniellos, with the aim of preserving its flora and fauna, including the building of a state-owned hotel (*parador*). In those years, Patrimonio Forestal del Estado had already finished an evaluation study of the forest, which finally, on 17 February 1973, was purchased by ICONA. Ten days afterwards, the partnership for reforestation was dissolved. In 1975, it was declared to be a public utility (MUP 339) and a royal decree of 15 October (3128/1982) created the Muniellos Biological Reserve. Years later, by means of a decree of 4 February (21/88), the Valdebueyes Forest, the Vallina de Abraedo (MUP 345) and the La Viliella Forest (MUP 349) were included in the Reserve. In

1994 it became the Muniellos Integral Natural Reserve, and it was included in the *Red de Espacios Protegidos de Asturias.*[22]

As an Integral Natural Reserve, its management, based on conservation and regeneration, is trying to bring these forests back to the free interaction of forces in nature. Today, Muniellos is a splendid forest, with a degree of preservation that represents the victory of conservationist over productive forces, in a battle that had started at the beginning of the 1960s.

Notes

1 Muniellos Integral Natural Reserve.
2 Fernandez Prieto, J.A. and Bueno Sanchez, A. (1996) *La Reserva Integral de Muniellos: Flora y Vegetación.* Servicio de Publicaciones del Principado de Asturias, Oviedo, + map, which includes a historical introduction, which this article is based on: Torrente, J.P. Breve historia de la Reserva Integral de Muniellos, pp. 23–39.
3 General Society for the Management of Mines and Forests Bosna Asturiana, SA.
4 The State's Forest Heritage.
5 National Institute for the Conservation of Nature.
6 Spanish Anonymous Company for Valledor Oak Management, SA.
7 *Archivo Histórico de Asturias*, box 12.730 and *Archivo General de la Administración del Principado de Asturias (AGAPA), Montes, MUP 345, Valdebueyes* and *Vallina de Abraedo.*
8 *AGAPA., Montes, MUP 349, La Viliella.*
9 *El Carbayón*, 16 February 1893.
10 Spanish National Railway Network.
11 Jefatura del Distrito Forestal del Oviedo (1962) *Informe sobre la Primera Revisión del Plan Dasocrático del Monte 'Muniellos'.* Oviedo, 19 February, 17 pp (*AGAPA, Montes, MUP 339, Muniellos, s. org.).*
12 *Proyecto de Ayuda Técnica.*
13 Terms of the partnership between Patrimonio Forestal del Estado and Muniellos SA: *Consorcio para la Repoblación Forestal, Conservación, Mejora y Aprovechamiento del Arbolado Existente en Muniellos*, April 1964 (copy issued in Oviedo, 5 August 1965, certified by the head of the Brigada de Asturias del Patrimonio Forestal del Estado, AGAPA, *Montes, MUP 339, Muniellos, s. org.).*
14 Regional Tourism Assembly.
15 *Asturias Semanal*, 15 and 22 November 1969.
16 Decree 2395/1964, 9 July.
17 Ministry of National Education.
18 General Director of Fine Arts.
19 *La Nueva España*, 27 June 1971.
20 Ministry of Information and Tourism.
21 Society for the Protection of Animals and Plants.
22 Network of Protected Areas in Asturias.

Une Illusion Française: la Pénurie des Ressources Ligneuses, 1814–1914

9

Andrée Corvol-Dessert

Groupe d'Histoire des Forêts Françaises (GHFF), Institut d'Histoire Moderne et Contemporaine (IHMC), 45, rue d'Ulm, 75005 Paris, France

Le mot 'illusion' résonne de manière un peu provocatrice, et sans doute serait-il préférable d'évoquer la 'spécificité' française, encore que cette illusion ait tellement marqué l'esprit de nos concitoyens qu'il semble aujourd'hui quasi impossible de l'en extirper. En effet, en France, depuis le Moyen Age, le thème 'crise forestière' paraît revenir sans cesse sous diverses appellations. Celles-ci recouvrent grosso modo l'idée d'une diminution rapide et continue des forêts et des bois. Si cette impression avait eu quelques fondements, il y aurait eu ruptures d'approvisionnement en temps de paix. Mais cette situation ne s'est jamais rencontrée, même dans le cas de la Marine, et bien que ses prélèvements aient souvent été incriminés. En fait, les importations furent motivées par la difficulté de transporter les bois autochtones et l'élévation des coûts qui en résultait; quant à l'abandon du bois par les architectes navals, il reflétait la difficulté d'obtenir suffisamment de bois courbes naturels et de résoudre la conservation des bois de marine avant et après emploi. Ainsi, la transition vers la marine de métal fut en relation avec des facteurs financiers et techniques, aucunement avec la pénurie ambiante: les forêts qui avaient fourni du matériau-bois aux arsenaux en livrèrent à la tonnellerie, et ce fut tout.

1 Les dimensions du problème

La période retenue, 1814–1914, diffère peu des précédentes, elles aussi caractérisées par quatre points: une reconstitution de la flotte militaire qui

(eds M. Agnoletti and S. Anderson)

reste une flotte de bois (1820–1840); des ponctions considérables pour les fortifications terrestres (1840–1850, 1870–1880); une demande croissante du secteur industriel, d'autant que les usines fonctionnent encore au combustible végétal (1820–1870) et que la première révolution industrielle (celle des mines et des transports) consomme énormément de bois de service (bois de soutènement, bois des traverses), révolution inachevée au lendemain de la défaite de Sedan qui oblige à repenser l'articulation du réseau ferroviaire dans les départements de l'Est et à rénover les installations sidérurgiques de la Normandie; une demande soutenue des foyers domestiques,[1] d'autant que les habitants des petites villes et des bourgs ruraux alignent leur comportement en matière de chauffage sur celui observé dans la capitale au milieu du 18e siècle et dans les métropoles à la fin du 18e siècle (1870–1914). Cette période, cent ans tout rond, voit cependant l'intrusion de deux nouvelles données. La première est la disparition quasi totale et subite des contraintes qui pesaient sur la propriété privée, contraintes théoriquement abolies en 1790 et ressuscitées lorsqu'il fut clair que le conflit ouvert en 1792 ne se résumerait pas à un face-à-face France–Angleterre et France–Autriche mais réclamerait des efforts prolongés dans le cadre d'une conflagration européenne. Voilà qui obligea les gouvernements successifs à maintenir leur contrôle sur l'ensemble public et privé des matériels ligneux sur pied mobilisables pour les arsenaux (1794–1814). La seconde donnée est le recul progressif et discuté des droits d'usage par la méthode des cantonnements. L'Empire y avait répugné car redoutant la protestation des ruraux, lesquels formaient ses troupes. La Restauration eut moins de scrupules, d'abord parce qu'elle n'avait pas cette clientèle à ménager, ensuite parce qu'il lui fallait bien régler la facture des guerres impériales, que le déficit du Trésor la forçait à vendre des forêts du Domaine et que l'expérience démontrait la médiocrité des enchères dès lors que les servitudes usagères n'étaient pas éteintes. La Monarchie de Juillet raisonna de même. Grâce à ces aliénations domaniales (1815–1848) qui faisaient suite à la vente des biens nationaux d'origine ecclésiastique (1794–1814) – rares furent les bois du Clergé (1792) et des Suspects (1793) dont les dossiers purent être prêts avant la Convention thermidorienne – une véritable propriété communale se constitua, tandis qu'augmentait brusquement la superficie sylvicole détenue par les particuliers. La conséquence en fut une progression très nette de la commercialisation des récoltes ligneuses, alors que celle-ci n'avait guère concerné jusque-là que les bois extraits des grands domaines royaux et seigneuriaux, l'autoconsommation et la consommation usagère caractérisant pour l'essentiel de la récolte les autres types de possession. Ce mouvement conféra une grande importance aux recherches entreprises à l'époque des Lumières. Les études s'organisaient selon deux axes: (i) calculer ce que fabrique une unité de surface en fonction des traitements sylvicoles appliqués afin de distinguer celui qui procure le maximum de marchandises; (ii) mesurer la qualité de leur volume, puisque le prix en dépend, ce qui implique d'en appréhender la résistance à la pression, à la torsion, à la flexion, etc. On avait jusque-là raisonné en termes de surface. On commença à penser que cela ne suffisait pas, qu'il fallait produire beaucoup, qu'il fallait produire mieux. Bref, on réfléchit enfin au rendement à l'ha et à la destination des bois. Ce qui signifie qu'auparavant on avait cru tout faire ou presque en soulignant les limites d'une propriété quelle qu'elle soit, et en structurant son intérieur de façon à ce que la première coupe faite porte sur des arbres atteignant l'âge fixé pour les récolter tandis que la dernière prévue correspondrait à des arbres qui, pour l'heure, étaient au stade de l'an I des recrûs.

Justement, comment repérer cette propriété quand elle ne relève pas de l'Etat directement (bois domaniaux) ou indirectement (bois communaux, autres bois soumis)? L'expérience que m'apporte la direction d'une lourde entreprise – la recension des Sources de l'histoire de l'environnement – cofinancée par les ministères de la Culture et de l'Environnement, les trois tomes couvrant les 18e, 19e et 20e siècles – renforce en moi le sentiment que les descriptifs présentés sur la forêt française dans sa globalité reposent sur un échantillonnage médiocre. Ce ne serait pas trop grave s'il était représentatif. Hélas, il ne l'est pas, il s'en faut même de beaucoup! Cette recension révèle en effet: un, des bois non signalés dans les séries forestières classiques (séries B, P, M, Q) mais mentionnés, par exemple, à l'occasion des tracés de lignes ferroviaires car figurant dans le dossier d'indemnisation monté par les syndicats intercommunaux une fois proclamée la déclaration d'utilité publique; et des bois nulle part signalés mais dont la présence contraignit les ingénieurs des ponts-et-chaussées à redessiner le tracé d'une route, dont la présence contraignit leurs collègues du génie militaire à les faire entrouvrir afin de ménager un angle de tir. Bien d'autres situations pourraient être citées, et pas seulement à propos des séries archivistiques que manipulent les spécialistes de l'histoire des infrastructures en transport[2] et des équipements de l'armée. Or le but de ceux-ci n'est pas d'examiner les éléments forestiers: ils les laissent donc de côté. Les spécialistes de l'histoire forestière seraient bien contents de les analyser, mais eux ne furètent pas dans ces papiers où l'objet de leur travail ne paraît qu'assez épisodiquement. Pourquoi? Parce que l'organisation documentaire n'y est pas celle que secrètent le fonctionnement des instances forestières publiques, telles les grueries royales absorbant progressivement les grueries seigneuriales, et les maîtrises particulières puis les conservations des eaux-et-forêts, et celui des instances forestières privées. Ajoutons que ces dernières ne furent mises en place qu'après la Seconde Guerre mondiale, qu'elles ne concernent que des bois dont la taille dépasse un certain seuil et qu'elles ont produit des archives pour l'heure interdites à la consultation! L'historien forestier n'est ni myope ni presbyte, non, mais porte des œillères qui l'empêchent de regarder ailleurs et de découvrir un océan de papiers inconnus.[3] Cela explique sinon ses silences du moins ses minorations. Il sous-estime assurément les surfaces en haies épaisses, les surfaces plantées qui étaient en friches, les surfaces boisées à partir d'une pépinière voisine, les surfaces reboisées parce que les premiers essais ratèrent.

2 L'éparpillement des sources

Le décalage s'avère particulièrement net pour tous les peuplements qu'un homme finit par remarquer mais dont la croissance, trop récente, empêche qu'ils soient jugés repères immuables dans le payage familier, trop récente aussi pour qu'ils soient décomptés par des observateurs officiels. Les deux réactions ont partie liée: les gens du pays n'indiquent aux enquêteurs que ce qu'ils apprécient comme étant 'vraye forest' de longue date,[4] désignation qui ne saurait évidemment pas caractériser des peuplements dont l'origine – la main de l'homme – est bien connue. D'ailleurs, nombre de ces bois que nul ne saurait qualifier 'd'immémoriaux' (c'est-à-dire ayant au moins 30 ans d'âge selon la jurisprudence d'autrefois) sont étiquetés dans le cadastre à la rubrique 'landes, vagues.' Pour quel motif? Uniquement parce qu'un tel classement évitait (et évite!) au propriétaire la difficulté d'en estimer et le revenu et le capital, et le désagrément d'acquitter la taxe foncière! Je n'ose dire après tout cela que la pénurie ligneuse ait été inconnue. Un, il existait des régions sans bois, que les pentes en fussent dégarnies de toute éternité ou presque, ou qu'elles eussent été rendues chauves par suite d'une dégradation ponctuelle et point trop éloignées dans le temps, ce qui accréditait la possibilité de les restaurer au moindre coût. Deux, il existait des régions manquant de bois d'oeuvre ou de bois de fente, ce qui amenait leurs habitants à recourir aux bois de réemploi, encore que cette attitude fût assez répandue, y compris là où la fourniture ne faisait pas défaut. Je prétends néanmoins pareilles configurations plutôt minoritaires[5] et sans doute bien plus que je ne le subodorais avant de conduire l'enquête sur les sources de l'histoire de l'environnement. Je prétends aussi qu'elles ont été montées en épingle par tous les acteurs qui relevaient de la filière-bois: leurs intentions différaient mais leurs plaintes et leurs rapports ont convergé. Les unes et les autres ont crédibilisé l'hypothèse d'une production de bois ne répondant pas à l'ampleur de la demande. Il était dès lors exclu de brider les particuliers qui s'engageaient dans la sylviculture. Il était même recommandé de les encourager dans cette voie. Cela leur octroya maints avantages fiscaux que les homologues européens enviaient fort et n'eurent jamais. Les eaux-et-forêts entretinrent de même cette illusion d'une production insuffisante. Le mythe confortait la nécessité de maintenir les effectifs admistratifs, voire de les accroître, d'augmenter leur formation technique et, surtout, de réviser à la hausse tous les crédits. Sans ce mythe, comment en effet promouvoir le reboisement tout azimut, celui orchestré en moyenne montagne sous la houlette des administrateurs qui ont *réellement* cru que 'leur' forêt de protection serait une forêt de production, celui effectué par les communes et par les particuliers aux dépens des terres *réellement* insalubres parce que mouilleuses ou déclarées insuffisamment rentabilisées par un élevage ovin extensif.

C'est dans cette perspective qu'il convient de s'interroger sur la manière dont les données chiffrées, publiées entre la fin du 18e siècle et la veille de

1914–1918, ont durablement alimenté la peur d'une France ayant perdu ou en train de perdre sa couverture forestière. Je ne reviens pas ici sur la démonstration de l'incroyable accumulation d'erreurs qui entachent ces données. Je l'ai fait par ailleurs et d'autres que moi, dont Messieurs Jean Boissière et Bruno Cinotti ont confirmé l'exactitude quant à leur caractère éminemment contestable. J'ajouterai seulement que la chose n'est pas surprenante car, au 19e siècle, les statistiques forestières n'en sont qu'à leurs premiers balbutiements, alors que les pré-statistiques agricoles remontent aux années 1720. Il fallait bien savoir après l'épouvantable hémorragie démographique causée par les guerres de Louis XIV et la désastreuse météorologie des années 1692–1696 (2–3 millions de morts) et 1709–1710 (600,000 morts), combien de producteurs agricoles demeuraient disponibles et ce qu'il advenait du potentiel cultivable: les villages transformés en déserts sont alors légion dans le Centre de la France et le Nord-Est frontalier. Le fait même que les opérateurs forestiers aient tant tardé à expédier les renseignements est en soi suspect. Les particuliers n'y avaient aucun intérêt, comme il a été dit. Les ressortissants du public étaient eux, écartelés entre l'envie de montrer qu'ils avaient bien protégé le patrimoine national et l'auraient même agrandi si les moyens financiers avaient été à la hauteur des besoins, et la crainte d'avouer la mauvaise tenue des archives locales, laquelle incitait à inventer ou à 'redresser' les chiffres . . .

On sait l'importance du cadastre pour la reconstitution des paysages et la définition du territoire. Traditionnellement, ses documents sont aux Archives départementales (sous-série 3 P). Cette opération cartographique, d'un volume sans précédent, fut imposée par l'arrêté du 12 brumaire an XI qui institua les plans par masses ou natures de cultures. L'élaboration des plans parcellaires dériva, quant à elle, de l'instruction du 1er décembre 1807. A ces deux premiers fonds, s'ajoutèrent les états de sections et les matrices, le tout dans l'ordre alphabétique des communes. Les textes de 1834 visèrent à compléter le maillage là où les opérations cadastrales étaient à peine amorcées. Dans la pratique, le directeur départemental des services du cadastre contactait son équivalent au service des eaux-et-forêts.[6] C'est ce dernier qui devait lui remettre les informations relatives à chaque commune, le tout classé par cantons, une Inspection en comprenant plusieurs. Bien entendu, ces informations étaient fort disparates. Elles étaient assez détaillées pour les espaces soumis à la vigilance de son personnel et passablement clairsemées en ce qui lui échappait: le domanial était mieux décrit que le communal, et le communal l'était infiniment mieux que le privatif. La distorsion était donc plus notable en France méridionale qu'en France septentrionale, le nord de la Loire concentrant les forêts héritées du domaine de la Couronne. La distorsion était également plus affirmée là où l'abandon des cultures se traduisait par l'enforestement des champs, là où la possession ecclésiastique étant conséquente, avait facilité l'accession à la propriété forestière des bourgeois aisés et des paysans cossus via la vente des biens nationaux.

Exemple: le département de la Haute-Marne (1839). Dans la liste qu'envoie le conservateur forestier à l'administration cadastrale, toutes les forêts égales ou supérieures à 100 ha sont citées. Leurs propriétaires avaient été contactés à l'occasion des réquisitions de 1812–1814, réquisitions dont la finalité était d'approvisionner les arsenaux terrestres et maritimes, alors que la lutte contre l'Europe coalisée se faisait plus âpre et prenait vilaine tournure. La grande majorité d'entre elles appartiennent soit au domaine, soit aux communes, soit à des particuliers alimentant le secteur sidérurgique, directement (à l'instar des maîtres de forges nantis de bois) ou indirectement (cas des sylviculteurs signataires d'un contrat d'affouage). Les gestionnaires publics fréquentent les uns et les autres suite aux adjudications qu'ils ont mission d'organiser. Les premiers parce qu'ils sont acheteurs de triages qu'ils feront exploiter, les seconds parce qu'ils servent volontiers de prête-nom à un parent ou à un ami et, dans tous les cas, assistent aux ventes automnales qui annoncent l'orientation des prix pour les prochains mois. A ces deux motifs s'ajoute le fait que les deux catégories recouvrent un même milieu, celui où se recrutent les munitionnaires dont l'activité suppose des contacts étroits tant avec les militaires (fourniture de l'armement, des munitions, de l'équipement, des rations, etc.) qu'avec les forestiers (fourniture des affûts de canon, des crosses de fusil, de la tonnellerie, du bois de chauffe, etc.). L'extraordinaire dans cette affaire, c'est que ladite liste de 1839 ait résisté durant plus de 20 ans! Ainsi, personne n'a songé à modifier, une fois apaisées les turbulences internationales, des données dont l'aspect lacunaire tenait aux nécessités du moment. Les étendues inférieures à 100 ha (qui sont toutes de droit privé), n'ont donc pas été intégrées, si bien qu'à considérer les chiffres, l'impression prévaut d'un département où le bois est rare, le couvert forestier fortement clairiéré. Ce sentiment est mensonger puisque les ruraux déplorent l'effet des lisières qui recèlent un nombre excessif de lapins et l'insécurité qui provient du refermement des bois. Leur profondeur a attiré les jeunes gens fuyant la conscription et les soldats dont la désertion s'amplifie une fois acquis l'échec de ce que les stratèges ont baptisé 'la campagne de France', la dernière que Napoléon eut à conduire.

3 Les biais début 19e siècle

On sait aussi comment le quidam de base a tenté de répondre aux demandes pressantes que lui adressaient sa hiérarchie. Certaines lettres sont restées accrochées aux dossiers transmis et montrent qu'à force d'avoir attendu l'annulation de l'ordre reçu, le malheureux dut l'exécuter hâtivement. La difficulté résidait assurément pour lui dans la connaissance des forêts particulières. Il fit donc ce qu'il put en prenant la documentation qui forme aujourd'hui le gros de la série Q dont les dates extrêmes oscillent entre 1810 et 1830. L'ennui, c'est que postulant le recensement préalable de toutes les superficies, la hiérarchie exigeait de lui des précisions quant à ce qu'elles contenaient en volume de bois sur pied et à ce qui en était annuellement ponctionné. Après tout, au chapitre agricole, des données analogues étaient maintenant systématiquement expédiées par les préfets au ministre de l'Intérieur, la quantité de grains récoltés influençant leur prix et, par voie de conséquence, l'attitude du cultivateur et l'humeur du consommateur, prêt à l'émeute si les cours crevaient un certain plafond: sans elles, comment prévoir un apport supplémentaire qui freinerait la hausse et calmerait les esprits? L'ère de la statistique avait commencé, de même que la notation météorologique, un orage ou un coup de froid pouvant renverser des prévisions qui paraissaient valables. On imaginait qu'avec un peu de bonne volonté, on pouvait en sylviculture collecter des informations tout à fait comparables. On découvrit lentement que la chose n'était pas aussi aisée.

En effet, pour avoir une idée du volume d'un arbre puis de tous les arbres, il fallait avoir la hauteur et la circonférence moyennes, le nombre d'arbres par unité de surface puis le nombre d'unités de surface. Naturellement, les deux premières (hauteur et circonférence) donc la troisième (le volume) varient fortement selon la nature du traitement et la classe d'âge. Qu'à cela ne tienne: l'échantillonnage par unité de surface en fonction de ces deux critères aplanissait l'obstacle et, si la paresse empêchait l'honorable fonctionnaire de satisfaire à sa mission, il pouvait toujours exhiber les chiffres imprimés dans les tables de cubage.[7] Les milieux négociants les ont adoptées sous l'Empire car elles facilitaient grandement leur estimation: y recourir, c'est ne pas porter trop haut les enchères pour une coupe qui ne le mérite pas car, si elle ne procure pas la quantité de bois espéré, le bénéfice sera moindre que prévu, pire: l'exploitation se soldera par un déficit. Il n'est pas besoin d'être devin pour savoir que la faillite guette tout marchand qui répète souvent une telle bourde! Ces erreurs sont le fait de négociants peu expérimentés, parce qu'ils se sont lancés dans le commerce ligneux sans avoir eu cette éducation familiale qui accélère l'apprentissage du métier, ou de négociants venus d'ailleurs et n'ayant aucun représentant dans le pays susceptibles de les avertir des spécificités qu'offrent les peuplements sylvicoles locaux. Ces erreurs, ils ne les reproduiront pas: la jeunesse est une maladie dont on guérit vite; et quant aux 'marchands horsains', la prochaine fois, ils iront de visu ou chercheront des conseillers éclairés. Non, le vrai problème est afférent à l'évaluation des unités de surface. Comment discerner l'étendue des sous-déclarations, voire des non déclarations? Comment deviner la nature de l'unité, question qui semble incongrue, surtout lorsqu'elle vise un siècle (le 19e) où le système métrique (mètre-étalon) et décimal (multiples et sous-multiples de 10) est supposé règle universelle?[8]

On sait que l'écart est grand entre ce que les gouvernants commandent et ce que les gouvernés pratiquent et, qu'en matière forestière, la distorsion est flagrante! Reprenons ce département de la Haute-Marne, mais le cas arriva des dizaines de fois. La région fait partie des confins Champagne-Ardenne. Elle lorgna tout au long du 18e siècle davantage vers la sidérurgie de l'Est de la France qu'en direction du Bassin Parisien, bien qu'elle envoyât des bois flottés jusqu'à Paris en temps de crise, jusqu'à Reims et Troyes en temps de presse. Cette position explique qu'y cohabitaient trois unités agraires: le journal de Choiseul, le journal de Sorneville et l'arpent de Lorraine qui, ramenés à la toise, équivalaient respectivement à 1135, 817 et 583 toises. On notera que cela oscille du simple au double . . . Manque de chance, l'habitude faisait que les particuliers et les marchands de bois aussi ne calculaient pas en hectares ou en ares, mais dans ces unités anciennes; qu'ils ignoraient superbement l'arpent forestier d'ordonnance, obligatoire pourtant depuis le code de 1669; qu'ils restèrent fidèles aux unités agraires variables d'un bourg à l'autre pour discuter de 'leurs' bois; que l'embrouille atteignait des sommets de par l'existence d'un arpent *forestier* de Lorraine qui n'avait pas la même valeur que l'arpent forestier d'ordonnance mais n'avait pas pour autant celle de l'arpent agraire!! On voit au travers des bordereaux survivants qu'il y eut un certain malaise devant les conversions, le rédacteur étant bien embarrassé pour distinguer à quelle unité il avait affaire. Il lui fallut trancher. La facilité l'emporta souvent. Il décidait alors que tous les arpents se valaient ou plutôt qu'ils vaudraient 2634 toises (soit un hectare) équation qui ne convenait que pour l'arpent forestier d'ordonnance. Il regretta sans doute de ne pouvoir s'en tirer aussi vite avec ces maudits journaux, d'autant qu'interférait quelquefois un journal ou une ouvrée utilisée dans les terroirs viticoles tombés en friche par suite des revers du temps, friche devenue bois grâce au dynamisme des essences pionnières! Au bout du compte, la propriété privée s'en trouva incroyablement minorée: les deux-cinquièmes sont passés à la trappe uniquement pour la part où les erreurs de calcul nous sont connues et auraient pu être corrigées. Or elles ne l'ont pas été: ces données de 1839 traînent dans les relevés réunis[9] (1892–1896) en vue de préparer la première statistique Daubrée (1908), du nom du directeur général chargé de la confection de l'Atlas des Forêts. Elles étaient fausses, colossalement fausses soixante-dix ans plus tôt. Elles n'avaient pas été rajustées depuis, sans compter que dans ce laps de temps – plus de deux générations . . . – maintes mutations foncières étaient intervenues, incitant qui, à planter des arbres, qui, à les ôter!

En fin de 18e siècle, le secteur des Forêts et du Bois est bon dernier au chapitre des statistiques. On connaît tout quant à la production des champs (on omet celle des vergers et des jardins). On connaît tout quant à celle des mines et de la sidérurgie (on a des incertitudes pour le textile et pas grand chose pour les autres industries). On connaît le nombre des habitants, le nombre des agriculteurs, celui des employés, celui des ouvriers, le nombre de logements édifiés. On connaît ce qui rentre dans le pays et ce qui en sort, ce qui rentre dans une ville et ce qui en part. Mais la brume flotte sur la superficie des forêts, la production de bois, sa répartition, sa transformation, sa commercialisation. Oh, on a des chiffres, certes, et même des chiffres publiés, mais ils sont à manier avec des pincettes.[10] Le haut commandement militaire accusa le choc en constatant leur manque de fiabilité pendant le premier conflit mondial. L'année 1917 marqua un tournant quand il lui fallut faire appel aux forêts du sud de la Loire, les autres étant soit occupées, soit ravagées, soit tellement ponctionnées qu'elles en étaient exsangues. Il s'aperçut alors que les statistiques avaient été plus que biaisées dans les séries domaniales afin que les normes de réussite en matière de reboisement eussent l'air atteintes: les arbres poussaient à peine là où ils avaient été déclarés installés depuis 20 ou 30 ans. Cette situation obligeait à demander plus aux forêts des particuliers, fort sollicitées déjà pour les besoins des civils, besoins accrus puisqu'ils ne recevaient plus le charbon de terre du Nord-Pas-de-Calais ainsi que des gisements du Massif Central, si tant est qu'ils eussent employé ce combustible fossile avant guerre. Celle-ci survenant, plus question d'importations anglaises, allemandes et russo-polonaises; quant à la production nationale, elle était affectée en priorité aux usines travaillant pour l'armement. Le chauffage des foyers domestiques venait donc des forêts. Mais, excepté les ménages parisiens, qui achetaient presqu'autant de charbons que de bûches à la Belle-Epoque, c'était le cas bien avant la déclaration des hostilités.[11] On oublie souvent, en effet, que cuire des aliments et réchauffer le logis s'effectuait grâce aux fourneaux (cuisine) et aux poêles (chambres) installés devant le rideau de la cheminée et raccordés aux anciens conduits, fourneaux et poêles emplis de bûches. La combustion des charbons garda longtemps mauvaise réputation, bien que ce soit le tirage, défectueux car mal réglé, qui apparaisse comme déterminant dans les enquêtes qu'ordonne la Préfecture après tout accident mortel. Au reste, l'axphyxie due au monoxyde de carbone aurait pu tout aussi bien survenir avec un poêle-à-bûches qu'avec un poêle-à-charbons. Si les Français n'ont pas retenu cette option, à la différence des voisins anglais et belges, c'est que le bois de feu était assez abondant pour demeurer une ressource énergétique compétitive.

4 Les biais de la fin de siècle

Sa ressource énergétique ne fut même pas amoindrie par les ponctions inhérentes à l'état de guerre! Le directeur général des Forêts admet ainsi que la production bois de feu n'a pas fléchi entre 1913 et 1923 et que cette stabilité fut indépendante des productions bois d'oeuvre et bois de merrain, puisque l'une et l'autre ont continué à se développer (3% et 4%). La progression de ces dernières se confirme sur le moyen terme: la première augmente de 19% entre 1913 et 1928; la seconde, de 14%. Ces pourcentages, diffusés,[12] mineraient la démarche commune entamée par les puissances alliées: exiger de la République de Weimar des livraisons de graines et de plants pour repeupler les forêts dégarnies voire massacrées par les opérations militaires et par l'occupation teutonne, lui imposer des versements en argent destinés à dédommager les particuliers lésés par leurs réquisitions. On évacua de justesse les problèmes que posait l'indemnisation des dégâts et des ponctions incombant à l'armée française. Au vu des chiffres, il était clair que ces propriétés-là (hors de la zone des combats, en arrière des lignes ou dans l'intérieur du pays) n'avaient guère souffert des pillages allemands, que leurs récoltes s'étaient négociées au cas par cas, ici avec des marchands travaillant pour les marchés civils, là avec des marchands en liaison avec les bureaux militaires, ceux de l'Hexagone ou ceux du Second Reich. Le directeur général des Forêts amenda donc son texte dactylographié[13] (les corrections sont manuscrites et sont essentiellement des biffages de données qui pourraient troubler) avant de le transmettre pour impression. Ne restèrent donc que les traits du commerce extérieur: 5.12 millions de m^3 à l'importation, 2.69 millions de m^3 à l'exportation, traits singulièrement constants dans toute la période, traits qui font de la France une exception en Occident puisque partout ailleurs (même aux Etats-Unis) les années 1913–1928 enregistrent un accroissement continu des importations ligneuses! Certes, les quatre-cinquièmes des volumes que la France réceptionne sont des bois de sciage: la grande presse évoque l'inquiétante carence en bois longs et forts qu'a provoquée le premier conflit mondial et qui entrave la reconstruction nationale; elle discute peu de ce qu'il en est pour le bois de feu, et encore moins de l'existence d'un courant exportateur. Sage précaution! Surtout que les deux-tiers de ce courant sont constitués de bois de . . . sciage! Comme l'avoue crûment un expert du Ministère de l'Agriculture, 'mieux vaut vendre à la Grande-Bretagne des bois de sciage de l'Ouest et du Sud-Ouest (via le cabotage atlantique), et importer (pour les départements du Nord) des bois de sciage scandinaves.'

Cette réflexion souligne la difficulté d'acheminer les grumes et non leur insuffisance tant en quantité qu'en qualité. Les grumes 'haut de gamme' ont même de la peine à trouver preneur dans la conjoncture des années 1920–1928, et seront bradées une fois terminée la reconstruction et installée la grande crise de 1929 (1931–1939). La faute en est aux tarifs de transport, prohibitifs pour ce type de pondéreux: la France d'avant 1914, sous-équipée en plateformes ferroviaires, déficiente en ce qui concerne les jonctions voies ferrées-entrepôts ligneux, l'est plus encore en 1918. Ils rendent du coup moins onéreuses les importations par voie maritime, le réseau ferroviaire assurant leur dispatching à partir des ports de la Manche (Rouen) et de l'Océan (Bordeaux et bientôt Nantes). Rares seront les journalistes à lever ce lièvre. On les comprend. Qui, dans l'euphorie financière que suscite le principe 'L'Allemagne paiera,' oserait déclarer que cette situation de déficit factice, qui reflète un réseau de canaux mal conçu et périmé, incombe aux pouvoirs publics, pas au gouvernement de 1923 (date du premier bilan d'après-guerre) mais à tous les gouvernements de la Troisième République qui n'ont débloqué des crédits pour la construction et la connection des canaux qu'en fonction de l'industrie lourde, et jamais des grands massifs? Qui oserait du reste proclamer que cette situation qu'on présente comme héritée de la guerre l'a très largement précédée? Cela ruinerait les espérances mises dans la 'loi de cadenas' que les amis belges reprirent à leur compte le 28 janvier 1921 et qu'un bel élan prorogea jusqu'au 1er janvier 1927. Les associations défendant la nature avaient durement bataillé pour l'obtenir, les circonstances les forçant momentanément à mettre en veilleuse leur programme sur l'apposition de 'réserves naturelles' en forêts périurbaines pour commencer, dans toutes les forêts domaniales ensuite, en attendant d'y songer pour les grandes forêts privées. Cette loi permettait aux autorités de 's'opposer à l'exploitation excessive de certains bois et de certaines forêts appartenant à des particuliers.'[14] Elle postulait leur avidité sans bornes, et désirait les empêcher d'anéantir par des coupes à blanc leur patrimoine afin de profiter des hauts prix qu'entraînait la reprise du secteur Bâtiments et Travaux Public (des maisons à rebâtir, à édifier, des ponts, des routes, des chemins de fer à rétablir). C'était manifester là une surprenante méconnaissance des réalités sylvicoles – une de plus – car les bois privés livraient surtout du bois de feu, étant traités en taillis plus souvent simples que composés!

En fait, les erreurs d'appréciation forment la constante la plus remarquable de l'histoire sylvicole française. On croyait en la fonte des forêts, en la pénurie du bois, en son aggravation par la guerre. Or tout se continuait comme par le passé, jusques et y compris la géographie du commerce ligneux. Elle était en tout point similaire à l'ex-cartographie des bois de marine! 84% des importations étaient originaires de l'Europe du nord, dont 29% pour la Finlande, 25% pour la Norvège, 25% pour la Suède. Mais les propriétaires privés firent du lobbying pour dénoncer l'intrusion des bois américains et obtenir contre eux des quotas – exactement comme les céréaliers l'avaient fait en 1885! 81% des exportations partaient vers l'Angleterre et la Belgique, cette dernière reprenant les parts que détenait l'Allemagne sur le marché français du dernier tiers du . . . 18e siècle! Les chiffres divulgués par le Ministère, les commentaires plus ou moins téléguidés qui parurent d'abord dans les revues professionnelles puis dans les médias de l'époque, lesquelles reprirent et grossirent les composantes de l'information, firent croire à une crise inouïe justifiant des mesures d'envergure: le boisement, le reboisement devenaient ainsi oeuvre pie, mobilisant les associations d'anciens combattants, les associations défendant le paysage forestier[15] et l'administration des eaux-et-forêts. Mais le soi-disant déboisement était une Arlésienne qui jamais ne fut au rendez-vous de l'histoire nationale. Quelques Cassandre le dirent dans cet Entre-deux-Guerres, notamment à propos des moyennes montagnes. D'autres Cassandre, moins nombreuses encore, l'avaient susurré au 18e comme au 19e siècle. On ne les crut pas. Au 18e siècle, on s'effrayait de la conjonction des demandes bois de feu–bois de marine, bien que cela n'affecta pas les mêmes types de peuplements et que les prélèvements Marine fussent peu de chose, comparés à ceux des foyers et des usines. Au 19e siècle, on s'effraya des ravages occasionnés par le paysan-défricheur, sans songer que les guerres révolutionnaires et impériales mobilisaient les jeunes gens disponibles et que l'augmentation de la consommation en bois que provoquait la guerre à outrance tant sur mer que sur terre était compensée par l'accroissement des pressions exercées sur les territoires étrangers (l'essentiel des combats se déroulant à l'est du Quiévain et du Rhin, à l'est de la ligne alpine, en Suisse, en Autriche et en Italie). Mais n'est-ce pas la définition même du talent de Cassandre qui condamna ses prévisions à être toujours exactes mais jamais admises?

Notes

1 La note remise au Ministre de l'Agriculture le 9 août 1886 dénonce le fait que les quatre-cinquièmes de la production forestière nationale, 'principalement dans les forêts des particuliers aménagées en taillis à très courte révolution' soient brûlés. Son auteur juge le morcellement des grands massifs privés responsable de ce gaspillage car la production de chêne en bois d'oeuvre implique une période très longue, insupportable pour qui ne dispose pas de très vastes surfaces où un revenu régulier est alors envisageable. (Fontainebleau. Centre des Archives contemporaines, versement 880470, carton 30 DF 163).

2 Fontainebleau. Centre des Archives contemporaines, versement 880470, carton 30 DF 206, pièce III 25 A h. Commerce et consommation des traverses de chemin de fer, 1905–1935.

3 C'est le cas notamment de la série J aux Archives départementales qui accueille les papiers de famille. Ainsi, dans ceux du château d'Ecot-la-Combe (Arch. dép. Haute-Marne, 5 J n° 107), une lettre du juge de paix adressée au comte d'Ecot nous apprend qu'il a créé un bois sur un terrain contigu à la forêt de la Bouloire. Il lui demande de faire élaguer les branches des arbres de lisière dont l'ombre nuit à la pousse de ses jeunes plants: 'J'affectionne d'une manière toute particulière ce petit coin où je passe un quart de mon existence. J'y vis de la vie de mes plantations. J'y souffre de leurs souffrances . . .' (8 mars 1866). Mais aucune trace, ni avant ni après, dudit bois ne figure au cadastre de la commune de Prez-sous-Lafauche (près d'Andelot, canton de Saint-Blin!).

4 Cette remarque mérite sans doute d'être étendue. Cf. Bruxelles. Archives générales du Royaume. Section Cartes et Plans manuscrits. Mémoires explicatifs et descriptifs en douze volumes, 4,108 pages, en rapport avec les cartes (275 fol.) dressées sous la direction du général comte Joseph de Ferraris (1775). Il signale que 'les bois de haute futaie y sont distingués de ceux de basse futaie par des arbres plus ronds et moins serrés, et le fond des bois de sapin (= pin) est teint en jaune' (Tome I, préface). Il doit pourtant reconnaître les limites de sa recension, mais n'a pu faire mieux, malgré les inconvénients que cela engendre lors d'opérations militaires. C'est le cas des petits bois et des haies, comme à Ypres, où elles 'rendent ce territoire particulièrement propre aux ruses et aux chicanes de la petite guerre (= guérilla paysanne)' (Tome VI, p. 365). C'est le cas des peuplements résineux, existant depuis la conquête romaine dans le nord de la province belge (étude des pollens) mais non présentés car ces pins sylvestres étaient des sujets isolés ou formaient de petits bouquets jusqu'au 16e siècle(= *vlieger*). Ils n'apparaîtront officiellement qu'au 18e siècle, soit un siècle après qu'Adrien Ghy (forestier) ait semé des graines de pin de Riga de façon à ce que cette essence constitue des massifs continus à Vosselaer-lez-Turnhout (1675). Il les avait eues grâce à ses

contacts en la cité de Breda (Hollande), mais déjà des plantations avaient été réalisées en Campine limbourgeoise (1575–1620), l'introducton remontant à . . . 1515! Au reste, les Primitifs flamands ne les ont jamais peints dans leurs aperçus paysagers (15e–16e siècles). Ceux des Ecoles de la nature (Dendermonde, notamment, en Campine) les prendront par contre en considération (milieu 19e siècle).

5 La Champagne par exemple est réputée pauvre en bois. Or les plantations et les reboisements furent si nombreux dans la seconde moitié du 18e siècle – idem en Gascogne et en Bretagne! – que les cahiers de doléances font écho aux protestation des cultivateurs. Dans le bailliage de Troyes (Aube), celui de Saint-Benoît-sur-Vanne propose même 'qu'il soit défendu aux seigneurs de planter en bois leurs terres labourables, bois qui, en grandissant, ruinent les terres estimées voisines par l'ombrage qu'ils causent et par les accrues autour de ces bois . . .' (extr. Pierre Fromont, op. cit. Cf. infra). Cette tendance resurgit durant la période 1859–1913: la statistique Agricole Départementale annuelle de 1920 mentionne de la sorte 691 ha enrésinés *déclarés* pour ces années-là dans le seul canton d'Aix-en-Othe.

6 Ainsi, dans le département de l'Aube, le taux de couverture forestière est de 20% en 1899, soit 122,578 ha (cf. Arbeltier de la Boullaye. *Les Forêts de l'Aube.* Troyes: Paul Nouel, 1899. 26 p.). A cette date, on intègre enfin les 23,189 ha reboisés ou boisés depuis l'instauration du cadastre, et jusque-là ignorés, peut-être parce que 18,000 d'entre eux avaient été enrésinés. Progrès notable. Il ne se renouvellera pas de sitôt. De fait, Pierre Fromont (*Le Pays d'Othe, sa Vie Économique.* Nancy: Imprimerie nancéïenne, 1923. 218 p.), étudiant au Chapitre XI Les Industries du bois, indique la production en bois et produits dérivés des forêts domaniales: rien à propos des forêts communales pourtant citées au Chapitre III (La forêt, pp. 20–27), et moins que rien pour les bois particuliers sinon qu'ils 'sont nombreux' (p. 156). Là aussi, il y a confusion entre récolte domaniale et récolte globale. On imagine ce que cela entraîne comme marge d'erreur une fois additionnées les productions sylvicoles de tous les départements, même si l'erreur n'en a touché qu'un sur deux, ce qui reste à prouver.

7 Corvol, A. (1989) La métrologie forestière. In: *Introduction à la Métrologie Historique,* sous la direction de Bernard Garnier et de Jean-Pierre Hocquet. Paris: Economica, 289–330.

8 Corvol, A. (1989) Le système métrique: une greffe difficile pour le commerce du bois. In: Actes du colloque Mesures et Révolution. Société de Métrologie Française. C.N.A.M. Paris. Publ. *Genèse et Diffusion du Système Métrique,* textes réunis et présentés par Bernard Garnier et Jean-Claude Hocquet. Editions du Lys, Caen, pp. 147–159.

9 Cf. *La Statistique Forestière de 1892,* p. 11. Elle n'a aucun mal à prouver la supériorité du rendement moyen des forêts domaniales eu égard à ce qui a été collecté au sujet des forêts particulières! (Fontainebleau. Centre des Archives Contemporaines, versement 771615, carton 158 ONF 906).

10 Les chiffres traînent plus longtemps encore. Ainsi, lorsque Egon Glesinger rédige *Le Bois en Europe: Origines et Étude de la Crise Actuelle* (Paris: Sirey, 1932. In 8°, 734 p.), il exploite dans son Chapitre II (La France, pp. 298–335), l'*Annuaire Statistique* qui prête à l'hexagone un taux de couverture de 19%, lequel satisferait 78% des besoins en bois de manière globale et 53% des besoins en bois d'oeuvre. Il s'étonne que les habitants puissent consommer encore 0.44 m^3 de bois de feu par tête. Il estime pour rendre compte de l'anomalie que l'emploi de la houille doit être délibérement freiné, ce qui est une contre-vérité. Elle n'a tout simplement pas atteint le seuil de compétitivité qu'elle aura après la Seconde Guerre mondiale, lorsque la France commencera à recueillir le

fruit des conversions: moins de bois de feu car davantage de bois d'oeuvre et d'industrie, donc un bois de feu relativement plus cher mais encore tout à fait vendable puisque, aujourd'hui encore, cette ressource se classe au deuxième rang des énergies domestiques, souvent combinée au chauffage électrique.

11 Arnould, A. (1925) Coupons-nous plus de bois qu'avant guerre? *Revue des Eaux et Forêts*, janvier–février.

12 Le détail de ces chiffres se trouve aux Archives départementales, série M. Prenons l'exemple de la Haute-Marne. La liasse 184 M n° 6 renferme les données fournies par la 31ème conservation au titre du Recensement agricole. L'administration ne s'est préoccupée que des forêts soumises, ce qui est normal, eu égard à ses attributions. On n'a donc rien à propos des autres. Mais, même pour ces forêts soumises, on demeure perplexe: le rapport cite leur production totale en bois d'oeuvre, en bois de mine, en bois de chauffage, en charbonnette et bois de boulange, en charbon de bois, en écorce à tan, et sa répartition par forêts domaniales et communales dans les quatre Inspections Chaumont-Ouest, Chaumont-Est, Joinville et Langres. Mais les superficies ne sont pas précisées. Pire, ces données sont confondues avec la production forestière départementale, ce qui l'amenuise terriblement!

13 Fontainebleau. Centre des Archives contemporaines, versement 880470, carton 30 DF 163. Le rapport du directeur général Léon Dabat porte sur le bilan de la production et de la consommation pendant la Première Guerre mondiale. Il l'a rédigé le 30 septembre 1917. Il y apporte des rectifications après consultation de l'autorité ministérielle le 15 mars 1918. Toutes visent le Titre III (Répercussions de la guerre sur la production des forêts françaises). Il y distingue le cas des forêts situées dans la zone des armées (pp. 8–11) et celui des forêts situées dans la zone de l'intérieur, arrière des armées inclus (pp. 12–16). Là, 'les bois d'oeuvre d'essences dures, propres aux constructions et à la fabrication du merrain (= bois de tonnellerie), le chêne notamment, se trouveront plutôt en excès.' Il ne prévoyait que deux difficultés dans sa première version pour l'après-guerre car, côté production, 'toute inquiétude doit donc être écartée:' la main d'oeuvre qui 'menace de rester insuffisante pendant de longues années;' la commercialisation qui suppose d'étendre '(le) rayon de diffusion (des marchandises en abaissant les tarifs de transport par voie ferrée et même, s'il se pouvait, d'arriver à l'unification de ces tarifs entre les diverses compagnies').

14 *Bulletin de la Société Centrale Forestière de Belgique*, n° 29, 1926. La coupe déclarée 'anormale' est définie négativement: s'il s'agit d'une futaie, ne pas laisser la moitié du matériel sur pied et au moins 75 m^3 par ha; s'il s'agit d'un taillis composé, ne pas laisser un tiers du matériel sur pied et au moins 25 m^3 par ha. C'est-à-dire toute coupe réduisant la quotité établie 10 ans auparavant (mais le texte de 1911 – belge – et de 1913 – français – ne fut jamais appliqué, et pour cause: la guerre intervint avant la rédaction des décrets). On n'échappe à cette contrainte de l'autorisation après déclaration notifiée en recommandé (la non-réponse dans les 30 jours équivaut à un consentement) qu'à condition de faire approuver son réglement d'exploitation par les eaux-et-forêts. Deux remarques à ce sujet. Un, cela annonce les 'plans simples de gestion' qui seront imposés dans les années 1960. Deux, cela démarque 'l'autorisation préalable,' introduite dans toutes les législations sur le bois de marine, dernier tiers du 17e siècle, période où se constituent ou se reconstituent les grandes flottes de guerre (en France: 1669, 1689).

15 Ces associations ne contestent pas alors l'enrésinement, contrairement à ce qu'elles faisaient avant guerre. Elles reprendront ce leitmotiv dans le contexte de 1930

en le rhabillant en quelque sorte: la chute des cours est imputée à l'administration forestière qui, en mettant du résineux dans le domanial, pratiquerait une concurrence déloyale. Le consensus repose donc sur l'idée que les résineux ne doivent servir qu'à enrichir le taillis des particuliers trop orienté vers la production de bois de feu. L'ennui, c'est que cette méthode est aussi largement usitée par les communes! cf. Arch. dép. Haute-Marne, 5 J n° 980. Création d'une pépinière centrale financée par plusieurs communes en 1932. Elle remplace les pépinières volantes de 1921 dépourvues de toute possibilité d'apport d'engrais et d'arrosage régulier; cf. Arch. dép. Haute-Marne, 5 J n° 981. Fourniture de plants gratuits repiqués à 3 ans par l'Association Nationale et Industrielle du Bois. La proportion est révélatrice: en accordant l'indice 100 au chêne, on dénombre 100 épicéas et mélèzes et . . . 2700 pins! Les demandeurs peuvent contacter l'ANIB par l'intermédiaire des Préfectures ou des Conservations des eaux-et-forêts.

Shortage of Wood? Towards a New Approach in Forest History: the Palatinate in the 19th Century

10

Bernd-Stefan Grewe

Universitaet Trier, Neuere u. Neueste Geschichte, D-54286 Trier, Germany

Many people are unaware that the idea of sustainability was born with forestry. The thesis that this idea was a reaction to the shortage of wood has often been repeated in the last 200 years. Today there is a new debate in Germany, mostly between professional foresters and some social historians, as to whether or not there has been such a scarcity. For a deeper insight, it is important to examine the actual German debate about the shortage of wood. In this chapter, a new model is developed to understand the problem of shortage of forest resources. It is important to include the agricultural uses as a factor of equal value to wood production and recognize the difference between three types of shortage: crisis of resources, crisis of environment and crisis of distribution. The Palatinate, in the middle of the 19th century, is used to illustrate these ideas. For the state forestry it was possible to avoid a shortage of wood by excluding agrarian users from the forests. But, if the agrarian uses of the forests are also considered, then the model shows there was a crisis of resources for the rural population. This case shows clearly why we can no longer focus only on the wood market.

1 Introduction

It is hard to open a German newspaper today without reading something about sustainable development. Many people do not realize this idea was born with forestry and that, as most forest historians claim, that this idea was a

(eds M. Agnoletti and S. Anderson)

reaction to the shortage of wood. This thesis was based on a wide public debate about the shortage of wood at the end of the 18th century. Today there is a new debate in Germany, mostly between professional foresters and some social historians, as to whether or not there has been such a scarcity. For a deeper insight, it is important to examine the actual German debate about the shortage of wood. This chapter attempts to develop a new model to understand the problem of shortage of forest resources. The case of the Palatinate in the middle of the 19th century is then used to illustrate the model.[1]

2 The debate in German forest history

The debate about the shortage of wood is as old as forestry. In Germany, the foresters and the forest history departments at the institutes of forestry defined it as a surplus demand for wood that represented danger for the forests. Different generations of forest historians, from Bernhardt (1872) and Schwappach (1886) to Hasel (1985), Mantel (1990) and recently Uwe-Eduard Schmidt (1997), have always presented the history of forestry as a successful struggle against this shortage and against the destruction of forests.[2] They were aware that we have to draw a distinction between different kinds of shortage. Was there an unsatisfied demand for firewood, for timber as material to build houses or for the handicrafts? I intend to use the expression 'wood shortage' only for a lack of firewood. If this wood was no longer available, we can be sure that there was no timber either.

Another perspective was chosen by some social historians. They are often considered as revisionists, but they enriched the debate with the question: Who claimed a shortage of wood? And who profits from this claim? Joachim Radkau, Ingrid Schäfer and Joachim Allmann denied a general shortage of wood. As Schäfer stresses for the Lippe region from the end of the 18th to the middle of the 19th century and Allmann for the Palatinate in early modern times, the rulers pretended that there was a scarcity of wood and a danger that all forests would be destroyed by the people, in order to put their hands on the forests and to exclude agrarian users from it. In doing this, they initiated a wide discussion in German history as to whether or not there has been a wood scarcity. They pointed out that, of course, there could have been a wood shortage in some regions, but that the question has to be studied and answered region by region.[3]

But we cannot judge the problem of wood shortage correctly only by comparing the supply and demand of wood, if we do not consider possible reactions to a shortage of supply. For instance, firewood and timber could also have been imported from other regions, where the demand was not as high. For this we need to study the wood trade and relations between different regions, as has been done for the Dutch timber trade with the Black Forest by Ebeling and Scheifele.[4] This kind of study is very complicated for the 19th century, because, after the foundation of the Zollverein (the German customs union, 1834), we rarely find records to measure the wood import or export of a region or state.

Secondly, historians of technology demonstrated that most of the pre-industrial energy systems contained a large potential for innovation and more efficient new techniques, such as new ovens or stoves. This innovating potential would have made it possible for a longer time to avoid the big change in the energy system from a basis of wood to basis of coal.[5] If we examine any shortage of wood, we also have to ask how the society used the technical possibilities of their age. How modern was the technology? Which technical standard was reached? Since the 1850s, we know exactly how many steam engines had been in use in every German region, but we do not know – and probably will never know – how much the rural society used the new ovens.

The third point, the main argument of the forest historians at the end of the 19th century (Bernhardt, Schwappach), which has often been repeated by their successors (Mantel, Hasel), is a higher productivity of the forests, managed by well-trained professionals – the foresters. For them, the exclusion of the damaging agrarian uses, such as wood-pasture, was inevitable in order to guarantee the survival of the forests. Due to the exclusions and to professional forestry, their struggle against the destruction of the forests was victorious. Nobody talks about the victims.[6]

Finally, timber as material to build houses and ships was more and more replaced by iron, and firewood by coal and peat. The main condition for these substitutes to be used in greater quantities was a well-developed transport system. In the middle of the 19th century, the construction of the railways made this possible and there was transport and exchange of burning material on a massive scale, independent of natural borders and rivers.

3 Shortage of wood or shortage of forest resources?

Arguments in current discussions that focus too much on the wood market ignore other users of forest products and see their specific needs only as a damaging element in the development of the forests. Recently, Bernward Selter (1995) and Winfried Schenk (1996) showed for the preindustrial era that forests were not only an area in which wood and timber were produced.[7] They insist on the fact that woodlands had also been a part of agriculture, providing not only another surface for pasture, but also a fertilizer to guarantee or even to reach a higher agricultural productivity. Agriculture was becoming more and more productive through use of farmyard manure. Therefore the farmers needed a lot of straw, which they could not produce to the extent needed because of missing unused surfaces. In most German regions, the farmers became more and more dependent on litter (dead leaves, twigs, branches) from the forests to keep up the productivity of their fields. There can be no doubt that the fertility of the forests was reduced by the removal of the litter, but this does not mean that this use and the pasture necessarily destroyed the forests. Of course, they influenced the ecosystem inside the forests; other species had to be introduced and the growth of the trees slowed down. In conclusion, the most important effect on the woodland concerned the production of timber and wood and thus the future financial output of forestry. Sometimes when we read the records in the archives, we forget that they are written by foresters and that the woodlands had many different functions for the rural society, not only the provision of timber and firewood.

Few will argue with the principle that we should no longer speak of a shortage of wood, but of a shortage of forest resources. Otherwise we continue to think in the same ways as contemporary foresters. This can be helpful in understanding how foresters made their decisions, but this approach makes it impossible for us to decide whether or not a society at a given time suffered from an access to the forests which was too limited. If we continue to analyse the shortage of wood without the demand for all other forest products, we make the same mistake as the contemporary foresters: we will exclude the agrarian users from the beginning and consider them as dangerous for the forests, without reflecting how much they depended on the forests for fuels and other products. The approach that includes agrarian uses as a factor of equal value as wood production offers the possibility to examine how much the socio-economic development of rural regions depended on the forests.

To complete the new approach, it is necessary to rethink the definition of a 'shortage of forest products'. Extending the term, it is possible to work with the former definition of wood shortage. To mark the stage when the regional demand for forest products was higher than the regional supply, I shall use the term 'crisis'. But, as we can find different factors influencing such a shortage, we can make a distinction between the following three types of crises:

1. A crisis of resources (Fig. 10.1) is a real unsatisfied demand for forest products, which cannot be satisfied because of missing reserves. But this demand could be satisfied with substitutes or imports from other regions, or the demand could be reduced by new technologies. There are still some ways out of this crisis.

2. A crisis of environment (Fig. 10.2) often follows a crisis of resources. This is the worst case for the forests. The survival of the natural cycles of life is in danger and the natural resources are increasingly destroyed. There are no imports and no substitutes or new technologies which can stop this crisis. All possible reactions have already been tried.

3. The third type can be considered a crisis of distribution (Fig. 10.3). Strictly speaking, this is not a real crisis. The demand for wood and other products is unsatisfied not because of missing reserves but because the access to the reserves is not possible – for example, because of lacking transport facilities or an overly conservative forest policy. Only part of the demand is satisfied, even if it would be possible to satisfy the whole. Some social groups or industries could be excluded from any access to the forest resources. Even if this is not

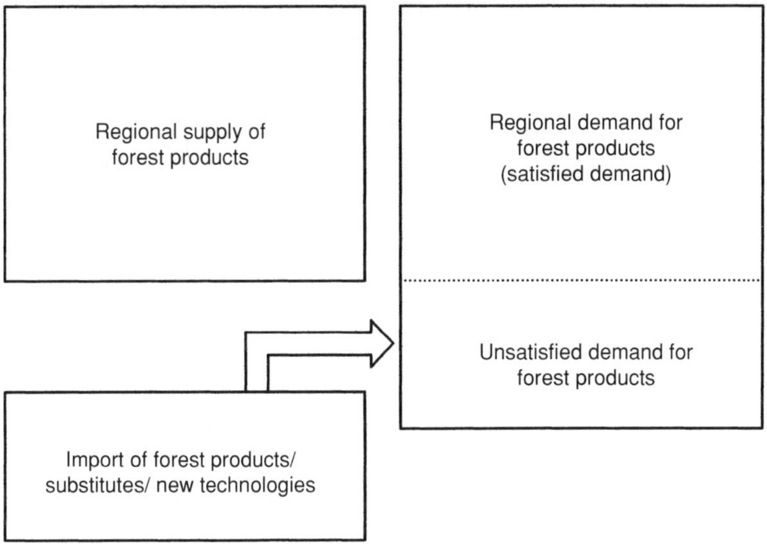

Fig. 10.1. Crisis of resources.

necessarily a shortage of forest products, for the people depending on the forests this is a severe situation.

4 The case of the Palatinate in the middle of the 19th century

The Palatinate lies in the south-west of Germany between the Rhine River and the French border (5928 km^2). In the 19th century, after a short period of French government, the Palatinate became part of the kingdom of Bavaria. When the French left the Palatinate, they left a very well-organized administration, a new system of justice and much debt. The fact that the Palatinate was separated from the rest of the state and that the government of Bavaria in Munich was quite far away created a difficult situation not only for the administration. Until the German Zollverein was founded, trade was also isolated from markets in Germany and France by customs barriers. Before the construction of the railway between the coalmines on the Saar River and the Rhine, we can consider the Palatinate a self-sufficient energy system. The Palatinate neither exported larger quantitites of firewood or timber nor imported substitutes for fire, such as coal.[8]

At the beginning of the 19th century, more than 80% of the Palatines were working in agriculture; until the 1860s there was little industry. Then the situation changed rapidly, the beginning of the industrial era in this region being marked by the foundation of the chemistry enterprise BASF in Ludwigshafen.

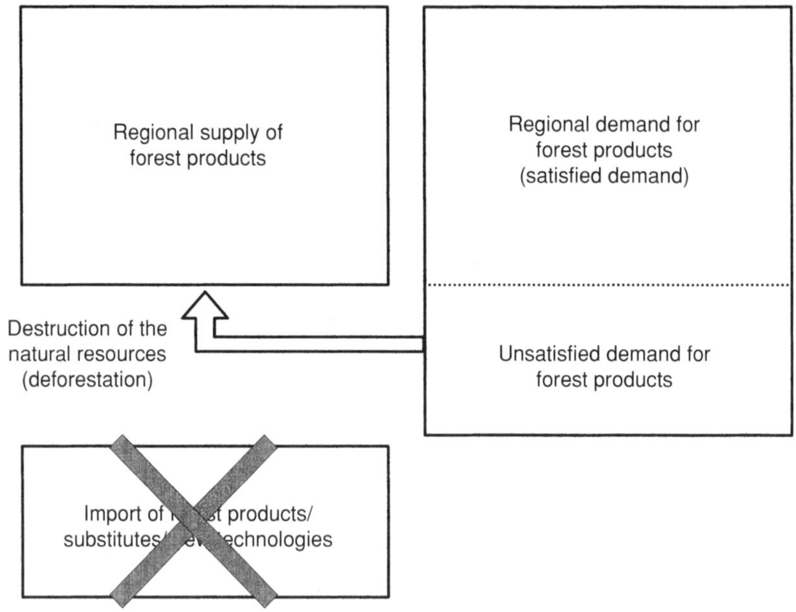

Fig. 10.2. Crisis of environment.

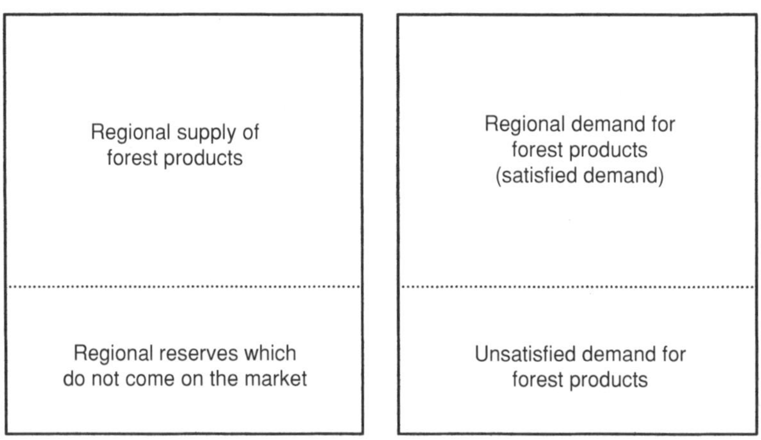

Fig. 10.3. Crisis of distribution.

More than 40% of the Palatinate's surface area was forestland. One-half belonged to the Bavarian state, 38% was the property of the communities and only 12% was owned by private persons. The forest administration was quite severe. Not only the state forests, but also the community forests were administered by the state forestry. Even the private forests were observed by the Bavarian forest administration to guarantee a sustainable forestry. For this reason, we have good records for most of the Palatine forests in the 19th century.[9]

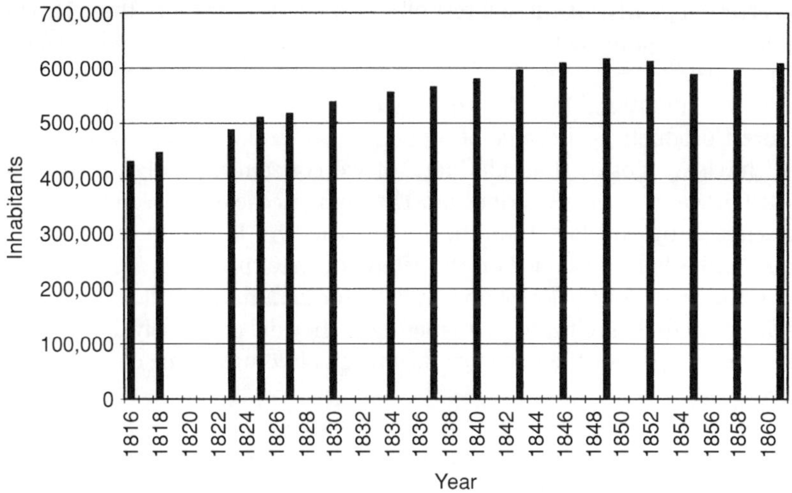

Fig. 10.4. Demographic movement in the Palatinate, 1816–1861.

The excellent sources are not the only reason for choosing the Palatinate as a favourite subject of study; this region is also particularly interesting for its demographic movement. Figure 10.4 shows the growth of population in the last century. The population grew in the period from 1816 to 1849 from 430,000 to 616,000 people. During the next 20 years, more and more people emigrated to other countries, most to the USA, so the population stopped increasing. This cessation of population growth could be taken to mean that a limit of growth had been reached.[10] The population density increased from 73 people per square kilometre to 104 inhabitants in 1849. This density was not reached by any other rural region in the first half of this century; in another non-industrialized region like Württemberg it was reached only more than 50 years later.[11] The pressure of the population on the forests was extremely high. There was only 0.5 ha of forest for each inhabitant. Where was there a shortage of forest products if not here?

There are several arguments in support of claiming a shortage of forest products. The prices for wood and timber were more than 50% higher than the average in the rest of Bavaria.[12] Apparently, every available surface in the woodlands was in use through the well-developed forestry, which had built nearly twice more trails across the woods than in the other parts of the state and also sustained 140 km of rafting channels.[13] Nevertheless, the rural population used peat as a surrogate for firewood wherever this was possible. In Munich, the applications for cheaper wood from the state forests in the second chamber of the Bavarian Landtag were very frequent during this time.[14] But the most significant indicator that a lot of people suffered a shortage of forest products was a very high rate of theft of wood and other offences against the forest laws. In the period from 1830 to 1845, more than 20% of the population was arrested yearly for these forest offences. Every year more than 100,000 Palatines were punished for these. The number of undetected cases must have been higher.[15]

These conditions indicate that there has been an unsatisfied demand for forest products, a shortage of supply. The question then is: which type of crisis have we got in the Palatinate? If we concentrate only on the wood market, it was a crisis of distribution. The high price level for wood and timber was caused by the fact that the finances of the Bavarian state and the communities both depended on the selling of forest products. The higher the prices, the better for the forest owner. Even the communities did not distribute the firewood to their citizens as frequently as they did previously, but sold it for high prices to pay for their school and church buildings or the construction of streets.

The well-developed forestry and transport systems through the forests contributed to the same goal; the more cheaply the wood and timber were transported, the higher was the gain.

The substitution of wood was a reaction of the public in order to make savings by using a cheaper burning material. The fact that there was no import of wood or timber, but even a small export to other regions, indicates that there was no crisis of resources.[16]

The very high rate of offences against the forest laws can be explained by the difficulties many poorer people experienced in paying the high prices for wood. They had no choice but to satisfy their undeniable need by other means.

It follows that, in the period from 1830 to 1850, there was no crisis of environment and there was no danger of deforestation, if we just consider the wood market. If 100 foresters were able to protect the forests against more than 100,000 people offending yearly against the forest laws, we can believe them when they claim that there was enough wood to satisfy the whole regional demand.

But this was only possible because they were able to exclude most of the agricultural uses from the forests. There still remained a very high demand for other pasture areas and an undeniable need for fertilizers to substitute for the missing straw. Indeed, the illegal removal of the litter from the forests heavily damaged their fertility and not only in the neighbourhood of the villages. Until artificial fertilizer was available at the end of the 1860s, we must speak of a heavy crisis of resources for the whole of the rural population. This case shows clearly why we can no longer focus only on the wood market when describing shortages.

Notes

1 This chapter is also based on the ample material in the Bayerisches Hauptstaatsarchiv München and Landearchiv Speyer. In Munich, I especially used the

records of the ministries in StR, M F, M Inn, M L, M A, M H and LVA. In Speyer: records of the Bavarian government of the Palatinate: H 1, H 3, H 5, H 5; records of different district administrations: H 39, H 42, court records: J 1, J 2, J 8; cadaster and tax records: L 56, L 57, L 82; records of forest offices: Q 24–40, Q 41, Q 55; communal archives: U 47, U 254, U 284, U 310. Further results will be published in a book.

2 Bernhardt, A. (1872–75, reprint 1966) *Geschichte des Waldeigenthums, der Waldwirtschaft und Forstwissenschaft in Deutschland*, 3 vols, Scientia Verlag, Aalen. Schwappach, A. (1886/88) *Handbuch der Forst- und Jagdgeschichte Deutschlands*, 2 vols, Julius Springer, Berlin; Hasel, K. (1985) *Forstgeschichte. Ein Grundriß für Studium und Praxis*, Paul Parey, Hamburg and Berlin; Mantel, K. (1990) *Wald und Forst in der Geschichte. Ein Lehr- und Handbuch*, M. and H. Schaper, Alfeld-Hanover; Rubner, H. (1967) *Forstgeschichte im Zeitalter der industriellen Revolution*, Duncker and Humblot, Berlin; Schmidt, U.-E. (1997) *Das Problem der Ressourcenknappheit dargestellt am Beispiel der Waldressourcenknappheit in Deutschland im 18. und 19. Jahrhundert. Eine historisch-politische Analyse*, Munich (in press).

3 Radkau, J. and Schäfer, I. (1987) *Holz. Ein Naturstoff in der Technikgeschichte*, Rowohlt, Reinbek bei Hamburg, pp. 16, 65; Radkau, J. (1983) Zur angeblichen Energiekrise des 18. Jahrhunderts. Revisionistische Betrachtungen über die 'Holznot', *Vierteljahrsschrift für Sozial- und Wirtschaftsgeschichte* 73, 1–37, 10, 15; Schäfer, I. (1992) *Ein Gespenst geht um. Politik mit der Holznot in Lippe 1750–1850. Eine Regionalstudie zur Wald- und Technikgeschichte*, Selbstverlag des Naturwiss. u. Historischen Vereins für das Land Lippele, Detmold, pp. 226–230; Allmann, J. (1989) *Der Wald in der frühen Neuzeit. Eine mentalitäts- und sozialgeschichtliche Untersuchung am Beispiel des Pfälzer Raumes 1500–1800*, Duncker and Humblot, Berlin, pp. 220–233.

4 Ebeling, D. (1992) *Der Holländerholzhandel in den Rheinlanden. Zu den Handelsbeziehungen zwischen den Niederlanden und dem westlichen Deutschland im 17. und 18. Jahrhundert*, Frans Steiner, Stuttgart; Scheifele, M. (1995) *Als die Wälder auf Reisen gingen. Wald-Holz-Flößerei in der Wirtschaftsgeschichte des Enz-Nagold-Gebietes*, Landesforstverwaltung. Baden-Württenberg, Suhrkaup, Stuttgart.

5 Radkau, J. (1989) *Technik in Deutschland. Vom 18. Jahrhundert bis zur Gegenwart*, Frankfurt-on-Main, pp. 59–114; Gleitsmann, Rolf-Jürgen (1985) Erfinderprivilegien auf holzsparende Technologien im 16. und frühen 17. Jahrhundert, *Technikgeschichte* 52, 217–232.

6 Bernhardt (1872) Vol. 1, pp. 178–179, 236–245; (1874), Vol. 2, pp. 289–297, 325–343; (1875), Vol. 3, pp. 122–126; Schwappach (1888), Vol. 2, pp. 678–755; Mantel (1990), pp. 423–476; Hasel (1985), pp. 109, 189–209.

7 Selter, B. (1995) *Waldnutzung und ländliche Gesellschaft. Landwirtschaftlicher 'Nährwald' und neue Holzökonomie im Sauerland des 18. und 19. Jahrhunderts*, Ferdinand Schöningh, Paderborn; Schenk, W. (1996) *Waldnutzung, Waldzustand und regionale Entwicklung in vorindustrieller Zeit im mittleren Deutschland*, Franz Steiner, Stuttgart.

8 Kolb, G.F. (1831/33), *Statistisch-topographische Schilderung von Rheinbayern*, 2 vols, Speyer; Alter, W. (1964–1994), *Pfalzatlas*, F.C. Kolb, Speyer.

9 Kermann, J. (1991) Wirtschaftliche und soziale Probleme in der bayerischen Pfalz an der Schwelle des Industriezeitalters, *Hefte zur Wirtschaftsgeschichte* 4, pp. 151–188; Keiper, J. (1930) *Pfälzische Forst- und Jagdgeschichte*, Speyer.

10 Heinz, J. (1989) *Bleibe im Lande und nähre dich redlich! Zur Geschichte der pfälzischen Auswanderung vom Ende des 17. bis zum Ausgang des 19. Jahrhunderts*, Institut für Pfälzische Geschichte und Volkskunde, Kaiserslautern, pp. 159–170.
11 Heinz (1989) 353.
12 *Die Forstverwaltung Bayerns beschrieben nach ihrem dermaligen Stande vom Königlich Bayerischen Ministerial-Forstbureau* (1861), Munich, p. 421.
13 *Die Forstverwaltung Bayerns* (1861) pp. 273, 385; Elsner, E. (1930) Die Trift im Pfälzerwald während des 19. Jahrhunderts, *Mitteilungen vom Verein der höheren Forstbeamten Bayerns* 25, 162.
14 *Verhandlungen der zweyten Kammer der Ständeversammlung des Königreichs Baiern im Jahre* (1825) vol. 8, Munich, pp. 439–484.
15 Mayr, G. (1867) *Statistik der Gerichtlichen Polizei im Königreiche Bayern und in einigen anderen Ländern*, K. Statistisches Bureau, Munich, pp. 75(86). Grewe, B.-S. (1996) Darum treibt hier Not und Verzweiflung zum Holzfrevel Ein Beitrag zur sozial-, Wirtschafts- und Umweltgeschichte der Pfalz 1816–1860, *Mitteilungen des Historischen Vereins der Pfalz* 94, 271–295.
16 *Die Forstverwaltung Bayerns* (1861), pp. 469–471.

11

Socio-economic Development and Changing Mental Concepts (Re)shaping the Woods in a German Region of Low Mountain Ranges: Contributions of a Study on a Regional Level to the Discussion of Major Issues of Forest History

Gerd Modert

Historiches Seminar Abteilung Neueste Geschichte und Zeitgeschichte, Ludwig-Maximilians-Universität München, Geschwister Scholl-Platz 1, D-80802 Munich, Germany

The study characterizes the development of the relation between a rural society and its natural environment in a geoclimatically disadvantaged region. The restriction to a very small political unit allows a closer look at the change of economic interests and material needs and demands affecting the character of the forest.

A clash of concepts characterizes the relationship of the rural society and the administration to the forest, that is, conflicting notions of how forests should look and what needs they have to provide for.

The analysis of the local sources giving evidence of the wood supply provides the basis for a factual comment on the question of wood shortage; in the given context the shortage of wood is more a social and political issue than a problem of natural resources and forest management.

In this chapter I should like to draw your attention to a small and rather remote area in the very west of the German low mountain ranges. But – as you will most certainly have inferred from the rather functional title of the chapter – I am more interested in making explicit basic social, economic and environmental processes than in giving a close description of a specific region.

(eds M. Agnoletti and S. Anderson)

How did the relationship develop between a vastly agrarian rural society and the forests, which were a central element of their natural environment? This is what I want to analyse and portray here by attempting to comment on the following three questions:

1. Which interests and which notions prevailed when the local population, the administration and the sovereign made use of the forests?
2. Which differences, antagonisms and conflicts resulted from the diversity of demands on the resources the forests could offer?
3. What can be said about the role of forest development in the social, economic and ecological conditions of a region and its society?

Let me start with a short introduction about the main characteristics of the region, the time in which this study is located and the sources upon which it is based.[1]

1. The former principality of Birkenfeld, located in the low mountain ranges between the River Rhine and the Franco-German border and south of the River Moselle, was a by-product of the Congress of Vienna of 1815. In this territorial and political reorganization of Europe, the 'artificial' principality of Birkenfeld was created as a compensation for war damages to the duchy of Oldenburg.
2. In the year 1815 about 20,000 people lived on this territory of approximately 500 km^2. The region stretches between heights above sea level of 400–700 m. A rather humid and cold climate brings about a relatively short vegetation period. Some 40% of the surface was covered with forests, consisting mainly of beeches mixed with some birches, ash-trees and alders.
3. The whole region was characterized by very traditional economic structures: that is, throughout the 19th century, the vast majority of people earned their living in the agricultural sector. In contrast to many other areas, the region experienced a time of deindustrialization, due to the decline of the traditional charcoal-based iron industry. The local inheritance laws had caused an extreme splitting up of land property. To compensate for the shortage of arable land, many farmers specialized in raising and dealing with fattened cattle.
4. Notwithstanding these poor economic prospects and massive surges of emigration, the population almost doubled in the period between 1815 and 1870.[2]

The historical period of time that this study is to cover is restricted to the decades between the founding of the state after the Congress of Vienna in 1815 and the 1860s. In these five decades – roughly speaking – the consequences of the implementation of the new techniques of forest usage and forest management became fully apparent.

The main sources comprise the correspondence between the various levels of the expanding forest service and the state administration, financial files, budget papers, statistical surveys, economic planning files and the papers documenting the almost endless discussion process about a new basic forest law.

I should now like to focus at this point on some characteristics of the usage of forest resources and the spectrum of legal, economic and political interests connected with the woods. Throughout the five decades analysed here, all social strata depended on the usage of forest resources and thus influenced the condition of their forest environment.

It was particularly the rural peasant population for whom it was not only the wood itself that was important – as building material and heating energy. They depended on the complete range of resources the forest system could offer. The generally critical economic, regional and historical context was influenced by the unfavourable geoclimatic framework. For the whole agrarian sector, this meant that people had to face a permanent shortage of most resources, such as litter and fodder for their cattle and manure for their fields. Most farmers also needed the forests as an additional pasture area, since their private land property was not sufficient to feed their cattle.

Life under this condition of general and permanent scarcity of resources made the woods an indispensable part of the rural economy. Among the peasantry it was an undisputed fact that the local forest had to serve as a natural extension of their private farming area.

A cheap and sufficient supply of firewood and building material was the primary interest of the rapidly growing part of the population who could no longer earn their living in agriculture or who had to become part-time farmers. Craftsmen and tradespeople who lived in the few small towns of the principality shared this interest in the wood. Quantitatively, this demand for firewood was absolutely dominant among all the other forms of forest usage, and it was not until the end of the 19th century that this massive demand started to decline, due to the slowly falling prices of coal, which could then become a substitute for firewood.

For the authorities, the forest was most attractive as serving their monetary interests. A closer look at the years from 1830 to 1860 shows that the annual state income from forestry came to a maximum of 31% (1846) and a minimum of 18% (1855) of the total revenue. Table 11.1 illustrates the changes in these figures over the three decades.[3] For the state budget the proceeds from the sale of wood were almost as important as the indirect taxes. These primarily financial interests could best be served by a forest management that focused on the production of wood. So a maximum output of wood – preferably timber because it was most profitable – was the financial and economic aim of the official forest policy. All the other forms of forest usage – that is, the whole concept of multifunctional forest usage – had thus become an obstacle to the fiscal interests of the state.

The forest service was the vital instrument for the enforcement of political guidelines and for putting the authority's financial interests into practice. On the one hand, it helped to suppress all forms of agrarian forest usage and, on the other hand, it organized the general implementation of new methods for the cultivation and management of the forests.

Table 11.1. Significance of forest revenue.

Budget item	1830	1839	1840	1846	1848	1855	1856	1859
Currency	Gulden	Gulden	Gulden	Gulden	Taler	Taler	Taler	Taler
Indirect taxes	49,000	52,000	48,000	49,000	28,000			
Forest revenue	43,000	50,000	50,000	62,000	34,000	32,370	32,030	33,320
Total revenue	180,000	181,980	184,120	197,530	124,580	178,850	145,000	173,600
Forest revenue (percentage of total)	24	27.5	27.2	31.4	27.3	18.1	22.1	20.2
Government spending	–	173,311	212,195	225,096	113,248	138,058	147,857	160,181

This role worked out not least for the benefit of the forest service itself. It made the service a central part of the public administration. The result of this can be read from the staff planning. I have assembled the available material about the quantitative development of the forest service and the result seems quite remarkable: between 1820 and 1850 the size of the service expanded by approximately 80% and in 1851 the expenditure on personnel alone amounted to 39% of the total revenue of the public wood trade.

So far, I have sketched a multitude of interests the forests had to face and the needs they had to provide for. Each group – farmers, tradesmen, craftsmen, the rural population in general, the authorities, the representatives of the iron industry, the timber merchants and the forest service – tried to safeguard its own interests. At some point, these interests could no longer be reconciled with each other. This point was finally reached when the forest service emphasized and generally enforced the new forest management methods.

In this economic system, which aimed at a monofunctional usage of the forest in order to maximize wood production, there was less or no room for most other traditional forms of forest usage – at least theoretically. It has to be considered that the new methods of cultivation, harvesting and management could only be introduced gradually, and the authority's sphere of influence was limited by the possessory rights. Although the forest administration had gained substantial rights to regulate the management of the local community's forest, it was always faced with the limits of its power. About 40% of all forests were private property and hence almost completely removed from state influence.

Here a significant feature of this system of relations becomes evident: the interests of liberal government politics were not always identical with the concerns and priorities of the forest service. The latter never succeeded in its efforts to extend its power to the forests in private ownership.

The rural population was entitled to make use of a diversity of forest resources by a subtle system of traditional rights. These legitimate claims represented a permanent obstacle to a fully effective enforcement of modern wood production methods. It took more than 150 years to settle these rights of forest usage. During the first half of the 19th century the complex process of redeeming these property rights gained some speed, but at the same time the growing population's material needs for forest resources increased.

In the politically explosive atmosphere of the years 1848 and 1849, these antagonistic interests turned into open conflict. Wood theft, which had always been a fact of everyday life in rural areas, now became an open demonstration of organized protest.[4] Aggression, which had been pent up over a long period of time, was now directed at the representatives of the power which had limited the freedom to make use of natural resources, and in many rural areas the members of the forest service were the almost ideal personification of a patronizing government.

Before concluding my analysis, I need to point to a problem which, since the 1980s, has caused a lively and long-lasting dispute among historians, historical geographers and forest historians. In the historical discussion about wood supply, the management of natural resources and energy the problem of a potential 'wood shortage' has become a popular catchword. Drastically simplified, this dispute can be summarized as a collision of two positions or interpretations:

1. On the one hand, there are those who proclaim the existence of a general wood shortage in the 18th century, which, in their opinion, played the role of a pacemaker in the process of using fossil sources of energy instead of traditional power sources, such as wood and water.
2. This opinion is opposed by those who doubt that there has ever been any kind of general wood shortage, that is, more than just local resource problems. They claim that the widespread lament about a threatening wood shortage was mainly an instrument of political rhetoric abused by the authorities to legitimize their economic interests.[5]

Let me make one thing clear from the outset: I have not made the attempt to construct an arithmetical work of art trying to give ultimate proof for an opinion. I think it is neither necessary nor realistic nor economically useful to try to collect the data required to quantify actual wood production and wood consumption. Even for a territory of rather limited size and even for a period of time when statistical sources gradually become more reliable, it seems to be somewhat naïve and unrealistic to hope for more than rough approximations.[6]

Contrary to some older estimations from the 18th century forecasting a serious wood scarcity for the Hunsrück area, the representatives of the contemporary forest administration hardly ever mentioned this issue. Some sweeping criticism can be read in the sources blaming the Napoleonic occupying power for exploiting the local forests. But, apart from these projections of contemporary problems to the former foreign authorities, there is almost no evidence of a lack of wood resources. As early as the 1820s, the public forest service even praised the good condition of private and communal forests. This is more than remarkable, since these non-governmental forests were under only limited public control, and traditionally the forest administration tried to enlarge its influence over these private areas. What is more, both the forest service and the authorities were integrated in the contemporary discourse about a wood shortage.

In 1830 the director of the public forest service gave an annual survey of the sustainable yield of the state forests and put it in relation to the popular demand. Summarizing his findings, he concluded that the situation was far from any kind of wood shortage, although the wood produce of the private forests was considerably lower than that of the public forests. He added that the supply situation even made it necessary to export wood to the neighbouring regions.

Judging from the material I have found about the wood supply – or at least the wood on offer – in the principality of Birkenfeld in the first half of the 19th century, at least one conclusion can be drawn: the situation in this specific region over this particular historical period gives no evidence of anything like a general wood shortage. However, what is certain is that the question of an effective and reliable wood supply for the whole population was a highly sensitive political and social issue. For the individual situation particularly of the socially and economically disadvantaged, soaring firewood prices could rapidly become a heavy burden. Such a situation of individual need could easily turn into a dangerous private 'wood shortage' for the poor, regardless of the actual wood reserves in the region and the condition of its forests.

To conclude my analysis, I should just like to emphasize that studies which focus on a distinct and compact region can offer quite substantial arguments to improve and enhance the general discussion of the central issues of forest history. Reliable research into the historical conditions that formed the specific relations of individual communities to their natural environment is not only a vital prerequisite for discussions on a more general level. It can also help widen our understanding of the central characteristics of everyday life.

Notes

1 See Eckardt and Schmidt (1987).
2 Based on my own research.
3 Based on Oldenburg budget files and contemporary state descriptions.
4 Blasius (1978), p. 56.
5 See Radkau (1983, 1986) and Schäfer (1992). Compare with Radkau (1996).
6 For evaluations of regional wood supply in Germany, see Schäfer (1992); Selter (1995) and Schenk (1996).

References

Blasius, D. (1978) *Kriminalität und Alltag. Zur Konfliktgeschichte des Alltagslebens im 19. Jahrhundert*. Vandenhoeck and Ruprecht, Göttingen.

Eckardt, A. and Schmidt, H. (1987) *Geschichte des Landes Oldenburg. Ein Handbuch*. Holzberg, Oldenburg

Radkau, J. (1983) Holzverknappung und Krisenbewußtsein im 18. Jahrhundert. *Geschichte und Gesellschaft* 9, S.513–S.543.

Radkau, J. (1986) Zur angeblichen Energiekrise des 18. Jahrhunderts: Revisionistische Betrachtungen über die Holznot. *Vierteljahrsschrift für Sozial- und Wirtschaftsgeschichte* 73, S.1–S.37.

Radkau, J. (1996) Wood and forestry in German history: in quest of an environmental approach. *Environment and History* 2, S.63–S.76.

Schäfer, I. (1992) *Ein Gespenst geht um. Politik mit der Holznot in Lippe 1750–1850. Ein Beitrag zur Wald- und Technikgeschichte*. Naturwissenschaftlicher und Historischer Verein für das Land Lippe, Detmold.

Schenk, W. (1996) *Waldnutzung, Waldzustand und regionale Entwicklung in vorindustrieller Zeit im mittleren Deutschland. Historisch-geographische Bezüge zur Erforschung von Kulturlandschaften in Mainfranken und Nordhessen*. Steiner, Stuttgart.

Selter, B. (1995) *Waldnutzung und ländliche Gesellschaft. Landwirtschaftlicher 'Nährwald' und neue Holzökonomie im Sauerland des 18. und 19. Jahrhunderts*. Schöningh, Paderborn.

Historical Interactions Between Agricultural and Forest Utilization in the Austrian Alps and Their Impact on the Present Forest Condition

12

Elisabeth Johann

Österreichischer Forstverein, Arbeitsgruppe Forstgeschichte, Bertoldstrasse 17, D-79085 Freiburg, Austria

No other section of human activity is connected more closely with the alpine region's population and its manifold needs than forestry. In the past, the forest was very much governed by other industrial activities. Very often, the forest had only been used as a helpful tool for another branch of industry, agriculture, certain trades or industries and hunting.[1]

The relationship between agriculture and forestry is an example of close interconnections and essential complementation, as well as much controversy. In the past, this assessment was valid, especially in the alpine regions of Central Europe, and even today it may still be observed in many countries of the Third World. For many centuries the forest served solely agricultural needs and only at the beginning of the 19th century, due to the development of modern farming, did this kind of dependency come to an end. Particularly with regard to forest utilization, there had been a very close connection between the rural population and the forest. The most important product was not wood. There were other products of the forest that served people's needs. Woodland grazing and forest litter utilization are examples for the forest in Austria, which was for many centuries dependent on agriculture, whose needs had to be met.

This chapter explores to what extent this kind of dependency really took place in the alpine region and how far the rural population's living could be ensured by means of forest utilization. It also investigates whether the agricultural utilization on its own was responsible for the evident degradation of alpine forests or whether there were additional deceptive factors and underlying conditions, making a decisive contribution to degradation.

(eds M. Agnoletti and S. Anderson)

1 Rural living conditions in dependence on forest utilization

In the pre-industrial period, the rural population's income was influenced by various factors. In the mountainous region of Austria the most important were: (i) the dimension (size) of available arable land per family (farmstead); (ii) local conditions of yield; (iii) the development of prices for agricultural products; and (iv) taxes and revenues (Table 12.1).

From the 16th century a shortage of arable land was noticeable in the alpine region of Austria, which led to a reduction in the size and extent of new settlements (Fig. 12.1). Local site conditions and climate further limited the extent of arable land. The importance of the size of farmsteads for the rural income and its impact on the present forest condition has seldom been discussed in economy history.[2]

To make a precise statement about the economic situation of the rural population, the economic and social facts of a selected unit, a domain of about 5000 ha, including 14 communities, have been investigated. This domain can be considered as representative, with regard to geographical site, topography, settlements and economic development, of the alpine region of Austria. Furthermore, the results have been compared with other samples and with the contemporary literature of the 19th century.[3]

At the beginning of this investigation, there was every reason to believe that the available local resources were sufficient to nourish the local population. It was assumed that the complaint about the shortage of yield and the obvious poverty of the rural population at the beginning of the 19th century arose from the increasing uses by the mining industry. But it was determined that the shortage of arable land and existing conditions for production for the

Table 12.1. Forest area in the Austrian Alps (19th century).

	Forest area as percentage of the total area (%)	Forest area as percentage of the cultivated land (%)	Forest area per capita (ha)
Salzburg	30	45	1.8
Styria	49	57	2.2
Tyrol	34	58	1.4
Carinthia	37	45	2.3

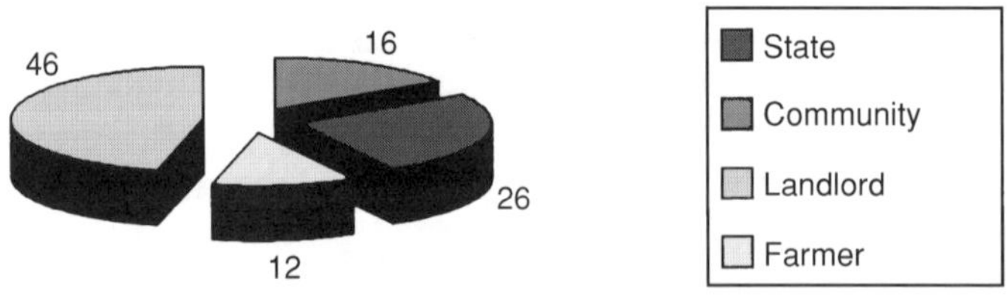

Fig. 12.1. Forest area in categories by ownership (%) (19th century).

rural economy had a greater effect on poverty than increasing interests in fuel wood and timber harvesting.

The most important factors which determined the living conditions of the rural population were: (i) size, extent and quality of cultivated land available for each family (farmstead); (ii) number of livestock being able to live during the winter period on the fodder grown on their own land; and (iii) possibilities to make up the expected actual lack of income.

The examination of yield and agricultural income of the rural population based on stock and arable farming followed the determinants listed below (Table 12.2):

1. Individually owned useful (effective) areas:
   - Garden.
   - Arable land.
   - Meadows.
   - Alternate fields.
   - Pasture.
2. Commonly owned useful areas:
   - Pasture.
   - Alpine pasture.
   - Forest pasturage.

1.1 Size and extent of farmsteads

Following from the assumption that each farmstead was originally created to an extent that enabled the family to support themselves adequately (private land and forest rights on common land), an increasing population put pressure on the little land available for agricultural utilization not later than the 16th century. As a result, the size of new farmsteads was reduced and existing farmsteads were divided into two to four parts. Small-scale agriculture forced people

Table 12.2. Landed property per farmstead (ha) (18th–19th century).

	Category 1 (ha)	Category 2 (ha)	Category 3 (ha)	Category 4 (ha)	Category 5 (ha)
Arable land	0	0.3	1.25	1.6	2.3
Meadow	0	0.5	1.6	3.8	6.7
Cultivated land	0	0.8	2.85	5.4	9
Pasture	0	0.05	0.25	2.2	5.3
Forest	1.5	1.4	0.85	2.7	2.1
Grazing ground	1.5	1.9	1.1	4.9	7.4
Total area	1.5	2.7	3.95	10.3	16.4
Considered minimum size of cultivated land				8.00	

to find jobs or part-time jobs in other professions, such as tradesman, miner, charcoal burner, transport and timber harvesting.

The definition of the existing rural classes (categories) was dependent on different factors, for instance age of settlement, structure of landscape or social relationship (Fig. 12.2 and Table 12.2). The available common land consisted of 1350 ha forest area, 115 ha pasture and 85 ha alpine pasture, altogether 1560 ha, situated on steep slopes on the southern side of the valley from the bottom of the valley (500 m above sea level) to the top of the mountain (2000 m above sea level).

1.2 Possibilities and limitations of making a living from agricultural utilization

To ensure the farmer's family nourishment (in consideration of the calorie content), which was composed of 70% cereals and their products, 10% crops from the garden and 20% livestock breeding, a family consisting of five to six members required a small garden, livestock consisting of two cows and a few pigs and a crop of 2.6–3.1 quintals per person. Taking into account the conditions of yield existing in the 18th/19th centuries and the local conditions of the mountainous region (crop harvest only three times the seed), an individual utilizable land of about 8 ha was necessary to supply a farmstead sufficiently (Tables 12.2 and 12.3).[4] When potatoes were included in the nourishment, the demand for arable land decreased. However, this crop was not introduced into the agricultural practices of the alpine region to a significant degree until the middle of the 19th century.

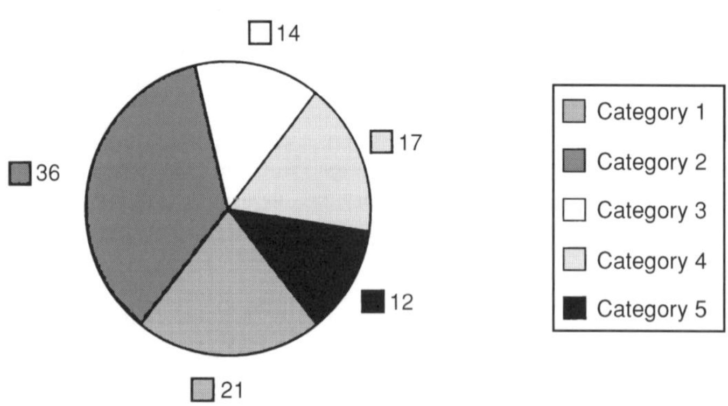

Fig. 12.2. Number of farmsteads in reference to dimension of private land (43 farmsteads = 100%). Category 1: less than 0.5 ha arable land and meadows; category 2: 0.5–2.5 ha; category 3: 2.5–5 ha; category 4: 5–7.5 ha; category 5: 7.5 ha and more.

Because of the lack of utilizable land, farmers were not able to make their living on it without an additional income in the mountainous region of Austria. During the heyday of the gold- and iron-mining industry, small-scale farmers were able to earn some money, but later additional jobs were limited by the gradual breakdown of this industry from the 17th up to the 19th century (depending on the site). Therefore, they were forced to fall back on the only reserves available in the mountainous region area until the end of the 19th century: cattle breeding supported by the forest (Fig. 12.3). This was the reason why livestock stocking levels per farmstead were remarkable considering the extent of the land owned (Table 12.3). The intended purpose of

Table 12.3. Number of cattle per farmstead (average 6.7 ha^{-1} $year^{-1}$) (18th–19th century).

	Horses	Cows/ oxen	Calves	Sheep/ goats	Number of cattle supplied by own fodder	Number of cattle supplied by forest grazing
Category 1	0	3.3	0.3	9.8	0.8	9.0
Category 2	0.2	6	0.8	10.6	2.5	8.1
Category 3	1	12	0	20	7.5	12.5
Category 4	0	14	0	24	11.8	12.2
Category 5	2.4	25.5	0	27.6	18.8	8.8

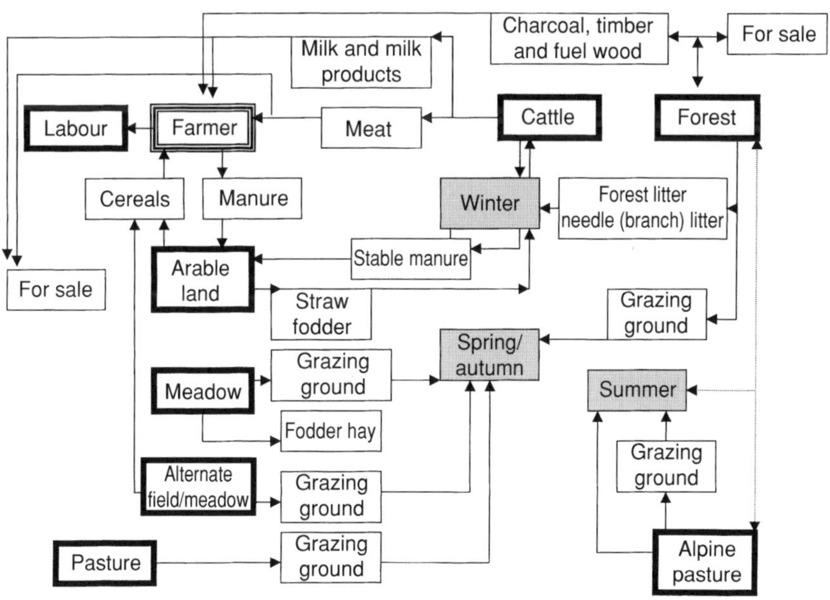

Fig. 12.3. Substance flow in a self-supplying farmstead in the alpine region (18th–19th century).

this increased number of cattle was, apart from the possibility of earning some money by selling (which in fact was forbidden by rules and orders), to produce a high amount of stable manure for agricultural use. Because straw served to feed the livestock, this manure could only be made up by application of forest litter to an extent considered to cause a severe impact on forest conditions.

2 Impact on present forest condition

2.1 Forest litter utilization

In the alpine region of Austria, forest litter utilization has not only been practised since the adoption of stable feeding, which came into being at the end of the 19th century. It was one of the most important forest rights already connected with the farmstead since the first settlement. The demand was constantly high and increased with increasing livestock. Several regulations and laws published by owners and the government and introduced by foresters tried to regulate this right by limiting amount, time and location, but without success. In the opinion of the rural population, forest litter utilization was considered to be a regular and natural utilization of forest areas. Especially in regions dominated by small-scale farming, forest litter utilization caused a severe conflict between the rural population and the forest management. Even when the forest rights were generally regulated and removed by law in 1853, forest litter utilization was still a political issue and played an important role in gaining votes in the first decades of the 20th century.

Ground litter utilization

The harvesting of ground litter led to severe degradation of forest soils through the loss of mineral and organic compounds, the decreased percentage of nitrogen and changes in the structure of the humus cover. In relation to site requirements, the degradation was estimated to be up to 75%. As a consequence, the increment of trees was reduced, for instance, of beech from 8 to 50% and of pine up to 30%. According to the frequency of harvest, the impact increased. In harvesting forest litter for the first time in the forest environment, the loss of mineral compounds was as high as the demand for nutrients of the growing trees on this area over 15 years.[5] It also prevented natural regeneration by destroying and injuring indigenous seedling plants and roots.

Branch litter utilization

Branch litter utilization was preferred by the farmers to ground litter utilization for two reasons. Even using iron rakes (which was forbidden in fact by various rules and orders), ground litter collection was difficult (because of the roots and stones) and could only be done during the summer period. However, branch litter utilization could also be done during the winter, depending on sufficient available time and weather conditions.

Usually branch litter utilization started at a tree age of about 30–40 years. It was achieved by climbing with irons high up to the crown of the trees while cutting the lower branches and was repeated every 10 years (if it was done carefully), in total five to six times until the final cutting. The use of climbing irons (already prohibited not later than the forest laws published by Maria Theresa in the middle of the 18th century) had an additional impact on the trees, injuring the bark and stems. The annual harvest from 1 ha forest resulted in the removal of nutrients of about 1000–1700 kg forest litter per year.[6]

The impact of forest litter utilization on the mountain forests of Austria included the removal of important nutrients, the disadvantageous changes in structure and quality of soil and the lack of natural regeneration. The decreasing increment and the impairment of health of the forest environment were remarkable. The impact on the quality of the forest has to be considered of long duration. In many mountainous regions even today, the forest has a reduced and unsatisfactory yield due to repeated forest litter utilization in the past.[7]

2.2 Forest pasturage

Forest grazing has always been thought to increase the damage done to the forest area. Because of the lack of fodder for livestock-breeding, farmers were totally dependent on forest pasturage (Table 12.3). In rural regions the impact on the forest increased in relation to the density of population (due to immigration by miners, workers and charcoal burners) and the increasing construction of roads. Forest grazing was also practised in coniferous forests. At the turn of the 19th century, almost every mountain forest area was utilized for grazing even at a considerable distance from the village. In the domain that has been investigated, the annual demand for forest grazing amounted to 6.7 cows per ha (compared with a compatible annual maximum of about 1.5 ha per cow).

Forest grazing has always been a very important forest right connected with the farmstead. According to this right, each farmer was allowed to drive the number of cows, horses, oxen, sheep and goats into the forest that he was able to feed during winter time. Forest grazing took place everywhere in the forest, but the farmers preferred the large clearings cut by the mining industry which offered good and sufficient fodder for 20–50 years. Therefore forest owners and forest authorities, as well as the mining industry, complained about the damage of forest grazing to natural regeneration from the 16th century onwards. To improve natural regeneration, forest pasturage was restricted and reduced by various regulations in all parts of Austria. But neither restrictions nor punishment were able to stop farmers from practising their right. As a consequence, natural regeneration failed and forests were degraded. In addition, forest grazing in the alpine regions was for many centuries the reason for the lowering of the timberline partly down to 400 m because of erosion and trampling.

It was mostly sheep and goats that were considered to damage natural and artificial regeneration and the grazing of goats in particular was strictly forbidden by law in the alpine region. Because of the lack of foresters and the needs of the rural population, these orders were never really followed. It becomes evident that the duration and increasing extent of forest grazing over several centuries put tremendous pressure on the forests and have contributed to the poor conditions in the forests noticeable in the middle of the 19th century.

Improved technology in agricultural management, for instance additional fodder from clover and the cropping of potatoes, at last made stable-rearing possible and was able to reduce the impact on the forest area. In 1853 forest grazing rights were removed by law but remained and still exist in some mountainous regions where farming is still dependent on forest grazing.

3 Conclusion

It is evident that it was not only forest utilization and forest grazing that were responsible for the unsatisfactory condition of forests in the middle of the 19th century. Additional determinants were overcutting for the supply of the mining industry, particularly the salt- and iron-mining industries, the financial interest of the government and the economic policy of mercantilism. The resulting impact on the forest has generally been of short duration. The implementation of improved silvicultural and management methods was able to compensate for damage within several decades. The decline of soil fertility caused by forest litter utilization and forest grazing resulted in permanent damage. The time needed for regeneration of the forest ecosystem impaired by forest grazing has been estimated by Kubiena to be between 300 and 500 years.[8]

It has been proved that long-lasting intensive forest grazing and forest litter utilization are responsible for soil compaction, rapid superficial run-off, perceptible soil erosion and decrease of soil organisms caused by the destruction of the soil's water-absorbing capacity. As a consequence, the danger that forest stands will be threatened by certain insects is increasing, leading to an increase in harmful insects on former forest pasturage. An unexpected correlation between agricultural management, for instance the structure and condition of rural farming (including forest grazing and forest litter utilization), and increase in insect numbers has been proved in the Austrian Alps for certain species, for instance the bark beetle. As an example of this development, the extent of bark beetle infestation should be mentioned as having occurred in the government- and privately owned mountainous forests of Upper Carinthia during the first decades of the 19th century.

Up to the middle of this century agriculture took a heavy toll from forest land, especially in the Austrian alpine regions. However, this has been by no means a particular feature of the forest history of Austria. Moreover, it is rather indispensable and in accordance with a law of nature that, anywhere on earth, subsistence agriculture is dependent on the benefits offered by the forest. It is only with advanced developmental conditions that the forests can get rid of their service-rendering connection with agriculture.

Notes

1 Hasel, K. (1968) Die Beziehungen zwischen Land- und Forstwirtschaft in der Sicht des Historikers. *Zeitschrift für Agrargeschichte und Agrarsoziologie* 16(2), 142–159.

2 Henning, F.W. (1969) Die Betriebsgrößenstruktur der mitteleuropäischen Landwirtschaft im 18. Jahrhundert und ihr Einfluß auf die ländlichen Einkommensverhältnisse, *Zeitschrift für Agrargeschichte und Agrarsoziologie* 17(2), 171–193.

3 Archiv FV Lainach: Forstamtsakten, Urbarium 1683, Protokollbuch, 1690 ff.

4 Konold, W. and Hackel, A. (1990) Beitrag zur Geschichte der Streuwiesen und der Streuwiesenkultur im Alpenvorland, *Zeitschrift für Agrargeschichte und Agrarsoziologie* 38(2), 180–181.

5 Rebel, K. (1920) *Streunutzung, insbesondere im Bayerischen Staatswald.* Diessen vor München, pp. 36–42.

6 Wessely, J. (1853) *Die Österreichischen Alpenländer und ihre Forste*, Vol. 1, Verlag Jos. C. Huber, Vienna, pp. 222–224.

7 Ebermayer, E. (1891) Die Waldstreufrage, *Forstlich-naturwissenschaftliche Zeitschrift. München* 2, 1–11.

8 Hasel, K. (1968) Die Beziehungen zwischen Land- und Forstwirtschaft aus der Sicht des Historikers, *Zeitschrift für Agrargeschichte und Agrarsoziologie* 16(2), 141–149.

Exploiting the Last Phase of the North European Timber Frontier for the International Market 1890–1914: an Economic–Historical Approach

13

Jörgen Björklund

Department of Economic History, Umeå University, S 90187 – Umeå, Sweden

During the three decades before the First World War, Sweden, the world's largest exporter of softwood, was exposed to very hard competition from Russia proper. Half of Russia's export went to Great Britain. Other important export markets included Germany, Holland, France and Belgium. By the outbreak of the First World War, Russian timber exports had surpassed Swedish exports. Especially the sawmill industry of northern Russia expanded quickly. The number of sawmills doubled, the workers increased eight times, while the production value grew five or six times during this period. Furthermore, about 20,000 people were employed in forest work and floating at the turn of the century.

The forests of northern Russia are enormous and were at this time regarded as the last large reserve of raw materials in Europe. The forests of the province of Archangel alone were almost twice the Swedish area. In addition it was possible for the north Russian sawmills to utilize the vast forests in the adjoining provinces. North pine and spruce predominated and were growing to very large dimensions. The main owner of the northern forests was the Russian state.

To a large extent the Russian sawmill industry was established with the support of foreign capital. This foreign engagement was the last phase of the so-called North European Timber Frontier movement, with new areas of virgin forests constantly being opened, and the timber frontier was based on both technological and institutional preconditions. With the October Revolution in 1917, the Russian sawmill industry and the capital assets of the foreigners were confiscated by the Bolsheviks.

1 Introduction

Economic history belongs to the social sciences. Economic history deals both with the use of natural resources, such as water power, and with economic and social change. Also, of great importance and interest in the research of economic history is the interplay between economic growth and its current influence on the forests. This research is focused both on forestry and on the forest industry. These studies can be directed either to the macro level or to the micro level.

The Swedish industrialization process started around 1870. Up to 1970, Sweden probably had the fastest economic growth in the world. However, Swedish industrial and economic growth accelerated substantially from the 1850s, so, during a period of 120 years, Sweden has developed from an agrarian, rather poor country to an industrial welfare state. Among the OECD countries in 1970, Sweden's GDP per capita was ranked as number four, which also meant that she was one of the richest countries in the world.

Sweden's transformation from an agrarian society to an industrial society can be divided into three periods. The first one, 1790–1850, was characterized by a commercialization of agriculture and the development of a domestic market. The period 1850–1890 saw an upswing in staples with a rapid expansion and an early industrial breakthrough, especially in the sawmill industry. The third period, 1890–1930, saw the breakthrough of the modern industrial society, when the Swedish engineering industry, together with the pulp and paper industry, developed quickly.[1] From 1930 to 1975 there was a fourth period, when Sweden entered the so-called mature industrial society. From the middle of the 1970s Sweden has been on her way from an industrial society to a post-industrial, or IT, society.[2] However, today Sweden is far from being top rank among the OECD members.[3]

To simplify, we can say that the Swedish economic growth and her development towards a welfare state to a large extent depended on the exploitation of forests, iron ore and power. For the international market particularly, the exploitation of the forests was of great importance and contributed to the creation of the Swedish welfare state (Table 13.1).

Table 13.1. Swedish export of timber and pulp and paper products.

	Percentage of the total export value						
	1881/85	1896/1900	1911/13	1924/25	1934/38	1951/54	1959
Timber products	40.4	42.8	26.1	22.0	13.2	11.8	8.3
Pulp and paper	4.6	8.1	17.6	27.2	28.3	30.5	22.4
Total forestry industry	45.0	50.9	43.7	49.2	41.5	42.3	30.7

Source: Fridlizius, G. (1963) Sweden's exports 1850–1960. *Economy and History* 6, 30, 55 and 76.

Before the turn of the 20th century sawn timber constituted the major part of the exports. At the beginning of the 20th century, there was a dramatic reduction in the export of timber products. Simultaneously there was a rapid increase in pulp and paper. This meant that, up to the middle of the 1950s, more than 40% of Swedish exports fell on the forestry industry. Thereafter the forestry industry has continued to diminish its proportion of the export value, but even today the forestry industry is the largest earner of foreign exchange in Sweden.

In the 1880s Sweden was the greatest exporter of sawn timber in the world, the Swedish proportion comprising 24%. Up to the beginning of the 20th century she almost kept this share. However, during the decade before the First World War, Sweden's share of the world market decreased more than that of many other countries. The main reason for this was a considerable diminishing in felling dimensions.[4] This meant, first, that the average sales price per unit decreased because the highest relative prices could be charged for greater dimensions, and second, production costs increased because the smaller dimensions required more work per unit volume at all stages of production: cutting, driving, floating and handling at the sawmill and in the timber-yard. This increase in production costs became all the more serious since the price of raw materials rose more rapidly than the sales price.[5] Thus there were indications that a slow degradation of the very foundation of the sawmill industry, the raw materials, was taking place. At this stage there were entrepreneurs, some of them from Sweden, who took advantage of the opportunities provided by the upswing in the business cycle at the end of the 1890s in order to exploit the forests of northern Russia.

From the turn of the century up to the outbreak of the First World War, Russia's export quantities more than tripled. In 1913 Russia was the greatest exporter of wood, well ahead of Sweden. In this process, the forests and the sawmill industries of northern Russia played an essential role.[6]

The development of the north Russian sawmill industry during the decades around the turn of the century took place within a market economy. After the October Revolution in 1917, the market economy was replaced, first by the war, then by the new economic policy and finally by Stalin's planned economy. After the collapse of the Soviet Union at the beginning of the 1990s, Russia is once again on her way towards a market economy. But, before getting there, many severe problems have to be solved, and these problems will be much more complicated since it is the first time in history when a state has had to transfer from a planned economy to a market economy. However, more than 80 years ago, the Russian industry was operating in a market economy. From that time there must be many important things to be learned. In northern Russia the contemporary problems are still more complicated. Besides the economic problems for the forest industry, the environmental pollution of the forest, the air and the Barents Sea have to be solved. To apply an economic-historical perspective to these problems is a relevant and challenging task for historians.

In this chapter, which is also a comparison between the province of Archangel in Russia and the sawmill district of Sundsvall in Sweden, the following questions will be discussed:

1. What characterized the technological preconditions and the institutional preconditions which took place to facilitate the development of the timber industry and the exploitation of timber in northern Russia during the decades before and after the turn of the 20th century?
2. How was export of sawn timber developed and to which markets were the Russian items shipped?
3. Where did the exploitation of timber take place and what characterized the felling dimensions?

2 The breakthrough of the Swedish sawmill industry

As a result of a sharp increase in demand in England and on the European continent, the Swedish sawmill industry was able to make a breakthrough in quantitative terms in the middle of the last century. The Swedish industry adjusted quickly to the new market situation. Earlier, timber production had been predominantly concentrated in western-central Sweden, particularly the county of Värmland. Timber exports went via harbours on the Swedish west coast, where the city of Göteborg was an important centre.

As the sawmill industry grew, the need for raw materials increased considerably. The old areas of production could not alone meet the increase in demand. One reason for this was that the forests of Värmland had been ravaged not only by the forestry industry, but also by the iron mills. Therefore, the sawmill owners began to turn to northern Sweden, Norrland, the forests of which had hitherto only been exploited to a relatively small extent, and which in many cases were of a primeval character.

Sawmills powered by water-wheels were already in existence, although the level of production was minimal, but, thanks to the newly invented steam engine, it was possible to establish much larger units on the coast. Timber was transported to the steam-driven mills on the rivers. In this way, more and more remote forest areas became accessible for industrial exploitation.

Sawmill entrepreneurs from Norway, Värmland and other places came to Norrland. The majority of entrepreneurs, however, had experience from the earlier water-sawmill epoch in Norrland. To simplify matters a little, we can say that the first steam sawmills were established in southern Norrland and that the industry gradually spread northwards. A so-called frontier movement of sawmills towards the north was established.[7] Up until the First World War, Västernorrland was the most important sawmill county in Sweden, with three dominating districts: Sundsvall, Härnösand and Örnsköldsvik. Of these districts Sundsvall was the largest. By the end of the 19th century, the Sundsvall district had the highest concentration of sawmills not only in Sweden but in all of Europe. There were no fewer than 43 steam sawmills in this area. In the Västernorrland county there were 106 steam sawmills. By this time, the water-driven sawmills in this county had practically disappeared.[8]

Within a 25-year period, Västernorrland took over the role that Göteborg had had as Sweden's most important timber product-exporting district. Parallel to that, Sweden's total export of timber products increased from about 500,000 m^3 in 1850 to 2.5 million m^3 in 1875, to reach 5 million m^3 by the turn of the century. Between 1850 and 1895, the export of timber products from the Sundsvall sawmill district increased from 31,000 m^3 to 790,000 m^3, or 25 times.[9]

Consequently, in the Sundsvall district the need for raw materials increased enormously. Between 1861 and 1865 the total yearly consumption of logs at these sawmills was 0.63 million pieces. Ten years later this amount had almost quadrupled to 2.35 million logs. The consumption of logs continued and, during the period 1896–1900, had reached no less than 6.88 million per year. It was impossible to provide the sawmills in Sundsvall with raw materials gathered only from within the county. The forestry companies had to turn to the forests of the adjacent counties of Jämtland and Västerbotten. But this was not enough either. The exploitation continued first to Norrbotten and later to northern Finland, from where the timber was floated across the Gulf of Bothnia to Sundsvall.[10]

From the middle of the 19th century up to the beginning of the 20th century, Sweden was the leading exporter of sawn timber. In 1880, when the world's timber exports were 13.5 million m^3, Sweden was responsible for a quarter of the total. During the decades up to the First World War, timber exports more than doubled. By this time, Sweden's share of the market had dropped to a sixth. Tsarist Russia, i.e. Russia proper plus the Grand Duchy of Finland, had a total export in 1913 that amounted to no less than one-third of the entire world timber trade. Thus, tsarist Russia had clearly overtaken the former market leader, Sweden.[11]

3 The North European Timber Frontier

Up to the middle of the 19th century Norway was the largest exporter in Europe of sawn timber. In an interesting essay, *Veien mot øst* (The road to the east), the Norwegian economic historian Francis Sejersted deals with the gradual changes in the fields of activity of the Norwegian sawmill entrepreneurs from *c.* 1850 to 1930. Sejersted shows that this process started around the middle of the 19th century and was directed eastwards – first across the Fjord of Christiania in southern Norway and then to parts of central Sweden. Thereafter the process went through northern Sweden and Finland and finally to

Russia. Sejersted characterizes the process as a sort of frontier movement, with new areas constantly being opened. The centre of gravity in this movement was, as time passed, moved constantly further to the east and north-east. The aim of the expansion was to utilize raw materials from hitherto unexploited forests, often of a virgin nature. This process is defined as the North European Timber Frontier and was based both on technological and institutional preconditions.[12]

The technological preconditions induced the so-called transport revolution, which began with better sailing ships, and was speeded up when the railway, the steamship and the telegraph were invented. It was when these means of transportation came into being that it became economically possible to fetch raw materials from distant areas.

The institutional preconditions were liberalism – the open society whose fundamental principle was free enterprise. The decreasing supply of raw materials, first in Norway and then in Sweden, raised the prices of timber. Because of this, the timber industry tried to get control over the supply of raw materials by purchasing forests of their own. After a while there was in both countries a reaction from farmers and politicians, which led to concession laws. There were particularly strict laws against the purchase of forests. It was prohibited for limited companies to buy forests of their own. The regulations were strict for private persons, too. This meant that, in a situation with a shortage of raw materials and with rising prices for these goods, it was impossible for sawmill entrepreneurs to control this supply. Borrowing a concept from the field of migration research, the situation for the Norwegian and Swedish entrepreneurs as regards raw materials was a push factor. In the large Russian forests, however, the entrepreneurs saw good possibilities of expansion–pull factors.

What, then, were the technological preconditions that allowed Sweden to exploit the forests of northern Russia? At this time there was no production of sawmill equipment in northern Russia. So these articles had to be purchased further to the south. Consequently, the Swedish entrepreneurs bought their equipment in Sweden. In the mid-1880s, the railway between Sundsvall in Sweden and Trondheim in Norway was completed. This railway could be used to transport equipment from Sweden to the Russian sawmills. The material was offloaded in Trondheim for further transportation by steamship. These steamships also enabled transportation of the finished products to their destination markets.[13]

In the past, it had been very difficult to arrange a supply of foodstuffs to the sawmills in northern Russia. Therefore the opening of the railway between Archangel and Vologda at the end of 1897 was of great importance. A railway connection was thus established between Archangel and, among other places, St Petersburg. The importance of this railway link was noted by the British consul in Archangel in his report of 1897: 'its chief use, exclusive of postal and passenger advantages will be to bring cheaper bread to the north . . . this passing from the ways of Middle Ages to something that borders on civilization'.[14] The use of the telegraph made it possible to make quick contacts with timber agents abroad. The telegraph was particularly important in this area because of the fact that the postal network was so bad. Both the Kola peninsula and the Rybatchy peninsula were provided with telegraph stations in 1897. When, at the end of the 1890s, the Swedish sawmill representatives arrived in northern Russia, they were able to take advantage of an infrastructure that was to some extent already established.[15]

4 Some perspectives on the forest resources of northern Russia

Exact details of the extent of the forests prior to the First World War are, in the case of Russia as for other countries, uncertain. According to contemporary calculations, the Russian forests consisted of 155 million hectares (Mha). In the district of Archangel there were approximately 35 Mha. The Swedish figure is 21 Mha. The forest area covered 45% of the total land surface, as against about 50% in Sweden. In addition, it was possible for the White Sea sawmill industry to exploit the approximately 64 Mha which were to be found in the adjacent provinces of Vologda, Olonets, Vyatka and Perm. The industries situated to the west also had the Finnish forests at their disposal. The most common timber sort was Scotch pine (*Pinus sylvestris*), which accounted for about two-thirds of the area, growing in stands with very large dimensions.

The majority of the forest was owned by the state – 96% in Archangel. In order to cover their need for timber, the sawmill companies were referred to the annual government auctions. In order to diminish competition at these auctions, many sawmills entered into a cartel, the Archangel Association of Saw Mill Owners and Timber Exporters. The most common mode of transport for timber was the rivers. From the point of view of floating, they were not as good as the Swedish rivers. The Russian rivers had shallow banks, which often led to flooding. The greatest difficulty, however, was the direction of the stream. Because the rivers flow to the north, the sources are the first to become ice-free. Blocks of ice then pile up in great ridges moving towards the mouth of the river, and this in turn causes flooding. The floated timber was thus easily forced up on to the land, which resulted in increased costs. The ice blockages delayed damming, which further reduced a flotation season which was already short.[16]

5 The Russian sawmill industry

Consequently the most important new element for the Swedish timber trade was the competition that it had to face from tsarist Russia during the three decades preceding the First World War. During this period, the value of Russian timber exports more than tripled. Finnish timber exports began to expand as early as the 1860s, but it was not until the 1880s and 1890s that exports from Russia itself began to play a more dominant role. It was at this time, too, that a rapid expansion took place in the White Sea area. Prior to this, the Russian timber trade had to a large extent been limited to the Baltic ports. The great importance of the White Sea area, however, would become apparent at the turn of the century, when it accounted for 18% of the Russian total. Before the First World War, this proportion had increased to 25%.[17]

The centre for the timber trade in northern Russia had long been the town of Archangel. From the middle of the 1880s this area saw a rapid expansion. During the 15 years up to the turn of the century, exports almost tripled, from 170,000 m^3 to 500,000 m^3. Expansion was particularly rapid during the latter half of the 1890s. In 1895, there were 21 sawmills in the district: in 1900 there were 33; in 1913 there were 44. Even outside the town itself, new sawmills were established and even before the turn of the century these industrial sawmill activities stretched from the Kandalasha Bay in the west to the River Pechora in the east. At this time the total export for the district was in excess of 700,000 m^3.[18]

Foreign interests in the Russian sawmill industry were considerable. Foreign investments in the Russian timber industry increased from 0.2 million roubles in 1890 to 7.8 million roubles in 1900 and reached 24.3 million roubles in 1915. By the turn of the century, 44% of all shares were in foreign hands, as against 6% in 1890. These foreign interests came from countries with a considerable import of timber products, such as England, France and Holland, but also from timber-exporting countries – Sweden, Norway and Finland.[19] It was only in a few of the north Russian sawmills that the foreign ownership was complete. Most of the foreign interests were in joint-stock companies.[20]

6 The timber exports from Archangel and Sundsvall

The value of Russian timber exports has been calculated to have risen from about US$16 million in 1892 to US$54 million in 1913. About 45% of the total export value came to Great Britain in the early 1890s. This figure changed very little before the First World War. Other important export markets for Russian wood products included Germany, Holland, France and Belgium. While the value of Russian timber exports was 3.3 times higher just before the war compared with 1892, the exports from the district of Archangel increased by no less than 7.6 times. Consequently the export development of northern Russia was much quicker than for the country as a whole. But to what countries were the sawn goods shipped (Table 13.2)?

Table 13.2. The export of sawn timber from the province of Archangel by country (%).

	1890	1901	1908	1909	1910	1911	1912	1913
Great Britain	73.3	66.9	61.9	59.8	59.0	60.0	55.6	61.6
France	11.8	9.0	5.8	7.5	8.0	8.4	8.6	6.1
Belgium	3.9	7.6	8.1	10.9	8.9	5.8	7.5	7.9
Holland	7.3	14.0	22.5	19.7	20.1	22.9	24.8	21.4
Germany	3.3	1.0	0.1	0.3	0.3	0.5	2.0	1.1
Spain	0.4							
Africa		0.9	0.8	1.3	2.7	1.4	0.8	1.2
Australia		0.4	0.6	0.5	0.7	0.8	0.7	0.6
Norway		0.2	0.2		0.3	0.2		0.1
Total standards	36,650	135,864	178,629	199,131	223,151	203,028	234,666	287,823

Source: Berättelser om handel och sjöfart. Archangelsk 1891–1914 (Consular and vice-consular reports for Archangel).

After the turn of the century Britain's share decreased considerably, but up to the war she was the predominant market for the North Russian sawmill industry. During the first decades of the 20th century, Holland's importance as an import market enlarged, while the French one diminished. After the turn of the century, two non-European markets, Africa and Australia, appeared. But these markets became insignificant.

With a timber export comprising 287,823 standards in 1913, the province of Archangel was one of the largest sawmill regions in the world. Since 1890 the shipped quantities had increased almost eight times. But what characterized the contemporary development in the sawmill district of Sundsvall?

In Sundsvall, the first part of the 1890s was not encouraging. It was impossible to press the price of timber to match the decrease in the price of the sawn products which took place after 1890. Shortly after the middle of the decade, the situation changed drastically and the sales prices began to rise. Generally speaking, we can say that companies with their own forest resources were placed in a more lucrative position but, for sawmills with limited supplies of forests, the situation was not as advantageous. In the latter category, we can find a number of large sawmill companies in the Sundsvall district that are particularly interesting in this context, since it was from Sundsvall that the sawmill entrepreneurs came who expanded into northern Russia. The development of the sawmill industry in the Sundsvall district should therefore be studied more closely. From the beginning of the 1870s, timber exports had increased almost continuously during two decades. By 1891, when the maximum was reached, the exports had tripled. During the latter part of the 1890s, no notable increase could be seen.

During the 1890s, there was a considerable decrease in the export of larger dimensions. The export of deals and battens decreased by almost a quarter. By the turn of the century, the export from the Archangel district was almost as great as that from Sundsvall. For Sweden as a whole, the value of exports increased almost every year during the 1890s. In 1890, Sundsvall's share was 22%, but this figure dropped steadily until 1900, when it was only 15%.[21] This decrease was a result of the fact that the sawmill had an insufficient supply of logs of large dimensions.

It was in order to obtain such logs that the sawmill companies in Sundsvall were forced to float timber across the sea from as far away as northern Sweden and Finland. But, despite these measures, the timber dimensions decreased considerably during the latter part of the 1890s. At the middle of the decade, an average of 25 logs was required for a volume standard product.[22] By 1900, this number had increased to 42.[23] As mentioned above, there was a degradation of the very foundation of the Swedish sawmill industry, the raw materials. Now, some Swedish sawmill entrepreneurs identified opportunities to exploit the forests of northern Russia.

7 The raw material situation

In the rapidly expanding sawmill industry of northern Russia, the need for raw material increased rapidly. For this reason, the industry was spread further and further away from the production centre, the city of Archangel. By the mid-1890s, attention was already being paid to the Urals and to the areas around the Pechora River, which were still completely unexploited. In this area both Swedish and Norwegian interests were engaged in forest exploitation. But what characterized the felling dimensions of the trees in the province of Archangel compared with those in the district of Sundsvall (Fig. 13.1)?

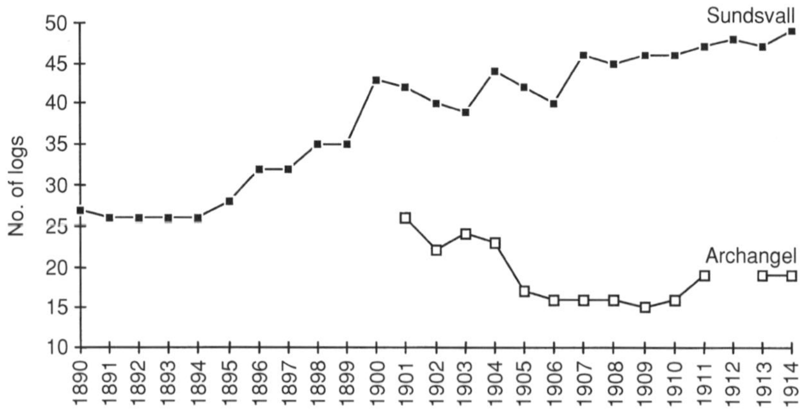

Fig. 13.1. Number of logs per standard in the district of Sundsvall and the province of Archangel. Source: Hammarland (1961), p. 249 (see n. 23) and Berättelse för år 1914 and 1915. Vicekonsulatet i Archangelsk (Annual report from vice-consul in Archangel 1914 and 1915).

Table 13.3. Sawmills in the province of Archangel in 1913. Number of frames (A), number of logs sawn (B), production of deals and boards in standards (C), number of workers (D) and number of logs per standard (E).

Name of the sawmill	A	B	C	D	E
M. Kirkaloff, Archangel	3	118,636	7,067	193	16.8
F.O. Vorobjeff, Archangel	2	8,000	300	22	26.7
J.E. Makaroff, Archangel	4	148,200	4,100	271	36.1
Amossoff, Gernet & Co., Archangel	5	214,292	11,200	715	19.1
S. Kirkaloff's, Majmaxa	4	97,576	5,581	323	17.5
N. Russanoff & Son, Majmaxa	7	193,902	13,161	500	14.7
Nord Trading Co., Majmaxa	6	303,939	16,900	861	18.0
Walneff Brs, Majmaxa	9	387,469	18,000	610	21.5
Pomor Trading Co. (Pryts & Co.), Majmaxa	2	28,258	1,398	55	20.6
Charles Stewart, Majmaxa	6	254,237	13,000	460	19.6
Economy Sawmill, Majmaxa	4	127,596	8,322	430	20.1
Archangel Sawmill Co., Majmaxa	6	187,853	9,500	416	19.8
Atltius & Co., Majmaxa	5	210,346	11,480	750	18.3
E. des Fontains, Majmaxa	8	313,769	21,234	915	14.8
Majmax Steam Saw Mill, Majmaxa	6	296,841	16,500	873	18.0
Appanage Saw Mill, Majmaxa	10	435,837	22,118	800	19.7
Olsen, Stampe & Co., Brebennik	9	305,203	16,917	381	18.0
H. Voelke, Brebennik	1	6,188	465	32	13.3
Petz Bros., Chavarka	3	148,747	9,100	312	16.3
A. Tjudinoff & Co., Zinglominka	5	286,245	17,662	495	16.2
F. Kirkaloff & Co., Kegostow	2	39,993	2,304	280	17.5
Surkoff & Shergold, 6th verst	10	290,000	14,500	865	20.0
Kolibin-Lund, 9th verst	3	119,819	5,851	193	20.5
Rinin Bros. Belogor, Koschka	2	107,340	4,460	190	
Volodin Bros., Pinega	3	86,368	4,455	280	19.4
A. Ruschnikoff, Kamenka, Mezen	5	146,574	8,739	378	16.8
N. Russanoff & Son, Mezen	5	161,768	9,299	516	17.4
Stella Polare, Pechora	10	183,335	14,317	650	12.8
Bache & Wig, Onega	6	126,000	5,800	299	21.7
Bache & Wager, Onega	3	104,416	5,655	294	18.5
Frischenbrüder Bros., Onega	6	121,664	4,866	479	25.0
Onega Wood Co., Ponga, Onega	8	272,872	11,093	666	24.6
Charles Stewart, Motjanoff, Kem	3	98,042	5,476	204	17.9
P. Bjeljajeff's Successors & Co., Soroka	14	460,599	26,170	942	17.6
Kem Sawmill Co., Kem	7	289,420	13,475	618	21.5
V.F. Sawin, Keret	3	115,800	4,426	237	26.1
N. Russanoff & Son, Kovda	4	105,532	5,679	382	18.6
Arendt Berggren, Kovda	6	51,928	2,745	376	18.9
Charles Stewart, Kovda	4	87,857	4,623	321	19.0
P. Bjeljajeff's Successors & Co., Umba	4	119,971	8,407	365	14.3
A. Berggren, Kola	3	32,215	1,700	80	19.0
Total		7,194,647	388,045	17,949	

NB. For three sawmills no figures are given: Golowin & Höst, Kegostow, Vitt & Co., Karitza, and E.R. Volgrin, Mogra, Mezen.
Source: Berättelse för år 1915 vicekonsulatet i Archangelsk. Kommersiella meddelanden. Årg 3, 1916, no 10 &16, bil (Annual Reports 1915 by Swedish Vice-Consul in Archangel).

Between 1894 and 1901, the average number of logs per standard in the district of Sundsvall increased from 26 to 42. This was an increase of no less than 60%. The number of logs per standard in this district continued to rise up to the outbreak of the First World War, which, for the Swedish sawmills, led to deteriorated sales prices and higher production costs. In 1901 the average number of logs per standard in the province of Archangel was 26, i.e. the same as when Sundsvall had entered in 1894. From 1902 to 1910 there was a remarkable increase in the dimensions of raw materials in Archangel, when the number of logs fluctuated between 15 and 19 per standard. In 1914 the average number for the province was still 19.

There were remarkable differences in dimensions between the separate sawmills in the province of Archangel (Table 13.3, column E). However, of the 40 sawmills in the province only 12 (30%) were booked for 20 or more logs per standard. Consequently, for the major part of the sawmills (70%) the number of logs was less than 20 per standard. Such large dimensions had not been seen at the Swedish sawmills since the middle of the 19th century. Furthermore, in 1914, the number of logs per standard in Sundsvall was more than double that in Archangel.

7 Some concluding remarks

From the 1890s until the outbreak of the First World War, the North Russian sawmill industry was in many respects prosperous. From 1895 to 1914 the number of sawmills more than doubled. Simultaneously, the number of sawmill workers increased from 2400 to 18,000 and the production value rose five times. The export quantities increased rapidly and new markets, even outside Europe, were opened. The foreign economic interests in the industry were considerable, while the enormous forests of northern Russia were almost exclusively in the possession of the tsar. In order to cover their need for timber, the sawmill companies were referred to the annual government auctions. The raw material situation was extremely advantageous, from the point of view of both prices and dimensions. Before the October Revolution, Russia was on her way to a market economy. But the Revolution led to the nationalization of the sawmill industry and the confiscation of the domestic and foreign private assets as well. The exploitation of the last phase of the north European timber frontier within a market economy was over.

Notes

1 Schön, L. (1997) Internal and external factors in Swedish industrialization. *Scandinavian Economic History Review* 3, 209–223.

2 Schön, L. (1993) Omvandling och internationellt beroende. In: *Aventyiet Sverige. En ekonomisk och social historia.* UR och Bra Böcker, Stockholm, pp. 266–268.

3 Lindbeck, A. (1997) *The Swedish Experiment*, SNS Förlag, Stockholm, pp. 31–37.
4 Streyffert, T. (1931) *Världens Barrskogstillgångar*. Stockholm, p. 20.
5 Björklund, J. (1976) Strejk-förhandling-avtal. Facklig aktivitet, arbets- och levnadsvillkor bland sågverksarbetare i Västernorrland 1875–1914. Dissertation, Umeå, pp. 32–34.
6 I am studying this development in a research project called 'Exploitation of Timber for the International Market: the Forests and the Saw Mill Industry of Northern Russia 1880–1917', which is financed by the Swedish Council for Forestry and Agriculture Research.
7 Söderlund, E.F. (1951) *Swedish Timber Exports 1850–1950*, Almqvist and Wicksells, Stockholm, pp. 3–115.
8 Wik, H. (1950) Norra Sveriges sågverksindustri från 1800-talets mitt till 1937. Uppsala, *passim*.
9 Söderlund (1951) pp. 156–157.
10 BiSOS H. Landshövdingarnas femårsberättelser för Västernorrlands län 1870–1905 (Quinquennial Reports by the County Governor of Västernorrland).
11 See note 4.
12 Sejersted, F. (1980) *Veien mot øst Vandringar*. Festskrift till Ingrid Semmingsen, Oslo, pp. 163–201. (Sejersted uses political preconditions, this chapter institutional preconditions.)
13 Björklund, J. (1984) From the Gulf of Bothnia to the White Sea. Swedish direct investments in the sawmill industry of Tsarist Russia. *Scandinavian Economic History Review* 2, 28–29.
14 Diplomatic and Consular Reports on Trade and Finance. RUSSIA. FO Annual Series – 1897, pp. 41–42.
15 This study is a part of an interdisciplinary project on cultural borders in time and space in progress at Umeå University, called 'The Dynamic Frontiers of the Forest Exploitation: Timber Frontiers in Global and Local Perspectives'. The purpose of this project is to explore and compare the timber frontier which moved across both North America and northern Europe during the 19th and the beginning of the 20th centuries, with close cooperation between the fields of economic history and forest vegetation ecology. Timber turned out to be a slowly weathering natural resource, which had to be continuously exploited in new areas. The forest which was exploited consisted of rather undisturbed natural coniferous forests, characterized as typical 'old-growth' stands, i.e. the trees were very old and there were a lot of dead trees. Regarding the movement of the advancing timber frontier, there is still a lack of essential information about the historical development, globally as well as locally, nor is there any all-encompassing synthesis and sophisticated comparative analysis of the economic and ecological forces that affect the natural coniferous forests. The creation of the timber frontiers also points at a very interesting aspect of the scientific approach to the borderland issue: the existence of dynamic and moving borders.
16 Lindberg, I. (1916) Norra Rysslands skogstillgångar och sågverksindustri, *Skogsvårdsföreningens tidskrift* pp. 139–156 and *Handbok för Nordisk Trävaruindustri Omfattande Sverige, Norge, Finland och Ryssland*, pp. 671–676.
17 Söderlund (1951) pp. 141–142; Zon, P. and Sparhawk, P. (1923) *Forest Resources of the World*, Vol. I, McGraw-Hill, New York, p. 294 and *Statistiska Tabeller över Skilda Länders Export och Import av Trävaror* 1911–1913, Almqvist and Wicksells, Uppsala.

18 Berättelse om handel och sjöfart. Archangel. Konsulns berättelse för år 1894. Stockholm, 1895, Diplomatic and Consular Reports on Trade and Finance. RUSSIA. Annual Series 1895–1900 and Konsulns årsberättelse för 1915. Archangel. Stockholm, 1917.

19 Björklund (1984) pp. 21–26.

20 Falkus, M.E. (1972) *The Industrialization of Russia 1700–1914*, Macmillan, London, p. 72. About 45% of the total value of the Russian timber exports fell to Great Britain. Other important export markets for Russian wood products included Germany, Holland, France and Belgium (see Zon and Sparhawk, 1923).

21 See note 19.

22 1 standard = *c*. 4.67 cubic metres.

23 Hammarland, T. (1961) Studier i svensk trävaruhandel och trävaruproduktion 1890–1913. Unprinted licentiate's dissertation, Uppsala, p. 249.

14

'Impossible' Roads and 'Inaccessible' Woods: Aspects and Problems of Wood Transport in the 18th Century, Southern Italy

Maurizio Gangemi

Facoltà di Economia, Istituto di Storia Economica, Università degli Studi di Bari, Bari, Italy

By examining the state activity for the recovery of trees suitable for the building of warships in the 18th century, it is possible to re-create the environmental situation in which military and civil engineers, drovers, pack animals and draught animals tried to penetrate the woods. The mountainous territory and the lack of government intervention to increase roadworks made the woods more difficult to reach and thus the transportation of wood was a difficult and risky enterprise. The conditions of the road network, the description of the Regia Carrea (the organization of the supplies of the Neapolitan dockyard), the attempts to introduce some equipment already in use in other areas in Italy to the transportation system, as well as the different types of trees used, are the main topics of this chapter, supported by material from archives and figures.

The decision of many 18th-century travellers to end their grand tour at Naples, or, at best, to proceed by sea in order to reach the continental provinces farthest from the capital of the Kingdom of Sicily, was not without foundation. The few who managed not to resort to such measures, mainly in the second half of the century, have left some disconcerting accounts of the difficulties and dangers associated with the decision to continue all, or most, of their journey overland: the appalling road conditions, where indeed roads existed, only a few dilapidated inns, brigands and malaria. Moreover, until the late 19th century, the often repeated, age-old statement that Neapolitans themselves would

(eds M. Agnoletti and S. Anderson)

never have undertaken a long journey without having first made a will is well known.[1]

It is also from these accounts, from this travel literature, that a careful description of the reality, as well as commonplaces and stereotypes, of the landscape and nature of the south, its customs and the characteristics of its inhabitants can be derived. However, the fact that a poor road network existed in the kingdom is indisputable.[2] Even the layout of the 'postal routes' in the early 18th century compared with those of northern Italy showed a significant difference (Fig. 14.1).

When considering the whole southern road network around the middle of the century, a greater concentration in the vast central area of the country as far as the Apulian lowlands existed, which then began to diminish, in one direction towards Abruzzo on the border with Church State lands and in the other, south of the Citra Principality, towards Basilicata and the provinces of Calabria.

In order to explain this particular development it is not sufficient to refer only to the morphology of the territory, which is increasingly dominated by the obstructive Apennine mountain range in the farthest continental regions. Much of the financial resources made available for the creation or improvement of communications was used by the Bourbon monarchy to facilitate the supply of foodstuffs to Naples. The capital, with its sudden population explosion (a demographic nightmare, comparable at that time only to London and Paris in Europe), its huge port, the seat of the court and the most important administrative, scientific and cultural institutions, was the political and commercial centre to which foodstuffs from the most productive areas, such as Apulia and what is now the Campania area, were conveyed. Equal importance was also given to the routes leading to the places of relaxation, enjoyment or entertainment of the monarchy, such as the numerous *cacce reali* (royal hunting lodges) around the capital or the royal palace at Caserta or the villa at Portici.

On the whole, however, the majority of roads, including the postal routes, were incomplete in many parts. It was impossible for carts and carriages to pass during winter. Paved stretches were rare, many bridges were missing over waterways and ferry crossings, the so-called *scafe*, on which feudal dues were levied, were not always provided. The building of new roads proceeded very slowly and with great difficulty and regular maintenance work was very rare.

Together with the constant complaints of the people to the sovereign, especially in Abruzzo, Basilicata and Calabria, about the damage and inconvenience caused by their extreme isolation, the minister Bernardo Tanucci's peremptory judgement, in 1762, left no doubt as to the real state of road communication in the south: the roads are impassable, 'horses, carts and loads are lost there; frightened passengers stay away; trade, the customs-house and even the reputation of the Kingdom are suffering'.[3]

Fig. 14.1. Postal routes in Italy *c.* 1720 (from Bortolotti, L. (1985) Viabilità e sistemi infrastrutturali, in: *Storia d'Italia*, Annali, 8, *Insediamenti e territorio*, Einaudi, Turin, p. 293).

In the last decades of the century there was a greater tendency to build and maintain a road network and it seems that an important role in influencing the monarchy in promoting the revival of this activity was played by the minister, John Acton. In 1778, having been commissioned by Ferdinand IV to reform the navy of the Kingdom of the Two Sicilies, he immediately showed a keen interest in the location of forests with trees that would be useful for shipbuilding and consequently in the conditions of the roads required for their transport to the dockyard in Naples.[4]

From the surviving documents of an inquiry on the forests of the kingdom proposed by Acton in 1779, much information on the possibility of wood transport in some provinces can be derived.[5] Also, in this particular sector the best roads were to be found in the areas surrounding the capital, where, apart from the main roads previously mentioned, a rather close network between the different towns had existed for some time, thus rendering the connection with the royal roads which led to Naples much easier.

The means of transport in inland areas of the kingdom were more problematic, not only due to the few main available routes but also because the forests were situated in mountainous areas, such as Basilicata or Calabria, which were often defined by the inhabitants themselves as 'horrid' and 'inaccessible'. The existence of *trattoj* or *tratturi*, which were the old tracks formed over centuries by transhumant breeding and which in some cases were also used by carts, must not be overlooked.

In general, the paths which connected the forests with the most important roads did not always permit wood transport driven by oxen and were usually barely visible, inaccessible and required much restoration work, especially bridges over ditches and valleys. Many forests could only be reached by pack animals or exploited only by hard and limiting manual labour. Unlike in other areas of Italy,[6] water transport could not substitute for or integrate with road transport. Due to the torrential nature of many of the southern waterways, neither navigation nor timber floating was feasible and only in a few cases were they used.[7]

Difficult as it was, the opportunity to enter the woods was of vital importance even for the people who lived in the mountains. Following the violent earthquakes that in 1783 and 1791 shocked central and southern Calabria, the population of Cortale asked the government authorities to provide for the repairs of two roads which led to the so-called Montagna de' faggi and Bosco di Carrà, on which the town of Cortale's economy was mainly based, thanks to agriculture and wood, with which it carried out trade with Catanzaro and Monteleone as far as Reggio. Therefore in 1795, in order to avoid further delays due to the slowness of the central governments, the local administrators proposed that all the inhabitants offered, free of charge, 2 days of excavation work and transportation of material and that the affluent classes contributed money to pay the *Maestri*, the master masons in charge of *inselciatura* (road paving with stones).[8]

Calabria was surely to be counted among the provinces which, for its dense forest area and for the quality of the wood, offered the greatest supplies for the construction of the new Royal Navy war fleet. Throughout the entire century, but at a faster pace after the great famine of the 1760s, there was a reduction in forest area in favour of the expansion of the cultivation of cereals, beech, oak, above all bay oak, pine-trees and firs, in particular *Pinus laricio*, locally called *zappino*. These were transported even by land as far as Naples or more often were 'lowered' to the coast before being loaded on to the dockyard of the capital and, after 1783, the new dockyard built in Castellamare.[9] In only 2 years, 1788–1789, hundreds of men and thousands of animals were used in numerous areas from the Sila up to the Aspromonte.[10] In various places in the Sila, they cut down and transported *zappini* and firs to the shores of Rossano, Crocchia and Bruca. From forests in the Aspromonte oak-trees were 'lowered' to the shores of Gioia (Fig. 14.2).[11]

Archive sources relative to these particular aspects of the raw material supplies give evidence of the complexity of wood-cutting and transportation defined as *Carrea de' legnami* and practised in many forests of the kingdom. From Naples, building engineers, *esperti di marina* (naval experts) and masters of shipbuilding brought the tools for chopping and the cordage needed for the transportation, whereas the local population had to supply oxen and drivers at their own expense. Workers were involved in cutting and arranging the trunks and in repairing the roads (excavation work, building bridges, etc.). When it seemed impossible or inconvenient to travel by land from the islands of the Parthenopean Gulf, the Amalfi coast or the coast of Sorrento, the ship-owners and crews of *martinganes* (a type of ship of that period) arrived who were entrusted with the task of completing the transportation by sea. Due to the scarce number of ports, the wood flowed to stacking areas and then to the dockyard. The local populations, which were obliged to supply pack animals or carts to pull the tree trunks when the road conditions allowed it, felt underpaid, and the work was hard, if not unbearable. The workload was also too heavy for the animals, who suffered from the difficulties – a scarce supply of cattle feed and animals often wounded as a result of the conditions of the impassable routes used for the removal of trunks. The loss of a part of the oxen was a real and threatening danger, which the local population tried to avoid in every way, by imploring the sovereign for exemption by default and paying for experts in the field themselves.[12]

Particularly in Calabria, in the attempt to render the wood transport system less difficult and more productive, some innovative proposals were made. There were some who hoped for a reform of the *carrea*, avoiding the excess and abuse which made it unpopular and burdensome for the state's finances,[13] others considered introducing technical improvements, such as the use of windlass,[14] and others suggested a common use of agricultural equipment, which they asserted would increase the wood transported by at least one-third (Fig. 14.3).[15]

The first two vehicles illustrated are H-shaped sledges, called *Leze*, to be used for the general transportation of weights. The first sledge needs two or more oxen to pull it, while the second (different in the front part of the runners and considered mainly useful on rough roads) required either one animal alone or more animals in a line. If used for transporting particularly unstable loads, the second type of sledge could be equipped with panels of intertwined

Fig. 14.2. Tree-cutting and loading areas in Calabria between 1788 and 1789 for the building of warships.

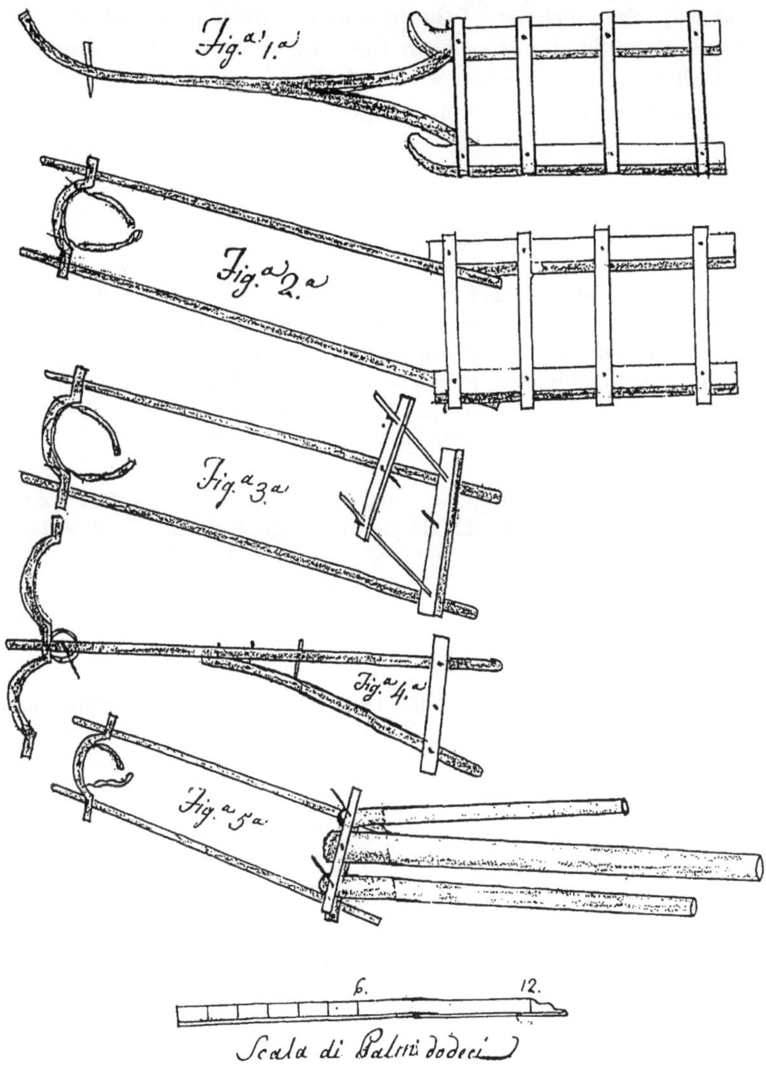

Fig. 14.3. Wood-cutting sledges (Archivio di Stato di Napoli, Sezione Militare, *Giunta dell'Arsenale,* Fascio 37. Year 1779).

twigs or a wooden loom. One or more oxen lined up would be ideal for pulling carts with parallel beams to transport one large tree trunk or three smaller ones. Finally, between these two illustrations, there is a Y-shaped cart, called a *Trazza*, with which two or more oxen can be used.

As Giuseppe Grimaldi, an expert in wood transport, maintained, the use of these carts was 'typical' of the mountains in Genoa, although unknown in

Calabria.[16] Even hypothesizing that the diffusion of these carts had improved, in some way, the efficiency of the practices of the removal of tree trunks, certainly this could not have been the only solution which would have resolved all the structural deficiencies that characterized this difficult but important sector[17] of the Calabrian economy.

Notes

1 This can be found, for example, in C.U. De Salis Marschlins (1906) *Nel Regno di Napoli. Viaggi attraverso varie province nel 1789*, Trani (reprinted Congedo, Galatina, 1979), p. 21, reproposed over the years from the late 17th century by Galanti up to Lenormant (1883) and referred to by Placanica, A. (1987) La capitale, il passato, il paesaggio: i viaggiatori come fonte della storia meridionale, *Meridiana* 1, 174.
2 For a longer period (14th–18th centuries), see De Rosa, L. (1982) Trasporti terrestri e marittimi nella storia dell'arretratezza meridionale, *Rassegna Economica* 3, 689–721. For the 18th century, see Lukacs, A.B. (1976) Le comunicazioni nel Mezzogiorno dall'arrivo di Carlo di Borbone al 1815: strade e poste, *Archivio Storico per le Province Napoletane*, pp. 1–27, and 1977, pp. 281–341; Ostuni, N. (1991) *Le Comunicazioni Stradali nel Settecento Meridionale*, ESI, Naples.
3 Mincuzzi, R. (1967) *Bernardo Tanucci ministro di Ferdinando di Borbone*, Dedalo, Bari, p. 117.
4 Ostuni, N., op. cit., p. 61.
5 For the provinces of Abruzzo Citeriore, Terra di Lavoro, Principato Ulteriore, Basilicata, Calabria Citra and Ultra, see Gangemi, M. (1996) Un'inchiesta sui boschi del Regno di Napoli nella seconda metà del Settecento. In: Cavaciocchi, S. (ed.) *L Uomo e la Foresta. Secc. XIII–XVIII* (Congress Acts, XXVII Settimana di Studi dell'Istituto Internazionale di Storia Economica 'F. Datini', Prato 8–13 May 1995), Le Monnier, Florence, pp. 551–566.
6 For more information on timber floating in regional areas, see Agnoletti, M. (1998) *Segherie e foreste nel trentino dal Medioevo ai giorni nostri*, Museo degli Usi e Costumi delle Genti Trentine, S. Michele all'Adige.
7 As well as Gangemi, M., *op. cit.*, pp. 560–561, see Gangemi, M. (1996) Il 'Paese de' Boschi': alcune osservazioni sull'Abruzzo del 1779, *Abruzzo Contemporaneo* 2, pp. 44–45. For timber floating in Basilicata, see also Demarco, D. (1988) *La 'Statistica' del Regno di Napoli nel 1811*, Vol. III, Accademia Nazionale dei Lincei, Rome, pp. 529, 558, 586, 598.
8 Archivio di Stato di Catanzaro, *Cassa Sacra, Segreteria Pagana*, Cartella 57, fascicolo 1243. The difficult access to the wood resources obviously involved different attitudes on behalf of the communities and each individual, positions which differed a great deal according to the interests at stake. Some people considered the trees 'useless', either due to the great logistic difficulties that had to be overcome to use or to sell the wood or due to the type of legislation issued for the protection of the forests, considered too binding in law, and so they preferred to resort to fires in order to obtain new pastures for the livestock or land for cultivation.
9 For a reference to the quantity and type of wood from some provinces of the Kingdom and used in the building of war fleets, see Gangemi, M. (1999) Des arbres pour

un arsenal royal: le cas du royaume de Naples (fin XVIII siècle). In: Corvol, A. (ed.) *Forêt et Marine*, L'Harmattan, Paris, pp. 41–61.
10 Archivio di Stato di Napoli, *Regia Sila*, Busta 24, fascicolo 135/II.
11 Archivio di Stato di Reggio Calabria, Sezione di Palmi, Fondo notarile, *Atti vari*, 2, *Libro degli Ordini*, fogli 42, 48, 70, 71.
12 Not only were the oxen heavily burdened but also other animals used for wood transport: the mules which pulled the tree trunks along the bad mountain paths of the Gargano, in Capitanata, 'vanno presto a morire per le piaghe, per dolori, e per istanchezza' (they soon die due to sores, pain and exhaustion). See Demarco, D. op. cit. Vol. 1, p. 427.
13 Lamanna, G. (1791) *Ricerche sugli affari della Regia Sila relativamente al dominio, ed alla pubblica economia*, Vincenzo Mazzolla Vocola, Naples, pp. 245 ff.
14 Jannucci, G.B.M. (1981) *Economia del commercio del Regno di Napoli*, part IV (ed. F. Assante), Giannini, Naples, p. 875.
15 Archivio di Stato di Napoli, Sezione militare, *Giunta dell'Arsenale*, Fascio 37.
16 F.S. Camilli confirms the use of sledges called *traglia* in the mountains in Abruzzo in the 18th century in *L'utilità delle pubbliche strade per tutto il regno di Napoli dimostrata col calcolo della spesa*, Naples, 1793, pp. 7–8. For the use of wheel-less carts in other southern regions during the 19th century, see Di Biasio, A. (1979) Gli 'ordegni rustici' nell'agricoltura napoletana, *Rivista di Storia dell'agricoltura* 2, 120–121. Regarding the ancientness and the diffusion in Italy and in the Indo-European area of similar vehicles, see Caselli, G. (1975) La treggia: nota preliminare per uno studio dei materiali delle culture non urbane in Italia. *Archeologia Medievale* 2, 440–455.
17 Gangemi, M. (1991) *Esportazioni calabresi nel XVIII secolo, Le tratte di 'Seccamenti' Salumi Tavole Legname ed Altro*. ESI, Naples.

Parks, Plebs and the Picturesque: Sherwood Forest as a Contested Landscape in Later Georgian England, 1770–1830

15

Ben Cowell

Department for Culture, Media and Sport, 2–4 Cockspur Street, London, SW16 1TL, UK

Towards the end of the 18th century, the remaining English royal forests were subject to a complex variety of claims to usage. Romantic tourists in search of the 'picturesque' were beginning to praise the sublime qualities of 'natural' forest territories. Landowners, on the other hand, generally sought to enclose such open forest land for the creation or expansion of parks and private woodlands as well as for the extension of agriculture during the Napoleonic period. They were encouraged by agents of the Crown, who proposed the enclosure and sale of a large number of royal forest properties. Both the celebration of forests as 'picturesque' landscapes and the large-scale enclosure of such forest wastes, however, were at odds with the survival, in many royal forests and open chase lands, of skeins of local and customary use-rights exercised by poorer plebeian inhabitants. Using the example of Sherwood forest in Nottinghamshire, this chapter examines the conflicts that such a clash of cultures generated. It concludes that the disappearance or survival of forest ecologies and landscapes in England was dependent in large part upon the social, economic and political factors that prevailed in the later Georgian period.

1 Introduction

This chapter considers the different 'ways of seeing' the English landscape park at the peak of its influence at the end of the 18th and beginning of the 19th centuries. It arises from research examining the different ways in which

(eds M. Agnoletti and S. Anderson)

landscape parks were valued and experienced by a range of social groups in the countryside, from the owners of parks and estates to those poorer members of rural society who bore the worst effects of agrarian change.[1] The chapter concentrates in particular on Sherwood forest, an area of north Nottinghamshire famous on the one hand for its 18th-century landscape parks, but on the other for its history as a royal forest and as an area of open commons and wastes. These different interpretations of the Sherwood landscape impinged upon the actual management of the landed estates and parklands of the region, so that for most of this period, until the final enclosure of the forest in the mid-19th century, Sherwood was a 'contested landscape'.

Recent theoretical work has taken seriously the idea that there are many 'ways of seeing' a landscape.[2] For example, in the case of landscape parks, studies have moved beyond the idea of parks as simply decorative adornments to country houses. Instead, parks can be also be seen as productive agricultural spaces or as expressions or even determinants of cultural and social power. Parks were more than just the two-dimensional views that landowners had from the windows of their country houses, since they were also the three-dimensional spaces in and around which everyday rural life was practised. Moreover, these were sometimes fiercely contested spaces. Bad feeling was sometimes created when landowners enclosed new areas of parkland against the wishes of local residents. This was especially the case in areas of the country where common or customary rights were most prevalent, such as the former royal forests. The history of such disputes, it might be argued, offers a rival account of the development of the landscape park in the period, and therefore has important repercussions for the contemporary management of such properties.[3]

2 Sherwood forest and the dukeries parks

In the technical meaning of the word, a forest implied territory over which the Crown exercised various rights of hunting and timber crop.[4] While not mentioned in the Domesday survey of 1086, the boundaries of Sherwood forest when they came to be formally recognized some time in the 12th century covered approximately a third of the entire county of Nottinghamshire, stretching almost the length of the western side as far as the town of Nottingham itself. Figure 15.1 shows the outlines of the parishes which were still subject in some way to forest law in the second half of the 18th century. The area was roughly coterminous with an outlying crop of bunter sandstone, the soils of which were naturally unsuited to intensive agriculture. Topographically, the medieval forest was predominantly open scrub heath, interspersed with areas of woodland, enclosed parks and cultivated arable. Much of this land was *terra regis*, the property of the Crown when the forest was laid out, but from the middle of the 16th century to the first half of the 19th century most of the Crown's interests in Sherwood were lost as a result of the breakdown of the forest courts and grants of soils and rights to private landlords. In 1798 Robert Lowe reported that '[t]he whole area of the soil of the forest is understood to have been granted off from the Crown to different lords of manors', although the Crown continued, theoretically, to hold rights over the game and timber

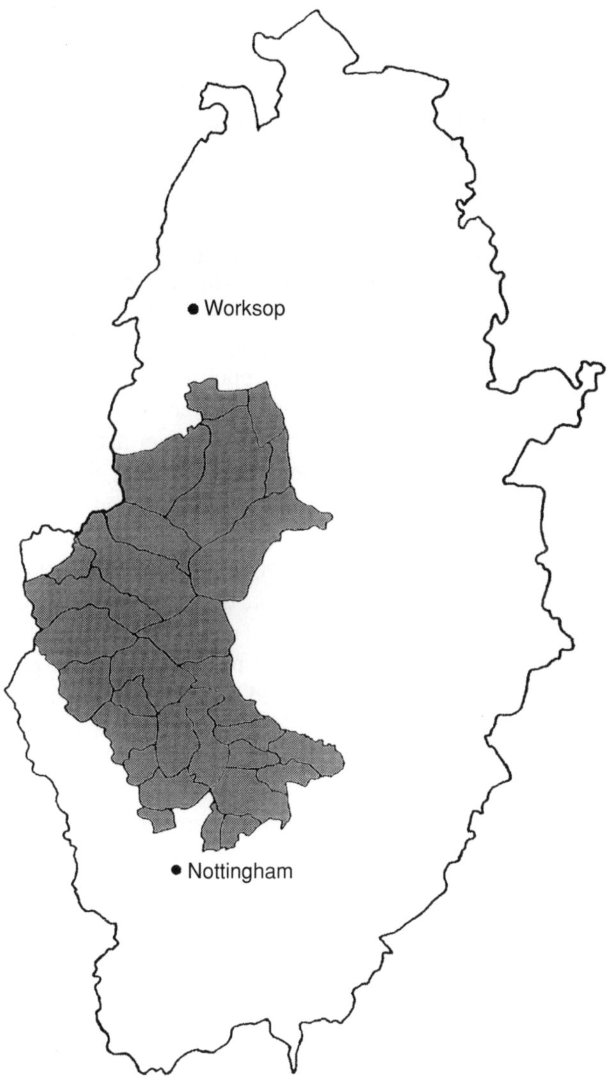

Fig. 15.1. Sherwood forest parish boundaries.

in many of these manors until their enclosure.[5] The last royal property in Sherwood, the woodlands of Birkland and Bilhagh, was sold to the Duke of Portland in 1826, and the enclosures of Mansfield and Mansfield Woodhouse in 1849 and 1850 saw the final end of a system of forestal rights which had prevailed since at least 1154.[6]

The principal distinguishing feature of a forest was its protection under special forest laws, which were designed to protect the venison (game) and vert (vegetation). These laws were administered by forest officers, who sat in a Swanimote court to hear offences and deal with the everyday management of the forest. Forest officers were answerable to a chief justice in Eyre, of whom there were originally two to serve the royal forests north and south of the River Trent. The inhabitants of certain forest districts might also be subject to the additional authority of the lord of the manor and his bailiffs and woodwards where the soil remained in private hands. Areas of forest had their own peculiar bureaucratic systems, which could often prove difficult to disentangle and which were prone to corruption and decay. In Sherwood, a Parliamentary report into the state of the royal forest in 1793 discovered that neither the justice in Eyre nor the Swanimote court continued to meet, except for the purpose of electing the forest officers. These officers included the lord warden (the Duke of Newcastle), a bowbearer or ranger (Lord Byron), four verderers, elected by the local gentry, and a steward, appointed by the lord chief justice.[7] The report was damning of the influence of these officers, claiming that they knew little of the forest territories and that they 'never acted, but in receiving their trifling salaries', which amounted to an annual fee tree out of the royal woods of Birkland and Bilhagh and a fee buck and fee doe from Lord Chesterfield's chase lands.[8]

The absence of a strong royal presence, coupled with the survival of archaic systems of forest law, explains partly why so much of Sherwood remained unenclosed into the later 18th century, relative to the rest of the county. As Table 15.1 shows, many forest parishes were not enclosed until the 1780s and 1790s, and some remained unenclosed well into the middle of the 19th century. While the Crown's interests in the soils of Sherwood had been almost entirely appropriated by the time of the 1793 report, the persistence of forestal rights over private manorial lands acted as a deterrent to their blanket enclosure and empalement, since compensation was required for all rights appendant on such properties. The Parliamentary commissioners of the 1790s were vocal in their complaints at the poor compensation that the Crown had been offered after the enclosures of Calverton, Arnold and Gedling parishes, and threatened further action to rectify the situation.[9]

By the middle of the 18th century, Sherwood was in a transitional phase, as it moved from being an area associated with royal authority and customary usage to one in which the principal feature of the region was the massive aristocratic estates, known as 'the dukeries', which were in the process of subsuming much of the original forest wastes. In 1756 the transformation had

Table 15.1. Sherwood enclosures.

Forest parishes enclosed by Act of Parliament, *c.* 1750–1850	
Calverton	1779
Arnold	1789
Basford	1792
Radford	1796
Blidworth	1805
Skegby	1808
Edwinstowe	1818
Mansfield	1849
Mansfield Woodhouse	1850
Forest parishes enclosed piecemeal *c.* 1750–1850	
Carburton	1770–1798
Rufford	1776–1798
Former forest parishes enclosed by Act of Parliament *c.* 1750–1850	
Gunthorp; Caythorp; Lowdham	1765
Burton Joyce, Bulcote	1768
Hucknall	1769
Ollerton	1778
Gedling; Lambley; Stoke Bardolph	1792
Sutton	1794
Woodborough; Kirkby	1795
Sneinton	1796
Annesley	1808
Warsop	1818
Linby	1850
Forest parishes enclosed piecemeal pre-*c.* 1750	
Fulwood; Colwick; Bestwood; Bulwell; Papplewick; Perlethorpe (part); Newstead (part)	

Sources: Tate, W.E. (1935) *Parliamentary Land Enclosures in Nottinghamshire*, Thoroton Society Record Series, V, Nottingham; Chambers, J.D. (1966) *Nottinghamshire in the Eighteenth Century*, Frank Cass, London.

proceeded to such an extent that Horace Walpole described the area as 'a very heptarchy of little kingdoms elbowing one another, and the barons of them want nothing more but small armies to make in-roads into one another's parks, murder deer and massacre park keepers'.[10] Nevertheless, despite frequent eulogies to the 'improved' and 'patriotic' plantings of the ducal landowners, there remained an uneasy tension between old and new within the bounds of Sherwood in this period, as the last vestiges of a customary way of life that had come to be associated with the forest region continued to make its presence felt at the borders of the dukeries parks.

3 Sherwood's customary economy

A range of common rights had been enjoyed on individual manors within Sherwood since at least the 17th century.[11] Some of these rights were already in decline by the middle of the 18th century; others, however, survived into the later part of the century, until they too were mostly extinguished as a result of the enclosure of most of the forest lands from the 1770s to the 1820s. They offer an alternative perspective or 'way of seeing' the Sherwood landscape at this time. Perhaps the most distinctive form of common right was the practice of breck agriculture.[12] Robert Lowe in his survey of the agriculture of the county of 1798 wrote that the laying out of brecks or temporary enclosures from the forest wastes and sheepwalks had been an 'immemorial custom for the inhabitants of the townships of the forests', although by this time the practice was under threat owing to the enclosure of the open sheepwalks upon which it depended.[13] Under the breck system, plots of up to 300 acres of rough pasture land could be temporarily fenced in and ploughed as arable for periods of up to 7 years, after which the fences were taken down and the land reverted to use as open forest. Where villages practised open-field agriculture, these brecks would also be farmed collectively, each farmer having a share of the breck proportionate to his or her allotment in the open fields. At Calverton, each messuage in the village was entitled to an acre each on the breck and each cottage to two roods, while the surplus was leased, sometimes to non-resident farmers, in order to raise money towards the upkeep of the poor. In other instances, such as at Rufford, brecks were enclosed for the benefit of individual farmers.[14]

Breck farming was considered by some outside commentators to be a primitive form of agriculture. Arthur Young wrote with some disapproval that the breck farmers of Sherwood parishes 'take as many successive crops of corn as the land will bear till at last they scarcely get their seed again'.[15] Nevertheless, while the taking in of brecks remained a customary practice rather than a legal right, there is evidence to suggest that the Sherwood brecks were farmed in a systematic and efficient manner by foresters, right up to the point of enclosure.[16] Typically, the rotation of crops to be sown in the breck was named in the petition which the villagers would have to make to the lord of the manor (or the chief justice in Eyre in the case of the remaining royal lands) in order to gain permission to lay out the enclosure. At Kirkby-in-Ashfield a sophisticated 9-year rotation was detailed in a petition for brecks of 1780, consisting of alternate years of oats, turnips, corn and ryegrass, with directions that only calves and sheep were to be allowed to eat the turnips.[17] At Edwinstowe the brecks were consolidated rather than scattered, indicating that they were farmed permanently or in regular rotation until the enclosure of the village in 1779.[18] Fowkes has suggested that brecks were central rather than peripheral to the agriculture of the forest villages, and as such 'it is unlikely that they bore much resemblance to forest in any physical sense, even allowing for the scrubby appearance of most of 18th-century Sherwood outside of the plantations of the great Dukery estates'.[19]

Other common rights enjoyed by inhabitants of the forest related to the more regular use of areas of wooded and scrubland waste for timber, food and fuel. Various forest inhabitants claimed a right to cut timber in the 17th century, although this seems not to have been the case in the 18th century, as evidenced by a number of presentments to the quarter sessions against labourers who had stolen firewood or damaged trees.[20] But a mid-19th-century local reported that as late as the 1790s residents of Edwinstowe 'could supply themselves plentifully with firewood, during the whole year' out of Birkland and Bilhagh, suggesting that this remained a customary claim, if not a legally prescribed common right.[21] The timber crop from the royal forests of Birkland and Bilhagh remained the privilege of the four forest verderers and the king's steward, who were each permitted to harvest a tree of their choice every year, and also to take 'fee trees' when they were approached in order to sanction a breck.[22]

Forest dwellers also held customary rights to grazing and the produce of the forest undergrowth. In the latter half of the 17th century, it was recorded that Sherwood villagers gathered bilberries from the forest, 'which every summer were wont to be an extraordinary great profit and pleasure to poor people who gathered them, and carried them all about the country to sell'.[23] Villagers continued to practise customary rights of grazing in the forest, as well as rights to certain undergrowth produce, such as ling (heather), in the 1790s and early 1800s. The Parliamentary commission of 1793 discovered that the villagers of Edwinstowe had a variety of claims in the neighbouring royal woodlands of Birkland and Bilhagh, including rights to acorns for their pigs and rights of grazing for sheep, but not for horses or 'horned cattle'.[24] Ling burning also continued in Sherwood, despite an Act of 1706 which had made it an offence by law.[25] A mid-18th-century observer remarked that the poor of the forest 'make a scanty livelihood, by gathering the fern and burning it, the ashes of which they sell for making soap'.[26] Much of this work was carried out by women and children. Some accounts of the practice were more positive, emphasizing the value of fern-gathering to the household economies of the Sherwood poor: 'sufficient was earned to pay off the year's shoe bill of the family, or the grocer's, tailor's, or other tradesmen's bills, and thus the comforts of the peasantry were advanced'.[27] The practice is known to have been carried out in Birkland and Bilhagh in the last decades of the 18th century, possibly as a temporary response to high prices and poverty. In 1788 the villagers of Edwinstowe petitioned the Duke of Portland, the lord of the manor, to allow them:

> to burn the weed called Bracken . . . pertickleory in that part of the wood called Birkland and Billow, or where your Grace will be pleased to order and appoint them, as has usually been done, in ye time of your Grace's noble ancestors.[28]

Such petitions were made in the years 1767/68, 1787–1790 and again in 1800/01, and each time the villagers stated clearly the circumstances of their application: 'all kinds of provision is verey dear, and all kinds of work being scarse' (1788); 'having nothing to depend upon for support but their own hard labour' (1789); 'cole, corn and all sorts of provision being extremely dear' (1790).[29] The petitioners also added that the Duke would himself benefit from the arrangement, since the cutting of the bracken would 'likewise prevent Many Dangers and actsedances that might happen to Gentlemen in persuit of their Game, by Discovering and laying open to view maney dangorus Holes and tree roots and stumps, wich might cause maney very bad misfortens' (1788).

In fact, the Duke of Portland was probably quite happy to recognize a right of common belonging to the villagers of Edwinstowe at this time, since such applications legitimized his claim as lord of the manor over the woodlands of Birkland and Bilhagh. By the later 18th century the royal interest in Sherwood was confined to the two areas of Birkland and Bilhagh, but as these petitions make clear the Duke of Portland was also claiming a right, as lord of the manor, to the common rights appendant to the woods. When Parliamentary commissioners approached Mr Gladwin, the Crown's agent in Sherwood, they were told that, while the Crown had the ownership of the wood and timber in Birkland and Bilhagh, the Duke of Portland was the proprietor of the soil, 'in consequence of a grant from the Crown to his ancestors'.[30] Moreover, when questioned about the rights of herbage in the forest, Gladwin answered that '[t]he Duke of Portland, his tenants and other copyholders within the manor of Edwinstowe solely exercise the right of common' in the property, and that 'the inhabitants of Edwinstowe insist that the Crown cannot enclose Birkland . . . having no other interest but in the wood standing thereon'.[31]

On other parts of the Portland estate, however, there was less accommodation towards the interests of the customary villagers. At Clipstone, a former medieval forest park which was managed by the Dukes of Portland purely for timber, a number of prosecutions were made in the 1760s against forest dwellers who illegally entered the park and caused 'great outrages and disorders'.[32] On one of these occasions, in 1767, the park was invaded by a riotous mob of labourers from the towns of Warsop and Worksop, led by James Newton, the constable of Warsop. Robert Bushaway has written of the persistence with which commoners defended their timber rights where these were perceived to be under threat, through forms of ritualized protest or through more covert methods of stealing and vandalism.[33] Whether oppositional motives lay behind the 1767 Clipstone riot is unknown, but it was certainly true that the process of agricultural improvement on the dukeries estates effectively put an end to the customary economy of the forest. John Throsby, visiting Clumber in the early 1790s, commented that:

> Ling and broom are natives of this part of the forest. Before it was so much enclosed the people used to burn acres together, and plant fresh to be eaten young as food for sheep; but now most of that ground is occupied by corn.[34]

Writing in 1847, Christopher Thomson, the self-styled 'artisan' of Edwinstowe, observed that some 34 parishes in the forest had been enclosed in the previous half-century, and that the Sherwood peasantry had suffered as a direct result. He articulated a somewhat different attitude to the open landscape of the forest and its subsequent 'mapping-out' by the aristocratic owners of the dukeries parks:

> Time was, when the humblest peasant enjoyed his right of 'focland,' – the land of the folks; but dare the peasantry of Bassetlaw set a foot upon one acre within those thirty-four parishes? Truly, some persons have been enriched by the mapping-out, and ring-fencing of these commons; but are they the 'common people?' . . . Doubtless, the peasantry are existing by thousands within these parishes; have these thousands been allowed to take the 'lion's share?[35]

4 Parklands, plebeians and the picturesque in Sherwood

Parkland enclosures had always been a feature of the Sherwood landscape, but only with the consent of the Crown, since emparkment implied restrictions on grazing for the king's deer. In the medieval period the Crown operated a number of deer parks, principally Clipstone, Bestwood, Bulwell and Nottingham, while grants were also made to monastic estates established within the forest boundaries, which also claimed a variety of common rights of grazing on the forest sheepwalks. The Premonstratensian Welbeck Abbey, established in the 1140s, was one of the biggest of the medieval monastic estates, while others included the Cistercian abbey at Rufford, monastic estates of the Augustinian order at Worksop, Newstead, Felley and Thugarton, and a Cluniac abbey at Lenton near Nottingham. When these establishments and their outlying properties were surrendered at the Dissolution, they passed into the hands of the larger gentry and aristocracy of the county and became the basis of substantial lay estates.[36] In a similar fashion, the royal parks had been entirely given up by the start of the 18th century. Clipstone was granted to the Duke of Devonshire in 1603, Nottingham to the Earl of Rutland in 1623, while Bestwood, the best maintained of the group, was alienated to Lady Elionora Gwynne in 1687.[37]

The aristocratic estates which comprised the 18th-century dukeries evolved partly through the appropriation of earlier monastic and royal territories, and partly as a result of the acquisition of the open forest lands through purchase, marriage, Crown grants and enclosures. The dukery landowners wielded great influence over Nottinghamshire as a whole, as landlords with estate properties throughout the county and as political leaders and patrons. The Welbeck estate of the Newcastle family was bought from its original post-dissolution owner in the 16th century and became one of the chief residences of the family. A map of the estate made in 1629 shows a substantial park of 1150 acres, surrounded by the enclosed fields, sheepwalks and open heath of the forest landscape.[38] The area of forest land subsumed by Clumber, too, was in part formerly owned by the monks of Welbeck, who originally ran it as an agricultural enterprise before it was eventually granted to the Duke of Newcastle in 1709. Newcastle used timbers from Welbeck and from the royal woods of Birkland to fence Clumber in the years immediately following the grant.[39] Thoresby Park, meanwhile, had been laid out some 30 years earlier by a royal grant of 1683, and a map of 1690 reveals it to have been some 1361 acres in extent at this time.[40]

By the later decades of the 18th century, the dukeries estates and parks had become famous for the 'improvements' that had been achieved within them at the hands of their aristocratic owners, in collaboration with several famous agricultural experts and landscape designers. Susanne Seymour has shown how this discourse of 'improvement' combined ideas of beauty with those of rational estate management, such that even areas of parkland were ploughed up, planted or used as grazing ground for intensive livestock rearing.[41] Substantial sums were spent on improving the landscapes of these parks, while simultaneously outlying areas of estates were converted from rough heaths to efficient farmlands and extensive plantations. Clumber, for instance, was used as a deer park until the accession of the second Duke of Newcastle, who transformed the estate into a highly productive agricultural unit. The deer were moved to nearby Haughton, and the park was improved by the addition of new plantations, a lake and cascade, a series of lodges and a new area of pleasure ground near the mansion house, which itself was rebuilt in 1770 by the architect Stephen Wright. Almost half of the area of the park was planted with a variety of native and exotic trees, while a geometric plantation of trees was laid out at Apley Head, formerly an area of open heath.[42] At Welbeck, the second Duke of Portland had developed a major home-farm complex between the 1760s and the 1790s, involving the enclosures of Carburton forest, Norton and Worksop forest, and Repton himself advised on the enclosure of Norton in 1793.[43] Seymour has noted how this process of reorganization had major repercussions on the nature of the forest communities. Where land had formerly been farmed collectively as brecks or areas of open grazing, it was now leased out to a small number of large tenant farmers. Sections of the forest population found themselves transformed from customary tenants claiming rights of common on the forest wastes, to landless labourers, solely dependent on the great estates for work.[44]

The framing of the forest region as a 'landscape' was an important aspect of these changes. Textual representations of the Sherwood landscape provided ideological vindication for the enclosure of the open forest wastes. Reports made by travellers to Sherwood in the 18th century made direct contrasts between the verdure and plantations of the parks and the decrepit nature of the open forest areas. Describing his journey from Mansfield to Welbeck in 1768,

Lord Grimston wrote that the surrounding countryside was famous only 'for the barrenness of its soil in the one part, & in the other for the size of its timber'.[45] Patching described the area as 'remarkable for nothing but a deep sandy road (through which our horses moved with the agility of tortoises) and a Desart [*sic*], overgrown with Heath and Fern'.[46] When the parks themselves failed to meet the approval of polite visitors, they were depicted as if the forest had seeped in through the pale. Grimston thought Welbeck 'very extensive', but, although in some parts it was 'much beholden to art for its Beauty', in others it was described as 'in the state that Nature has left it, very dreary and disagreeable'. A visitor to the region in 1750 noted that 'the parks in this country seem for the most part originally to be large commons inclosed, being a sandy poor soil, which naturally produces little but fern'.[47] At Clumber Throsby saw patches of the native forest broom and gorse in the park, 'left, it may be presumed, for ornament'.[48] In contrast, Grimston praised Thoresby for its 'very good turf, which in this country is very much wanting'.[49]

Such writers were drawing on established and widely held ideas of forest landscapes. The most common mode of representation of forests was to stress their infertility, their unenclosed and therefore inefficient state and, by direct implication, the debased and immoral nature of their indigenous inhabitants.[50] Towards the end of the 18th century, however, forests began to receive some attention as landscapes in their own right. William Gilpin's *Remarks on Forest Scenery* (1791) was an early work which praised the picturesque qualities of areas of forest and woodland. Although Gilpin recognized that 'the capricious nature of *picturesque ideas*' often produced landscapes that ran counter to 'utility', *Forest Scenery* was an extended discussion on the value of blasted oaks, withered branches and moss-covered bark in landscape scenery.[51] Gilpin's aim was partly to draw unproductive forest scenes into the same picturesque frame as parklands. Nevertheless, he continued to make a sharp distinction between parks and forest, between culture and decay. 'Beauty . . . is not the characteristic of the forest,' he wrote, '[i]ts particular distinction is grandeur and dignity.'[52] Repton, too, was careful to draw a clear distinction between park and forest in his designs for Welbeck Abbey. While Repton was keen to promote the value of a large, unbounded park in the vicinity of the mansion, he also voiced concern that the absence of clear boundaries would lead to confusion between land which was park and that which was forest. The fundamental differences between the two types of landscape were readily apparent to him:

> a park has a character distinct from that of a forest; for while we admire, and even imitate, the romantic wildness of nature, we ought never to forget that a park is the habitation of men, and not solely devoted to the beasts of the forest.[53]

Repton's dilemma was that parks and forests remained very similar in appearance, since 'the trees, the water, the lawns, and the deer, are alike common to both'.[54] Consequently he advocated the strategic placing of buildings among the park woods, as well as the floating of pleasure boats on the lake, in order to accentuate the patrician character of Welbeck Park over the plebeian forests at its border. Similarly, at Worksop the steward of the Duke of Norfolk, Mr Wake, made a number of proposals, in relation to the landscaping of the park, that were designed to accentuate the differences between the park and the surrounding open forest land. Should the 'proper course of husbandry' not be followed, warned Wake, the greatest part of the park would inevitably 'return to its original forest state, and be overrun with heath, fern, whins [furze], moss and pernicious weeds; consequently it will lose its beauty'.[55] Correct modes of cultivation were therefore crucial to the maintenance of Sherwood's internal hierarchies.

The presence of a nexus of customary claims to the forest lands, however, threatened the sanctity of the dukeries parks in more direct ways. On a few occasions, the creation of massive areas of parklands from former commons and wastes was the cause of open conflict between the aristocratic estate owners and the resident forest communities. At Bestwood Park near Nottingham, a new tenant of the Duke of St Albans, Mr Barton, ran into problems in the mid-1770s when he attempted to repair the pales of the ancient park in order to turn the park over to agricultural uses. The wooden pales had been in a bad state for many years prior to Barton's arrival, partly owing to their removal by 'the poor people in the neighbourhood' for fuel. When Barton rebuilt the pale, local residents who had formerly used the park as open forest, pulled it down at night time in order to let their cows and sheep back in. Barton took several of these residents to court, to contest their claim that 'time had established a right' for them to use the park as if it were a legal forest common right. A survey of the park in 1775 reported that Barton had won his case, but that he still suffered 'extremely by his neighbours, pulling the hedges &c in various parts of the estate for firing but he has conquered them respecting their pretended right of commonage'.[56]

Rival interpretations of what was meant by customary and common rights came to light during a mid-18th-century dispute over an area of common that bordered Welbeck Park to the north. During the course of the dispute, a number of local people were interviewed in order to determine who enjoyed customary rights over the common. It emerged that, towards the end of the 17th century, the parishioners of Worksop had actually used Welbeck Park as a customary resource, both for gathering wood and also as part of the route of the annual 'perambulation' of the parish boundaries. The parish perambulation was an important symbolic and ritual occasion in the customary calendar which affirmed the attachment of the community as a whole to the land of the parish. However, by the time of the mid-century dispute, the parishioners had long ceased to perambulate through the park. Different reasons were advanced by witnesses as to why this was so. John Nelson said that the principal cause was that Mr Calton, the priest who led the perambulation and a local figure of some standing, was given another living in Staffordshire by the Duke of Newcastle, owner of Welbeck. But another witness, Michael Wittaker, suggested that the Duke of Newcastle had been more proactive in stopping up the way through the park. He said that the Duke had sued a number of the parishioners for entering the park, and that after this the officers and parishioners 'were afraid to venture into or thro' such part of the said park as was known to be in the said parish of Worksop, for fear of incurring the said Duke's displeasure'.[57]

5 Conclusion

The coexistence of large private parks enclosed from and laid out alongside the surviving areas of open and commonable forest wastes was perhaps always bound to provoke such disputes, and these disputes in turn threw into sharp relief the social and cultural cleft between those who looked upon Sherwood as a picturesque ruin, capable only of emparkment, enclosure or romantic abandonment, and those who lived and worked within its bounds on a daily basis and whose very livelihood was threatened as a result of emparkments and enclosures. For native residents of Sherwood forest, areas of commonable land were important everyday resources. Matthew Johnson has argued that the practice and performance of customary rights implied a more corporeal and ecological relationship to the land: 'it [custom] was an idea associated with everyday practice and the subjective experience of the landscape rather than with abstract, legal concepts of outside observers'.[58] Unless common forest rights could be proved in law, the Sherwood landlords were in a strong position to deny all forms of customary access to the forest resources. At the time of enclosure, only those who held 'ancient copyhold messuages' or cottages were taken into consideration, despite the fact that many others made claims to common rights based on possession of freehold land.[59]

Freeman has recently argued that the existence of conflict between different groups over the resources of a forest did not automatically lead to ecological instability, provided there was a sufficiently active regulatory body to assess rival claims and enforce sustainable management practices.[60] In Sherwood, as in many other forests, the internal bureaucracy had long been in decline by the later 18th century. Unlike in other areas of open, unenclosed wastes, however, the Sherwood commoners were limited in number and political power. By the mid-19th century, the final Sherwood parish had been enclosed and the region had ceased to be a 'forest' in any legal sense. Instead, it had become an area noted for its large aristocratic estates. The 'patrician' view of the forest was therefore the final victor in the struggle over the Sherwood landscape, but we need not assume that other 'ways of seeing' the Sherwood landscape either had never existed or were not in themselves important. The effacement of the traditional landscape through enclosure and emparkment was at times directly threatened by the residual claims of local inhabitants to common right or customary uses of park territories. Ironically, the very making and management of a landscape park could bring to light exactly those customary relationships to landscape that park owners were seeking to eradicate through their enclosures and emparkments.

Notes

1 Cowell, B. (1998) Patrician Landscapes, Plebeian Cultures: Parks and Society in Two English Counties, *c.* 1750–1850, PhD thesis, University of Nottingham.
2 See, for example, Mitchell, W.J.T. (1994) *Landscape and Power*, University of Chicago Press, Chicago; Bender, B. (1993) *Landscape: Politics and Perspectives*, Berg Publishers, Oxford; Hirsch, E. and O'Hanlon, M. (1995) *The Anthropology of Landscape: Perspectives on Place and Space*, Clarendon Press, Oxford.
3 See also Cowell, B. (1997) The politics of park management in Nottinghamshire, *c.* 1750–1850, *Transactions of the Thoroton Society* 101, 133–143.
4 See James, N.D.G. (1990) *A History of English Forestry*, Basil Blackwell, Oxford.
5 Lowe, R. (1798) *General View of the Agriculture of Nottinghamshire*, G. Nicol, London, p. 97.
6 Cox, J.C. and Whitworth, R.H. (1906) Forestry. In: *The Victoria County History of Nottinghamshire*, Archibald Constable, Folkestone, pp. 365–381.
7 The Fourteenth Report of the Commissioners Appointed to Enquire into the State and Cultivation of the Woods, Forests, and Land Revenues of the Crown, and to Sell or Alienate Fee Farm and Other Unimprovable Rents (1793) *House of Commons Journal* 48, 467–511.
8 Ibid., p. 469.
9 Ibid., pp. 471–472.
10 To Richard Bentley, August 1756. Lewis, W.S. (1973) *Horace Walpole's Correspondence*, Vol. 35, Yale University Press, New Haven, p. 266.
11 Seddon, P.R. (1978) The application of forest law in Sherwood forest, *c.* 1630–1680, *Transactions of the Thoroton Society* 82, 37–44.
12 Mastoris, S.N. and Groves, S.M. (1992–1993) Sherwood forest in 1609: a Crown survey by Richard Bankes, *Thoroton Society Record Series* 40, xxxii.
13 Lowe, *General View*, p. 21.
14 Fowkes, D.V. (1977) The breck system of Sherwood forest, *Transactions of the Thoroton Society* 81, 55–61.
15 Young, A. (1771) *A Farmer's Tour Through the East of England*, Vol. 1, W. Strahan, London, p. 427.
16 Fowkes, D.V. (1971) The process of agrarian change in Nottinghamshire *c.* 1720–1830, PhD thesis, University of Liverpool, p. 42.
17 N[ottingham] A[rchive] O[ffice] DDP 15/73.
18 *Study on the Enclosure and Clearing of Sherwood Forest in the Seventeenth and Eighteenth Centuries*, n.d., NAO DD 148, p. 17.
19 Fowkes, Breck system, p. 59.
20 Meaby, K.T. (1947) *Nottinghamshire. Extracts from the County Records of the 18th Century*, Thos Forman & Sons, Nottingham, p. 237.
21 Thomson, C. (1847) *The Autobiography of an Artisan*, J. Chapman, London, p. 301.
22 The Fourteenth Report of Commissioners, p. 481.

23 Thoroton, R. *The Antiquities of Nottinghamshire* (ed. J. Throsby, Nottingham, 1790–1796; repr. 1972; hereafter 'Throsby's Thoroton'), Vol. ii, p. 161.
24 The Fourteenth Report of Commissioners, p. 481.
25 Meaby, *Nottinghamshire*, p. 237.
26 Patching, R. (1757) *Four Topographical Letters, Written in July 1755, upon a Journey Thro' Bedfordshire, Northamptonshire, Leicestershire, Nottinghamshire, Derbyshire, Warwickshire etc. from a Gentleman of London to his Brother and Sister in Town* . . ., Newcastle, p. 10.
27 Thomson, *Autobiography*, p. 301.
28 N[ottingham] U[niversity] M[anuscripts] D[epartment] PwF 9925.
29 NUMD PwF 9925, 10034, 9926. Original spelling has been retained.
30 NAO DD4P 75/57.
31 Ibid.
32 Meaby, *Nottinghamshire*, p. 207.
33 Bushaway, R. (1982) *By Rite: Customs, Ceremony and Community in England, 1700–1880*, Junction Books, London, Chs 5 and 6.
34 Throsby's Thoroton, Vol. iii, p. 405.
35 Thomson, *Autobiography*, p. 263.
36 Cameron, A. (1975) Some social consequences of the dissolution of the monasteries in Nottinghamshire, *Transactions of the Thoroton Society* 79, 50–59.
37 Gillott, S. (1985) Bestwood: a Sherwood forest park in the seventeenth century, *Transactions of the Thoroton Society* 89, 57–74.
38 Seymour, S. (1988) Eighteenth-century parkland 'improvement' on the dukeries estates of north Nottinghamshire, PhD thesis, University of Nottingham, p. 141.
39 NAO DD 4P 75/74/1–20.
40 NUMD M 4P 19.
41 Seymour, Eighteenth-century parkland, *passim*.
42 Ibid., p. 300.
43 Ibid., p. 394, n. 36.
44 Ibid., p. 338.
45 H[ertfordshire] C[ounty] R[ecord] O[ffice] D/EV F13.
46 Patching, *Four Letters*, p. 10.
47 Pococke quoted in: Baddeley, V. (1996) The early eighteenth century landscape gardens of Worksop Manor, Nottinghamshire, *Transactions of the Thoroton Society* 101, 123–136.
48 Throsby's Thoroton, Vol. iii, p. 405.
49 HCRO D/EV F13.
50 Thomas, K. (1983) *Man and the Natural World. Changing Attitudes in England 1500–1800*, Allen Lane, London, p. 195.
51 Gilpin, W. (1791) *Remarks on Forest Scenery*, R. Blamire, London, p. 7.
52 Ibid., p. 209.
53 Loudon, J.C. (1840) *The Landscape Gardening and Landscape Architecture of the Late Humphry Repton Esq.*, Longman, London, p. 78.
54 Ibid., p. 79.
55 S[heffield] A[rchives], A[rundel] C[astle] M[anuscripts] W138/3.
56 NUMD BP 4.
57 SA, ACM W156/3.
58 Johnson, M. (1996) *The Archaeology of Capitalism*, Blackwell, Oxford, p. 73.

59 NAO DD WD 30, dispute over Blidworth enclosure 1805.
60 Freeman, M. (1997) Whichwood forest, Oxfordshire: an episode in its recent environmental history, *Agricultural History Review* 45, 137–148.

16

Social and Environmental Determining Factors of Forest Fire History: the Case of Eastern Andalusia, 1840–1890

Salvador Cruz Artacho,[1] Manuel González de Molina,[2] Monserrat Núñez Delgado[2] and Antonio Ortega[2]

[1]Departamento Territorio y Patrimonio Histórico, Universidad de Jaén, Paraje de las Lagunillas, S/N CP 23071, Jaén, Spain; [2]Departamento Historia Contemporánea, Universidad de Granada, Campus de Cartuja S/N CP 18071, Granada, Spain

In eastern Andalusia, forest fires have historically been used by rural communities to protest against the process of privatizing the ownership and use of public woodlands since the 19th century. This is demonstrated by the close correlation between the incidence of the privatization process of woodland areas and the increase in the number of fires.

1 Myths and realities of the historical presence of fires in the high Andalusian woodlands during the second half of the 19th century

It is well known that perhaps one of the most devastating effects traditionally suffered by Andalusian wooded areas, both in the recent past and today, is that caused by fire. The disorders caused in the ecology, territory and landscape of the affected area and the economic costs that arise are only some of the more common arguments used when justifying the importance of forest fire, the cause of which has been found to be of great complexity.

In this regard, leaving aside all the fortuitous, accidental or natural causes and focusing on those which could be described as intentional or criminal, their interpretations are and always have been of a wide and varied nature. This diversity, however, does not hide the fact that, in the historical presence of fires in the Andalusian woodlands, the intentional cases easily exceeded the merely fortuitous ones, as has recently been demonstrated by Eduardo Arque Jiménez (1997)[1] in a study carried out on the Jaén woodlands in the second half of the 19th century. The 1693 reports of fires in woodlands and pastures and those due to charcoal-making in the provinces of eastern Andalusia (Almería, Granada, Jaén and Málaga) received in the Salas de la Audiencia Territorial de la Real Chancillería de Granada (Territorial Court of Granada Royal Chancery Division) in the period 1840 to 1889 follow a similar pattern (Table 16.1).

These cases indicate the intentional nature of the fires. If they had only been fortuitous or a result of mere chance or accident, the cases would never have reached the higher courts of justice. In many cases, these reports even include the names of the offender(s). The intentional or criminal nature of many of these fires is apparent. Moreover, animal and agricultural farmers ended up being the object of many criticisms and proceedings. The repeated examples in this respect appear to confirm the excessive use of fire as a means of regenerating grazing land or as a mere tool for gaining additional space that could be cultivated from the woodlands. If to this is added a long history of fires caused by charcoal-making, by widely varying types of negligence and other reasons which were aimed at hiding or propitiating administrative practices surrounded by obvious elements of illegality, a considerably complex panorama arises, in which the problem that historically surrounded the intentional forest fires in Andalusia was found to be rooted in socio-economic and institutional factors which were directly related to the consolidation of the capitalist market model of agriculture in 19th-century Andalusia and its process of 'agriculturization'.

Numerous studies on the subject make repeated references to forest fires, especially those related to the practices of ploughing up the woodlands. A line of argument whose main reasoning lies in the evolution of cereal prices and others in which no less evident connections with strategies clearly aimed at

Table 16.1. Reports of fires in woodlands and pastures and due to charcoal-making in eastern Andalusia, 1840–1889 (global figures).

Time interval	No. of fires in woodlands	No. of fires in pastures	No. of reports due to charcoal-making	Total figures
1840–1855	161	18	13	192
1856–1867	402	48	17	467
1868–1874	436	48	3	490
1875–1889	433	55	56	544
1840–1889	1432	169	92	1693

Source: files of the Real Chancillería de Granada (Records of legal proceedings, Granada Royal Chancery), Partes de Prevención de Causas, 1840–1890.

provoking a devaluation in the price of wood are pointed out. This occurred especially when wood was in high demand during the last decades of the 19th century as the railways were being constructed and mining activities were at their peak.

The intentional nature of these fires and their apparently clear association with the market obviously had the effect of relating and integrating the phenomenon of the forest fire within a broader, more significant historical framework, i.e. the processes of privatization and change in the use and management of the Andalusian woodlands throughout the second half of the 19th century. This highlights the low value which should be attributed to the explanations for these fires that describe the problem of fire as an ahistorical fact inherent in the Mediterranean woodlands and their particular meteorological conditions. Various studies have already clearly shown how the privatizing processes, originating in the 19th-century sales of institutional lands, ended up modifying the traditional patterns of use of the woodlands of eastern Andalusia. The most important consequences of this were the very evident intensification of human pressure on increasingly scarce wooded areas and the transformation of the woodlands and their use into an object of conflict within the Andalusian rural communities themselves.

The figure of 452,990 cases recorded by the police in the whole country between 1844 and 1900 of woodland damage and the 352,009 forestry offences would appear to be evidence for this (Domínguez Rodríguez, 1986).[2] This circumstance again appears if we consider the 128,593 arrests for offences in the Andalusian woodlands between 1875 and 1920. The high and ever-increasing number of these cases reported throughout the second half of the 19th century included a significant amount of intentionally caused forest fires.

It is precisely this circumstance, together with the fact that clear parallels can be observed between the general increase in rural criminality associated with the woodlands in eastern Andalusia and the evolution of the number of reports of forest fires (Table 16.2), which leads us, first, to situate the problem of the forest fires within the historical context of the Andalusian woodlands during those decades of the 19th century and, secondly, to try to find possible connections which could have existed between many of the intentionally provoked fires and social practices of struggle or resistance offered by the peasants from the communities in which they took place.

This exercise of putting woodland fires into context obviously has one of its clearest points of departure in the procedures of transfer and privatization of the Andalusian public woodlands, particularly if one bears in mind the actual dimensions of wooded areas which were finally affected. A total of 264,136.55 ha were included for disposal in the four provinces of eastern Andalusia in the General Classification of 1859, which represented 34.48% of the total public forest area existing in these provinces at that time. This loss, in global terms, became even greater over the years in some particular cases.

In Málaga, the 144,129 ha of wooded area exempted from disposal in 1859 became only 106,663 ha in the *Catálogo de Montes Publicos de 1901* (Catalogue of Public Woodlands of 1901) (Table 16.3). In Jaén, the evolution was fairly similar; over 200,000 ha were declared for disposal in 1859 and to this a new loss of 29,861 ha must be added between 1859 and 1901, if we consider the data referred to in the Catalogue of 1901. Only the province of Almería experienced a different evolution, recovering, in global terms, 20,861 ha of public wooded area by the year 1901. The public wooded areas of Granada, for their part, remained practically unaltered between these two dates.

The reductions in surface area observed between the data provided by the General Classification of 1859 and the Catalogues of 1864 and 1901 can be explained, for many of the woodlands in eastern Andalusia at least, by the fact that the privatization of the woodlands mentioned previously was not exclusively channelled through the decrees for disposing of land previously owned by the church and other institutions. Other possibilities must also be added, such as expropriations which occurred either as a result of errors or carelessness in the elaboration of the corresponding catalogues of woodlands for public

Table 16.2. Number of reports of forest fires and of any other damage to woodlands or fruits: eastern Andalusia, 1875–1889.

Time period	Woodland fires	Pilferage of wood	Tree-felling	Theft of fruits	Illegal ploughing
1875–1879	192	1461	1472	2678	806
1880–1884	224	2072	2285	3293	688
1885–1889	17	1244	1340	1006	469
1875–1889	433	4777	5097	6977	1963

Source: *Partes de Prevención de Causas* (Records of Court proceedings) and statistics of the cases of the Civil Guard (police force) in their role as forest wardens.

Table 16.3. Evolution of the public wooded area in eastern Andalusia, 1859–1901 (ha).

	General classification of 1859			
Provinces	For disposal	Exempt	Catalogue of 1864	Catalogue of 1901
Almería	7,652.00	53,786.00	13,092.00	74,647.00
Granada	36,244.55	114,393.00	80,931.00	114,934.00
Jaén	212,249.00	189,410.00	184,453.00	159,932.00
Málaga	7,991.00	144,129.00	122,556.00	106,663.00
Total	264,136.55	501,718.00	401,032.00	456,176.00

Source: Clasificación General de 1859 and Catálogos de Montes Públicos of 1864 and 1901.

use, either through irregularities in the delimitation process or as a result of the illegal ploughing up of woodlands, which was later ratified.

By the beginning of the 20th century, these processes had privatized 59.56% of the total public woodlands originally existing in eastern Andalusia and used by its rural communities in the mid-19th century. This percentage speaks for itself as regards not only the actual magnitude of the phenomenon, but also the importance that this process must have had in the daily life of many of these rural Andalusian communities, especially if we bear in mind that the traditional woodlands constituted complementary resources on which many peasant families relied for their subsistence.

The sharp increase observed throughout the second half of the 19th century in so-called woodland offences (theft of wood, fruits, charcoal-making . . .) in these rural communities and the fact that in the republican period of the 1930s the recovery of the common lands became a persuasive argument for recruiting the people to their cause and to that of many trade unions in eastern Andalusia seem to confirm our hypothesis. We believe that this hypothesis is strengthened even further if one considers the intervention of forestry administration, which led to a considerable reduction in the area that could be exploited, and the multiplication of the so-called extraordinary auctions, which were closely related to chronic shortages in the local government treasuries. The final result was obvious: the separation of the woodlands from their use or the right to their use by the peasant communities. The only alternative left open to them was a progressive change in their economies to a dependence on salaries and the market.

Andalusian peasants did not always respond in a negative way to this process in which the communities became salaried, commercialized and privatized. However, resistance did often form part of their reaction. These strategies of resistance did not always constitute practices which excluded adaptation and did not always occur in the form of collective protest. Apart from the latter, which evidently did exist, individual actions also emerged, which had the aim of mitigating or rejecting the privatization and commercialization mentioned above and which attempted to optimize the so-called 'weapons of the weak', with which the peasants traditionally defended their interests both against the non-peasant élite and against the state. Among these 'weapons' and forms of everyday resistance, those of particular note were false submission, pilfering, poaching, pretended ignorance, calumny, sabotage, desertion, fraudulent ploughing up and also arson, and among these were the forest fires.

The inclusion of the forest fires within the peasant resistance strategies should not be interpreted as an attempt to suggest that the cause of all the forest fires that occurred in eastern Andalusia came within this framework and logic. It is obvious that some were the result of this and others were not. The important and difficult problem is to be able to distinguish the former from the latter. This difficulty arises, to a large extent, from the fragmentary nature of the historical information conserved for this period and by its scattered nature.

Indeed, records of the criminal proceedings resulting from many of these intentional fires are conserved only rarely in the archives of the justice system. At the same time, also with a certain predictability, in those cases in which some records are conserved, the documentation is incomplete. There exist hardly any statements made by the defendants in which they explain the reasons for their action, or at least their point of view of the facts. Furthermore, the judicial sources (mainly the records of police proceedings) that were used in this study to quantify the evolution of the forest fires occurring in eastern Andalusia between 1840 and 1889 also suffered from some of these problems, particularly the lack of defendants' statements mentioned above. Also, specific data are not given regarding the surface area affected by the fire in question.

These difficulties and gaps in the sources could well make one believe it impossible to prove or disprove the hypothesis suggested that forest fires should be considered as a means of resistance used, on occasions, by the Andalusian peasant communities against the privatizing thrusts of the second half of the 19th century. This would indeed be so if we were trying to accurately quantify the principles of the hypothesis. However, the case is made through qualitative analysis of the presence of fire in the Andalusian woodlands during the 19th century from the number of reports of forest fires and the coincidence of this quantification with other sources, information and historical processes which were contemporaneous in these rural communities throughout this period.

This exercise of quantification, combination and qualification allows us to elicit connections, tendencies and indications – and even facts in some particular cases – which concur with the interpretation given by the hypothesis. For example, one can see how, in some particular cases, a very close correlation appears to exist between changes in the evolution of the number of forest fires reported as intentional and the arrival of the privatizing process. This is so in the case of the Málaga courts of Gaucin, where the important increase in the number of forest fires from the end of the 1850s coincides with an equally progressive decrease in the woodlands available for public use, as shown in the Catalogues of both 1862 and 1901.

Similarly, we could also cite the case of the Málaga district of Marbella, in which the sharp rise observed in forest fires during the 1850s and 1860s coincides precisely with a period in which the public woodlands in the area were reduced to a minimum, as reflected in the Catalogue of 1862. Similarly, the rapid fall in the number of fires recorded in the final decades of the 19th century coincides with an ostensible recovery of woodlands for public use, as can be seen in the Catalogue of 1901, especially if these data are compared with those of 1862. This argument could also be extended to a certain degree to the Jaén districts of Andújar, La Carolina and Villacarrillo, etc.

2 Numerical and territorial size of forest fires

Prior to beginning the numerical quantification of forest fires, some critical observations should be made with respect to the sources used. These sources have been assessed by the *Secretaría de la Audiencia Territorial* (Secretariat of the Territorial Courts), using information provided by different public order posts throughout the geographical area covered by the Granada Territorial Courts.[3]

The *Libros de Partes de Prevención de Causas* (Police Proceedings Records) contain information collected every year and classified by district court areas and in chronological order, listing the cases reported in each area, although these might not yet have been heard in court.[4] However, the level of information provided on the offence committed and the accused is less detailed than that provided by other sources, although the following data are included:

1. Name of the person arrested.
2. Reason for his detention.
3. Date of detention.
4. District court covering the area.[5]

From an analysis of these sources, a study was made of the number of reports of forest fires between 1840 and 1890. Table 16.4 shows that the total reports of fires in this period were 1432 for the whole of eastern Andalusia (an average of 33 fires per year), a figure that is comparable with the 525 fires in the period 1980 to 1994 in the same geographical area and with a similar area of woodland.[6]

We can therefore state that the integrated woodlands, i.e. woodlands used for forestry, agriculture and livestock, and populated with autochthonous species with slow growth, suffered fewer fires than the currently segregated woodlands (with a basically commercial or environmental orientation, e.g. to fight against erosion in basins, etc.) and in which monospecific and non-autochthonous forest masses predominate. Over the 50 years considered, the distribution of the fires was not homogeneous. For this reason a series of

Table 16.4. Evolution of the number of reports of fires in woodlands, eastern Andalusia (1840–1889).

Year	Unspecified	Private	Common	Own	State	Total
1840–1855	146	11	–	2	2	161
1856–1867	347	35	4	7	9	402
1868–1874	396	26	–	2	12	436
1875–1889	320	75	1	8	29	433
Total	1209	147	5	19	52	1432

Source: records of police proceedings and statistics of the cases of the police in their role as forest wardens.

periods have been taken (see Table 16.4), which enable us to establish as far as possible whether the fire reported corresponded to a criminal act of a common and universal nature, to a change in the management of the woodlands that had made traditional practices there illegal or whether, on the other hand, it was a conscious act of reprisal by the locals who disagreed with the implementation of the liberal legislation.

During the period 1840–1855, the agrarian measures of the liberal revolution were proclaimed (*Desamortización Eclesiástica, Decreto de 1834* (Sale of Church Lands Decree of 1834), which allowed the local councils to sell their rural heritage). This was at a time when the municipal institutions themselves obtained a large part of their income from the exploitation of the woodlands (thus reducing ordinary tax pressure on the local community), based on the provisions of the *Ordenanzas de Montes de 1833* (Woodland Regulations of 1833) – consecrating private property, transforming common land into municipal property, fencing private woodlands and eliminating communal rights.

The next period (1855–1868) involved the first systematic attack on the municipal and communal property of Andalusian villages (*Ley General de Desamortización* (Sale of Institutional Land General Act) of 5 May 1855), both through the conversion of public woodlands into private property and the introduction of the market as a central element in the management of the public woodlands.

The period 1868–1874 – *Sexenio Revolucionario* (Revolutionary Six Years) – is used to evaluate the incidents of these agitated years, which involved a conservative and statist slant, backing private interests in the agrarian legislation.

The last period tries to evaluate the incidence of fires in private and public woodlands, supported by a liberal legislation and protected by the stability of the recently created Canovas conservative regime, implemented in the period 1875–1889: disposal of the institutionally owned woodlands susceptible to public sale, auctions of forest produce, control and repression of the woodlands by the police, who impeded activities contrary to the commercial use promoted by the local and state authorities, the project and execution of the first plans for regulating the woodlands, etc. This was the time when access by locals to the public woodlands was restricted, due to the introduction of market dynamics in a society in which property owners controlled the majority of the local councils,[7] a phenomenon which we have called forestry despotism.[8]

Within the time sequence established, most of the forest fires were concentrated at the beginning of the 1870s and 1880s, when a confrontation arose between the local people, defending traditional uses, and the forestry administration, backing state intervention and privatization of both ownership and use of the woodlands (Table 16.5).

There exists a high correlation between the attempts to privatize the use and ownership of the public woodlands and the reports of fires, implying that

Table 16.5. Average number of annual reports and their relationship with area of woodlands.

Year	Annual average	Report per 100,000 ha
1840–1855	10.0	0.4
1856–1867	33.5	1.4
1868–1874	62.3	2.5
1875–1882	52.0	2.1
Total	32.9	1.3

Source: records of police proceedings and statistics of the cases of the police in their role as forest wardens.

forest fires were a tool used by the poorest locals against this privatization. This is supported by the low incidence of fires prior to the *Ley General de Desamortización de 1855* (General Act for the Disposal of Institutional Lands of 1855), when the local population had relatively recognized rights to use the public or private woodlands (common rights).

However, it is undoubtedly in the period of the *Sexenio Democrático* (Democratic Six Years) (1868–1874) that the most forest fires occurred, which was due to various factors: an increase in the strength of the peasant movement opposed to the privatization of the common woodlands, the passing of the Regulation of 17 May 1865, which assimilated woodlands arson as an offence within the criminal law, and the application of the forestry exploitation plans, which fixed local quotas for use of the woodlands and the volume of products destined for public auction.

The fall in the number of fires in the following period can be explained by the beginning of the restoration of the monarchy, accompanied by the implantation of the conservative regime of Canovas del Castillo, which led to an intensification in prosecution for this kind of offence, the vigilance and custody of the woodlands being made the responsibility of the police in 1876.

From 1855 (when according to the Woodlands Catalogue of 1859 there was 1 ha of public woodland for every 3 ha of private woodland), fires were mainly concentrated in the public woodlands, as they were subject to a process of privatizing ownership and use, as opposed to those which were recognized as communal by the administration, which suffered fewer attacks. This confirms the relationship between privatization of the municipal or communal woodlands and the increase in intentionally caused fires.

3 Spatial distribution of fires

Table 16.6 shows the incidence of forest fires according to time periods and provinces, and demonstrates that the woodlands of the provinces of Málaga and Jaén suffered more fires in the periods under consideration. However, no correlation exists between the woodland surface area and the number of fires – in this period, Granada had the largest area of woodlands but not the highest number of fires. It seems that neither the vegetation, climate nor extension of the woodlands was a determining factor in the number of forest fires, but rather they were a result of the changes in ownership and ways of managing and administering forest resources provided for by forestry legislation.

To confirm this hypothesis, an evaluation of the public woodlands in each province and their relationship with the number of fires reported was necessary. This comparison was made using the Catalogue of 1859, in spite of numerous doubts about its reliability, and the data available for 1910, and this enabled us to assess the importance of the privatization of the ownership of public woodlands in each province (but not of their use, which is a question that requires an analysis at micro level for its quantification).

Table 16.7 sets out the correlation between the number of fires and the amount of privatized public woodlands. The greater the area that passed into private hands, the greater was the number of reported fires (cases of Jaén and Málaga). In public woodlands subject to an increasing restriction in use,

Table 16.6. Distribution by province of the number of reports of woodland and pasture fires, eastern Andalusia, 1840–1882.

Year	Almería	Granada	Jaén	Málaga	Total
1840–1855	10	49	70	64	193
1856–1867	12	68	202	187	469
1868–1874	10	68	175	229	482
1875–1882	5	79	182	255	521
Total	37	264	629	735	1665

Source: records of police proceedings and statistics of the cases of the police in their role as forest wardens.

Table 16.7. Relationship between the number of and average annual number of reports of fires and woodland surface areas for each province, 1840–1882.

Concept	Almería	Granada	Jaén	Málaga
Woodland surface area in 1900	563,470	857,248	656,459	351,734
Public woodlands in 1859	61,438	150,637	401,659	152,120
Public woodlands in 1910	119,903	132,397	196,765	98,140
Variation	+40,465	–18,240	–204,894	–53,980
No. of reports	37	264	629	735
Average annual no. of reports	0.8	6.1	14.6	17.1

Source: Clasificación General de 1859 and Catálogos de Montes Públicos of 1864 and 1901 and records of police proceedings.

through public auction of part of the yield, an intensification of these peasant practices occurred in the public woodlands.

The importance of fires in the 19th century makes it possible to conclude that the woodlands did not catch fire on their own, or, rather, that fires tended to have fairly clear social causes, according to each historical moment. We therefore reject the suggestion of the forestry administration, who argued that the proliferation of these fires was due to natural causes, negligence or pyromaniacs, their reasoning being that fire was inherent in Mediterranean woodlands and therefore inevitable.

The immediate causes of the fires can be found in the interrelationship between humans and the environment, although the composition, density and disposition of the forest masses influence their extent; thus the lower number of fires during the 19th century, with respect to the present day, is due to the botanical composition and disposition of the masses, but the exponential increase shown during the 19th century can be explained by the growing breach between local communities and woodlands, encouraged by the liberal legislation.

Having established that fire was a form of social protest, a study was made on the conflict associated with the privatization of the woodlands between 1830 and 1920 in the provinces of Jaén and Granada, using judicial sources and police records published in the *Boletín Oficial del Estado* (Official State Bulletin). This research concluded that the forest fire was a form of protest or a way of obtaining certain fraudulent benefits (the term used for the traditional exploitation made illegal by the liberal legislation relating to the woodlands). The relative frequency of arson with respect to these actions (pilfering or wood-cutting, theft of fruits, ploughing up) gradually fell over the period considered, from one fire for every eight offences to one fire for every 22 other offences in the province of Granada, as opposed to the stability found in the case of Jaén (four offences for each fire reported).

4 Assessment of the causes of forest fires

The privatization of the woodlands is not the only cause of the increase in forest fires, but is one of the factors resulting from the separation of the local communities from the woodlands (meaning a loss in energy and food resources and a lack of the economic profit from their exploitation to outside the community) encouraged by the liberal legislation.

In the study of the immediate causes of forest fires during the 19th century, the social context should be taken into account. This was characterized by a progressive privatization of the use and legal ownership of the public, local and communal woodlands, which were all originally of common use. This process must have had a severe effect on the poorest people and those who had fewest resources (who lost the right to collect wood and esparto, make charcoal and graze their livestock), and it was therefore these social sectors which promoted the systems of protest and who were principally responsible for the fires in the last third of the 19th century (see the case of Zújar in the District of Baza, a municipality in the north of the Province of Granada, whose esparto resources became subject to public auction from 1870).

Thus fire was used as a form of reprisal by some locals against the privatization of harvests (e.g. esparto) or woodlands which they were no longer allowed to use. They also burnt other areas of woodland to improve communal grazing land (a practice which was prohibited by the woodlands legislation). The state and private intervention in the communal woodlands created a situation which was bound to increase the commission of certain forestry offences by outlawing practices which until then had been considered legal.

The case of Zújar is only one example of the decided reaction of the rural communities against the privatization process of the use of the public woodlands, developed by the liberal legislation since the mid-19th century, in which the multiple action of the locals in causing forest fires destroyed the woodlands subject to commercial exploitation for public auction.

Notes

1 Araque Jimenez, E.A. (1997) Privatización y agresión a los montes públicos jienenses durante la segunda mitad del siglo XIX. Désputación de Jaen.

2 Dominguez Rodriguez, F. (1986) Los Servicios. In: *Revista de Estudios Históricos de la Guardia Civil*, Vol. XIX, pp. 126–176.

3 This includes the provinces of Almería, Granada and Málaga.

4 Sometimes, the offence was recorded even before it was known who had committed the crime. In any event, the judicial process had not yet begun.

5 On some occasions, the place where the offence was committed was included with the crime, but this was additional information and not the norm.

6 Casero, F. and González de Molina, M. (1997) Mito y realidades de los incendios forestales en Andalucía [Myth and reality of forest fires in Andalusia], in: González Alcantud, J.A. and Jesus Buxo Rey, M. (eds) *El Fuego. Mito, Rito y Realidades*, Anthropos, Barcelona, pp. 377–410.

7 See Grupo de Estudios Agrarios (Agrarian Studies Group) (1995) Transformaciones Agrarias y Cambios en la funcionalidad de los poderes locales en la Alta Andalucía (1820–1935) (Agrarian transformations and changes in the functionality of local power in high Andalusia), *Noticiario de Historia Agraria* 10, 35–66.

8 Cobo, F., Cruz Artacho, S. and González de Molina, M. (1992) Privatización del Monte y Protesta Campesina en Andalucía Oriental (1836–1920) (Privatization of the woodlands and peasant protest in eastern Andalusia), *Agricultura y Sociedad* 65, 253–302.

17

Common Woodlands in Mediterranean Societies: Commercial Management Versus Forms of Peasant Resistance in Andalusia (Spain), 1750–1930

Antonio Ortega Santos

Departamento Historia Contemporánea, Universidad de Granada, Campus de Cartuja S/N CP 18071, Granada, Spain

The common woodlands of Andalusia (southern Spain) ceased their socio-economic function for the surrounding communities in the mid-18th century, when they became subject to forestry legislation whose objective was market production (associated with national or international industry), thus limiting traditional uses of this land and provoking differing responses from each community.

1 Geographical–climatic characterization

The District of Baza (where the municipalities of Zújar and Cúllar studied in this research are situated) is in the north of the Province of Granada (eastern Andalusia, Spain), between the mountain ranges of Sierra Nevada and Cazorla-Segura, on a plateau at an altitude of between 1000 and 2000 m. This region is formed by a continuous stretch of tectonic depressions (originating from alpine orogeny) with a continental cold climate – continental Mediterranean mountain. The mean annual temperatures oscillate between harsh winters (January between 4 and 6°C) and hot summers (24–26°C). This climate defines a precipitation regime of rainy winters (35–40% of the total annual rainfall) and similar levels in autumn and spring (25–30%), with a totally dry summer. Total rainfall, however, often does not exceed 350 mm per year.

It is evident that the result of this climate is an arid, semi-desert landscape, due to a historical erosion process of these plateaus, with irregular watercourses which define small irrigated valleys (both in Zújar and Cúllar),

(eds M. Agnoletti and S. Anderson)

together with an unirrigated area dominated by cereals, bushes and desert areas running between a succession of gulleys. It is a large area of badlands evolved over clays, marly limestones, metamorphic reliefs and enormous semi-fossilized limestone reliefs (as in the case of Jabalcón hill, 1500 m, the main esparto-producing area of the region). For this reason, the type of vegetation existing there is dominated by Mediterranean autochthonous woody species (*Quercus ilex* or *Quercus rotundifolia, Pinus nigra*), together with a wide range of bushy species, such as esparto (*Stipa tenaccissima*), broom (*Retama sphaerocarpa*), gorse (*Genista scorpius*) or thyme (*Thymus orospedamus*).

2 Socio-economic functionality of public woodlands in Mediterranean economies

The state forestry policy in Spain after the *Ordenanzas de Montes de Marina, 1746* (Marine Woodland Regulations), *Ordenanzas Generales de Montes de 1833* (General Woodland Regulations of 1833), *Ley de Montes de 1863* (Woodlands Act of 1863) and *Ley de Repoblación Forestal de 1877* (Forestry Repopulation Act of 1877), led to a progressive subjection of the management of woodlands to internal market needs through greater state and municipal interventionism in the systems of forestry exploitation of the woodlands. This was just one more element in the liberal agrarian reform, which involved a liberalization of production factors[1] and an extension of the dynamics of the market in an agrarian economy dependent on the cereals system.

This legislative system involved both potentiating the commercial management of woodlands and progressively restricting traditional uses, i.e. production and consumption needs of the rural communities, needs which were secondary in market strategy considerations and in the implantation of the capitalist system in all realms of production. This system of commercial exploitation of the woodlands was only possible in the communities in the south-eastern peninsula once the public woodlands were defined and delimited, as another element in the construction and consolidation of the municipal districts since 1860.[2] Only after the municipalities – including the common woodlands – were delimited and defined could the commercial exploitation of the public woodlands be developed (a system controlled by the local council) in Mediterranean woodlands of scarce silvicultural character and dominated by bushy species (esparto, grazing, aromatic plants, etc.). However, the imposition of this homogenizing legislation of the forestry exploitation systems led to a confrontation between the two patterns of use of the woodlands:[3] an industrial–commercial pattern and a peasant use of the common land, related to the level of energy and food consumption.

The public woodlands in the Mediterranean rural economies represented the key element in meeting the demand for energy (wood for heating or building houses) and food (grazing, forest fruits and products, poaching, etc.) until the mid-19th century.[4] This socio-economic functionality of the common land – involving a multiple and integrated use of the ecosystem through an agricultural/forestry/grazing exploitation of the woodlands – was dismantled by the action of commercial forestry policy, generating differing responses in each community. This action involved:

1. Cultivation of the public woodlands ('agriculturization') by large property owners, a process only possible after deforestation (as occurred in Cúllar from 1770), to enable the introduction of large herds and increase the extensive cultivation of cereals.
2. Commercial exploitation of common land from the mid-19th century,[5] which led to the reduction in traditional uses of the woodlands by the peasants and the implantation of a forestry option subject to the demands of the national and international market (in the case of Zújar, this was orientated towards the extraction of esparto as a basic raw material for the paper and packaging industry).

In the mid-19th century, in order to intensify their commercial exploitation, it was necessary to catalogue and quantify the Spanish woodlands[6] – although these data are of little validity – and this was supported by the attitude of the municipalities, who saw a way of financing the local government treasury through this system (from 1877, 20% of the products obtained in the auctions of forestry products were handed over to the state to finance forestry repopulation or to improve the public woodlands).[7]

Although the communities of Zújar and Cúllar responded differently to the state/municipal interventionism in the management and exploitation of the common lands (one potentiating the agricultural and the other the forestry option), in both cases the traditional uses of the public woodlands were privatized, by changes in property rights (by ploughing up) or by restricting the peasant pattern of use by introducing public auctions for the sale of esparto to regional or national merchants.[8]

A conflict thus arose as a result of the commercialization (= silviculture) of the public woodlands, together with an expansion of the agricultural area to meet market demand, which was to lead to an end to the practice of multiple use of the woodlands and to generate a new cycle of environmental destruction in the ecosystems,[9] which was of particular significance in the fragile Mediterranean ecosystems – all this in search of a supposed productive efficiency in these ecosystems.[10]

The 'tragedy of the commons' can thus be seen in the context of changing property rights and a historical evolution of the patterns of use of the woodlands[11] in rural economies with regeneration difficulties due to interventionist and commercial practices of state forestry policy, which limited the rights of use previously exercised by these communities. The productive option chosen by each community depended on the confluence of various factors peculiar to each: an increase in the population, demand for energy (firewood, charcoal, etc.), the soil characteristics of each area, which limited (due to the predominance of lime soils in Zújar) or favoured (as in Cúllar) the expansion of the agricultural area, etc. Zújar, taking into account the demand for products in relation to the strategy of consumption and the scarce viability of cultivating its woodlands due to the soil characteristics existing there, opted for the commercialization of its woodland products, particularly esparto. On the other hand, Cúllar opted for agriculture instead of woodlands, and this was encouraged by a group of large property owners anxious to extend their control over any land suitable for agricultural exploitation, from the mid-18th century.

This paints a picture in which each community reacted differently to state–municipal control of the management of the public woodlands, either subordinating traditional uses to the introduction of systems of commercial exploitation or expanding the area of cultivation as a means of extending the cereal system and increasing grazing lands, to the detriment of silvicultural exploitation.

3 Agriculturization of the commons: extension of the agrarian system and limitation of the option of forestry management

Dating from 1770, an intensification of the deforestation and cultivation process of the public woodlands took place, mainly carried out by large property owners with a view to extending grazing land for their herds and, at a later stage, increasing cereal cultivation to meet the demands of a higher population with higher levels of food consumption. It is evident that the imposition of this agrarian system in the areas containing public woodlands led to the limitation of the possibilities for providing energy supplies for the community (introduction of grazing or crops inevitably meant limiting the collection of wood, charcoal, etc.). A confrontation thus arose between the various patterns of use of the woodlands: peasant uses (collection of wood, etc.), which became an ‘offence’ and illegal, versus agricultural use of the woodlands, leading to the destruction of the tree cover of the commons[12] – ploughing up, introduction of herds of livestock, forest fires, etc. (Table 17.1).

From the study of social conflict (Table 17.1), the confrontation between the three ways of using the public woodlands in the last third of the 18th century in the south-eastern peninsula can be defined:

1. Industrial pattern of use, postulated by the Naval Department in those woodlands under its authority since 1746 (*Ordenanzas de Montes de Marina* (Naval Woodlands Regulations)), which aimed to maintain the levels of wood production in these areas, limiting human presence there (by penalizing the collection of wood, etc.).

Table 17.1. Typology of forestry offence, woodlands of Cúllar, 1790–1830.

Type of offence	Percentage
Pilfering of wood	61.9
Deforestation	9.4
Theft of wood	8.8
Invasion by herds	8
Ilegal grazing	7.5
Forestry fire	3.6
Ploughing up	0.6

Source: *Instrucción por Causas de Daños en Montes* (Prosecution for Damage to Woodlands), AHMC.

2. Agricultural pattern of use, argued by the large property owners of the region as a means of increasing productivity of woodlands with 'scarce economic viability', especially in a community which had to satisfy higher levels of demand for food as the population increased; this is supported by the extent of offences of deforestation, invasion of herds, etc.
3. Peasant pattern of use: defined as an integrated and multiple use of the public woodlands, related to energy and food needs (wood, grazing, charcoal, poaching, sale of wood, etc.), which were extracted from these woodlands, in a community with a very unequal agrarian property structure.

Thus the action of the large property owners in the areas was focused on privately acquiring the public woodlands,[13] by moving the boundary marks which delimited them (*mojones*) to increase the extent of their private property. They then initiated a process of deforestation (by cutting down trees and invading the land with their flocks) and then, at a later stage, extended the area for cereal cultivation in these areas of ploughed common land, particularly as, if they could prove the land had been cultivated for an uninterrupted period of time, they could then consolidate private ownership rights over this land.[14]

This process of extending the cultivated areas is evidenced in Table 17.2 by the sharp increase in unirrigated cultivated land (almost exclusively cereals) in the period 1750–1870, together with a certain reduction in 'uncultivated' land – essentially public woodlands. On the other hand, in the case of Zújar, the option for forestry – gathering of esparto, public grazing, etc. – meant that the structure of the cultivation systems remained stable, the area of recognized woodlands increasing during the period under consideration (due to the process of delimitation of the municipality of Zújar in 1877, which facilitated a better knowledge of the area of woodlands). This shows the different response of the communities studied with regard to the management of the public woodlands; in Cúllar, the cereal system was potentiated (at the height of the liberalization of the production conditions initiated by the liberal agrarian reform, which favoured the maintenance of private ownership of land and the creation of markets for a specialized and expanding agrarian

Table 17.2. Percentage of cultivated land with respect to total of the municipality, Cúllar and Zújar, 1750–1870.

Municipality	Unirrigated	Irrigated	Uncultivated	Total
1752				
Cúllar	32.31	7.24	60.45	100
Zújar	24.4	8	67.6	100
1860–1870				
Cúllar	42.22	0.47	57.31	100
Zújar	24.3	3.09	72.61	100

Source: *Catastro Ensenada, Respuestas Generales* (a kind of land register reflecting ownership and use), Amillaramientos Municipales, AHRChG; Cano García, G. (1974) *La Comarca de Baza*, Disputación Provincial de Granada/Instituto Juan Sebastián Elcano Ed; A. Ortega, own research.

sector), whilst Zújar potentiated the commercial forestry option in the management of the woodlands.

From 1850, these processes of ploughing up and appropriating the woodlands by large property owners were intensified, in order to circumvent the cataloguing and delimitation of the public woodlands, which were being initiated at that time and which could evidence the illegality of later appropriations of forest land (Table 17.3). But these were not the only factors which led to the ploughing up of common land in the mid-19th century. The beginning of the process of selling public lands (1855) was a factor that directly affected the pre-emptive action of these large property owners, as this project was aimed at disposing of a considerable amount of the country's public woodlands – which had previously been in the hands of public institutions, nobles or the church – at public auction. The only exceptions were those woodlands with a proven silvicultural character or which were essential in the subsistence levels of the rural communities (common lands).

The incidence of the expropriation of these church and other lands in Cúllar was small, both because of their relative silvicultural nature there (dominated by *Quercus* and pines) and because the woodland area had been reduced by the ploughing up carried out by the large property owners since the end of the 18th century. Indeed, ironically, these large property owners – supported in the control that they exercised by the local council during the 19th century – defended the socio-economic importance of the public woodlands for the community and, for this reason, these woodlands were not disposed of at public auction.

A historical cycle was thus completed, in which the public woodlands had become a key element in the extensive agrarian system, generating a sharp reduction in the forested area as the unirrigated cultivated area increased (Table 17.2). This privatization process of the common lands was incompatible with the commercial system of exploiting the woodlands (esparto, grazing,

Table 17.3. Ploughing up of common woodlands in Cúllar, 1850–1872.

Year	Situation	Identity of farmer	Area
1852	Cortijo Bermejo y Madroñal	Conde de Jiraldeli	151 ha, 44 a.
1855	Montes de Matián, Pililla, Bermeja y Santaolalla	Ramón Galera Galera José Burrueco	403 ha, 84 a.
1857	Monte de Llano de Abad	Ramón Galera Galera Andrés Galera Galera José Martínez Torres Pedro Bautista Torres	252 ha, 40 a.
1857	Montes de Pililla	Antonio Galera Mtnez.	30 ha
1872	Cortijo D. Pedro García	Pedro Bautista Reche	160 ha, 20 a.

Source: *Acciones Judiciales contra Particulares* (Legal actions against private individuals), Files 15–20. AHMC, *Sección Montes* (Woodlands Section), AHDPG, File 447, Items 1–10.

wood, etc.) from the end of the 19th century, especially when, at this point in history, private property rights over the public woodlands – by means of ploughing up – were being consolidated through judicial proceedings,[15] in which court decisions certified the legality of this ploughing up, arguing the possession of these woodlands by the large property owners since the mid-17th century.[16] The final result meant the consolidation of a historic process of transferring forestry land to private property and agricultural production, to the detriment of the common usage.

4 Commercial exploitation of the common lands of eastern Andalusia: Zújar, 1860–1930

As opposed to the case of Cúllar, the public woodlands of Zújar were submitted to an intense process of commercial exploitation from 1860, centred on the public auction of the esparto produced there.[17] This system of commercial exploitation of the woodlands meant the following:

1. Limiting the collection of esparto by the peasants for making farm tools or for sale in local or regional markets (peasant pattern of use of the woodlands).
2. 'Commercialization' of the management of the woodlands according to mercantile practices, which privatized the use of the woodlands by assigning the control and management to private individuals. In this sense, a direct confrontation arose between the peasant and commercial patterns of using the woodlands, which caused all sorts of social conflict.
3. Intensification of the extraction of forest products, due to the demands of the national and international market, which not only privatized the rights of use of the land, but led to a productive crisis of these woodlands in the long term, underlining the semi-desert character of the ecosystem.

The system involved an ever-increasing extraction of esparto, over and above the legal levels permitted, and a non-compliance by the winning bidders in the auction to hand over a share of the esparto for local villagers' use, a fact which is demonstrated by the 'illegal' extraction carried out by the locals and the low international market prices of the product from 1870 (pilfering and theft of esparto from the woodlands was a mechanism of protest by the local people against the public auction system). This public auction system of esparto from the woodlands, which attracted the main national industries related to the production of paper fibre, led to a dependence by the peasant community on seasonal work during the periods in which this product was harvested (and thus peasant economies became 'salaried').

In this way, the common land was robbed of its socio-economic role within the agrarian system and became just one more element in the state forestry policy, its levels of extraction and harvesting (which became the economic mainstay of the community by providing casual labour and salaries) being determined according to market interests (the logic of subsistence versus the logic of profit).[18]

Between 1870 and 1930, the collection of esparto from the common lands of Zújar dropped sharply, both in volume and price, due to the abusive and uncontrolled extraction of this woodland product for commercial purposes, which impeded the productive regeneration of the woodland; nor was it possible to maintain the level of forestry exploitation, as the grazing rights over the public woodlands of Zújar had also been auctioned (Table 7.4). It is evident that the effect of the crisis in the world market of esparto (attributed to the greater presence of esparto produced in Algeria, which, although of lower quality, had a higher annual production)[19] was a significant fall in market prices and bidding at the auctions, which did not, however, put an end to this system of managing the woodlands.

The tight municipal hold over the commercial exploitation of the woodlands involved controlling the labour force and its seasonal nature in the community (which was dependent on the collection of esparto during one part of the year to ensure a minimum number of paid working days) and financing the local treasury through the public auction system (and, by extension, controlling the level of taxation in the community, as the increase in income by exploiting the woodlands reduced indirect taxes).

From these circumstances, it can be deduced that the continuity and stability of the public auctions were very important for the municipalities and the local oligarchs who controlled the local councils (*Municipalización–Oligarquización de la Gestión Forestal* (Municipalization-oligarchization of forestry management)).[20] The fall in esparto prices led to a reduction in the capacity to fund the local councils, which depended on the income from the exploitation of the woodlands to cover their annual budgets. In this sense, one could talk of an intensification of municipal control over the systems of commercial exploitation of the woodlands in the interests of maintaining

Table 17.4. Commercial exploitation of esparto: total volume and local* exploitation, woodlands of Zújar and Cúllar, 1870–1930.

Time period	Total volume esparto of Zújar	Total volume esparto of Cúllar	Local exploitation of esparto of Zújar
1870–1875	13,000	–	2,900
1876–1880	15,000	–	2,800
1881–1885	20,000	–	2,100
1886–1890	17,000	–	2,000
1891–1895	15,000	–	2,200
1896–1900	14,000	–	2,200
1901–1905	14,500	4,000	1,700
1906–1910	10,000	3,800	1,650
1911–1915	14,500	3,150	1,700
1916–1920	10,000	2,400	1,500
1921–1925	10,000	1,120	1,000
1926–1930	7,000	350	780

*No. of quintales métricos thousands per year on average per 5-year period (1 QM = 46 kg).
Source: Auction records, AHMZ and AHMC.

this income and thus reducing tax pressure on the local community (taxes on consumer goods, land-ownership taxes, etc.).

The case of Zújar, which had no space for extending the agricultural area, presents a model of intensification of the commercial exploitation of the woodlands, subordinating the traditional regenerative strategies of the community to a dependence on seasonal work in the collection of esparto (which occurred between February and May). Thus, the system of public auction meant a privatization of the woodlands, not through the transformation of property rights but simply by privatizing the use of the woodlands, granting the control of its management to individuals or companies, which, in order to carry out this exploitation, limited the multiple use strategy of the peasant groups in the public woodlands (Table 17.5).

This confrontation between the two patterns of use of the woodlands (peasant and industrial) and the protest of the community are evidenced by the pilferage of esparto, theft of wood and forest fires, a conflictual kind of opposition to the privatization of common usage (Table 17.5). This system of peasant protest is proof of the regenerative importance of the woodlands in the peasant economies and how the commercial pattern of use, through public auction, limited their reproductive possibilities (by restricting grazing, collection of esparto or wood for heating, charcoal, etc.). Indeed, Table 17.5 illustrates the fact that forestry offences show an evolution parallel to that followed by the dynamics of the market for the product (Table 17.6), meaning that the theft of esparto or wood became the symbol of protest against the

Table 17.5. Forestry offences: pilfering of esparto and wood*, woodlands of Zújar, 1850–1930.

Time period	No. of pilferings of esparto	No. of pilferings of wood
1850–1860	17	4
1861–1870	13	4
1871–1880	8	6
1881–1890	4	4
1890–1900	4	2
1901–1910	7	3
1911–1920	5	2
1921–1930	8	–

*No. of cases reported per year on average per 10-year period.
Source: Town Council Records, AHMZ.

Table 17.6. Forestry exploitation of esparto: reserve price and highest bid*, woodlands of Zújar and Cúllar, 1870–1940.

Time period	Reserve price of Zújar	Highest bid Zújar	Reserve price of Cúllar	Highest bid Cúllar
1870–1875	4,500	19,800	–	–
1876–1880	4,500	76,180	–	–
1881–1885	24,000	70,000	–	–
1886–1890	22,000	60,000	–	–
1891–1895	25,000	58,000	–	–
1896–1900	24,000	55,000	–	–
1901–1905	21,750	29,120	–	–
1906–1910	19,000	30,700	6,400	2,600
1911–1915	15,000	19,300	1,960	2,150
1916–1920	9,000	44,000	1,400	1,300
1921–1925	43,200	33,255	599	610
1926–1930	15,200	23,250	422	422
1931–1935	16,060	16,060	431	410
1936–1940	16,000	18,500	371	391

*Reserve price and highest bid per year on average per 5-year period.
Source: Auction records, AHMZ and AHMC.

intensification of the commercial exploitation of the woodlands and a defence of the peasant pattern of use.

As traditional uses became illegal, due to their incompatibility with the commercial exploitation of the woodlands, the rural communities opted for various modes of resistance, in the form of petty crimes in the woodlands,[21] as a way of responding to the imposition of the *Planes de Aprovechamiento Forestal* (Forestry Exploitation Plans) for the various public woodlands (*Ley de Montes*

1863 (Woodlands Act)), which did not provide any answers to the consumption and energy needs of these communities. The state forestry policies of the last third of the 19th century, which opted for the commercial management of the woodlands, therefore led to a violent rupture with the regenerative bases of organic economies, potentiating industrial use as opposed to seeing the woodlands as an integral factor in the agrarian system.

Notes

AHDPG: *Archivo Histórico de la Diputación Provincial de Granada* (Historical Archives of the Provincial Government of Granada).
AHMC: *Archivo Histórico Municipal de Cúllar* (Municipal Historical Archives of Cúllar).
AHMZ: *Archivo Histórico Municipal de Zújar* (Municipal Historical Archives of Zújar).
AHRChG: *Archivo Histórico de la Real Chancillería de Granada* (Historical Archives of the Royal Chancery of Granada).
RD: *Real Decreto* (Royal Decree).
RO: *Real Orden* (Royal Order).

1 Garrabou, R. and Sanz Fernáncez, J. (1985) La agricultura Española durante el siglo XIX: ¿Inmovilismo o Cambio? In: Garrabou, R. and Sanz Fernáncez, J. (eds) *Historia Agraria de la España Contemporánea*, Vol. 2. *Expansión y Crisis*, Crítica/Grijalbo, Barcelona, pp. 7–93. González de Molina, M. (1993) Nuevas Hipótesis sobre el Campesinado y la Revolución Liberal en los Campos de Andalucía. In: Sevilla Guzmán, E. and González de Molina, M. (eds) *Ecología, Historia y Campesinado*. Colección Genealogía del Poder no. 12, Ediciones Libertarias, La Piqueta, pp. 267–309.

2 The process of delimiting the public woodlands was already mentioned in the *Ordenanzas Generales de Montes de 1833* (Title II, art. no. 24), although this system of delimiting the woodlands was based on RD 1/4/1846; RO 12/5/1857; 11/12/1857; and only with the *Ley de Montes* of 24/5/1863 (art. no. 20–39) were these delimitations of public woodlands begun, in a process supervised by the local councils and civil governors, and those who modified these delimitations were penalized with a criminal sanction according to RD 8/5/1884, *Para la Reforma de la Parte Penal de las Ordenanzas de Montes*, art. no. 3 (RD 8/5/1884 for the Reform of the Penal Part of the Woodlands Regulations).

3 Guha, R. and Gadgil, M. (1993) Los hábitats en la historia de la humanidad. *Ayer* 11, 49–111.

4 Toledo, V. (1993) La racionalidad ecológica de la producción campesina. In: Sevilla Guzmán, E. and González de Molina, M. (eds) *Ecología, Campesinado e Historia Colección Genealogía del Poder*, Ediciones Libertarias, La Piqueta, pp. 197–219; Cazzola, F. (1996) Diboscamento e riforestazione Ordinata nella Pianura del Po: La Piantata di Alberi Nell'Economia Agraria Padana (Secoli XV–XIX). In *Storia Urbana*, anno XX, no. 76–77, July–December, Franco Agnelli Editori, Milan, pp. 35–65.

5 The system of commercial exploitation of the woodlands through public auction was consolidated in the *Ley de Montes 24 de mayo 1863*, which, in chapter VII, sets out the conditions and time periods for these auctions; complemented by the *Instrucción para Formación de Planes Provisionales de Aprovechamiento según art. 86 del Reglamento de Ejecución de la Ley de Monte 24/5/1863* (Instructions for making provisional plans of

exploitation according to art. 86 of the Regulations for Implementing the Woodlands Act), which defined the forestry year and the types of exploitation.

6 *Clasificaciones Generales de Montes Públicos* (General Classifications of Public Woodlands) of 1859, 1862, 1901 and 1931.

7 *Ley 11 de Julio de 1877 de Disposiciones sobre Repoblación, Fomento y Mejora de Montes Públicos* (Act 11 of July 1877 for Provisions for Repopulation, Promotion and Improvement of Public Woodlands), arts 1–6

8 Sánchez Picón, A. (1992) *La Integración de la Economía Almeriense en el Mercado Mundial (1778–1936). Cambios Económicos y Negocios de Exportación.* Instituto de Estudios Almerienses, Almería.

9 Sivaramkrishnan, K. (1995) Colonialism and forestry in India: imagining the past in present politics, *Comparative Studies in Society and History* 37, 3–40; Mann, M. (1995) Ecological change in north India: deforestation and agrarian distress in the Ganga–Jamna Doab 1800–1850, *Environment and History* 1, 201–220.

10 McNeill, J.R. (1992) *The Mountains in the Mediterranean World: An Environmental History.* Cambridge University Press, Cambridge; Corona, G. (1995) Il posseso collettivo della terra nell'Italia contemporánea trastoria e storiografia. In: Vicedo, E. (ed.) *Bens Comunals als Països Catalans i a L'Europa Contemporània.* Institut d'Estudis Ilerdencs, Fundació Pública de la Diputació de Lleida, Spain, pp. 529–553.

11 Fenny, D., Berkes, F., McCay, B.J. and Acheson, J.M. (1990) The Tragedy of the Commons: twenty-two years later, *Human Ecology* 18, 1–19; Martínez Alier, J. (1992) Pobreza y medio ambiente. A propósito del informe Brundtland, in: González de Molina, M. and González Alcantud, J.A. (eds) *La Tierra. Mitos, Ritos y Realidades.* Anthropos/Diputación Provincial de Granada, Granada, pp. 295–332.

12 The only significant obstacle that this process of massive destruction of the woodlands came up against was the formal cataloguing and quantification of the woodlands (distribution and number of species in each area of common woodlands) developed by the Marine Department in 1804 to assess the possibilities of producing wood destined for the domestic market and the construction of boats. *Noticia del Estado actual de los Arbolados que tiene la Subdelegación de Montes de la Villa de Cúllar . . . 1804* (News of the Current State of the Woods contained in the Subdelegation of Woodlands of the Township of Cúllar . . . 1804). AHMC, Unnumbered file.

13 Between 1780 and 1830, the Galera and Burrueco Masegosa families were repeatedly reported by the woodland wardens (over 50 complaints exist) for forest fires or invasions of the common land with their flocks to try and take over the use of the woodland and, by extension, become the owners of the common land by incorporating it into their *Cortijos* (estates) (Cortijos D. Pedro, Colorado and Bermeja), AHDPG, Woodlands Section, Files 434–449.

14 The consolidation of private property rights over the common land, obtained through ploughing up and usurping the woodlands, was facilitated by the fact that, by proving occupation and use of this land for a period of more than 30 years, proceedings could be initiated to claim private ownership, *Ordenanza General de Montes*, RD 22/12/1833, Title II, Section 1, art. 24. (General Woodlands Regulations).

15 Sentence of the Civil Court of Granada 8/6/1878 in AHDPG, *Sección Montes* (Woodlands section), File 447, Item 9.

16 The structure of land ownership in these municipalities originated from the expulsion of the Moors at the end of the 16th century, at which time the Crown distributed their land among the groups who repopulated the region and cultivated them,

including in each property an area of woodland. This generated some confusion – in the mid-18th century – between the common land and the privately owned land, which was resolved by the inclusion of all woodlands as common land (later public woodlands managed by the state). This formed the main argument for the appropriation and ploughing up of the woodlands by the Cortijos del Madroñal, Bermejo, D. Pedro and Colorado, arguing that they were private lands originating from the royal concession at the end of the 16th century. See *Pleito Judicial entre el Ayuntamiento de Cúllar y la Familia Galera y Burrueco Masegosa* (Case between the Cúllar local council and the Galera Family and Burrueco Masegosa), AHDPG, *Sección Montes* (Woodlands Section), Files 434–449; *Sección Fomento* (Promotion Section), Files 80–90.

17 The esparto from the public woodlands of Zújar was auctioned annually to merchants from throughout the country, who had to offer a predetermined minimum total price for the esparto (reserve price), although a much higher economic figure was offered to win the auction (highest bid). This auction process was governed by a series of previously fixed economic and administrative conditions, which obliged the merchant to hand over, free of charge, a certain amount of esparto to the local council for use by the local community (Ortega Santos, A. (1995)[20]).

18 Iturra, R. (1993) Letrados y campesinos. El método experimental en la antropología económica. In: González de Molina, M. and Sevilla Guzmán, E. *Ecología, Historia y Campesinado*, pp. 131–153.

19 Jiménez Blanco, J.I. (1986) *La Producción Agraria en Andalucía Oriental, 1874–1914*, Universidad Complutense, Madrid; Sánchez Picón, A. (1992) op. cit.

20 Ortega Santos, A. (1995) Los montes comunales en Andalucía Oriental: el ciclo del esparto en Zújar. In: Vicedo, E. (ed.) *Béns Comunals als Paisos Catalans a l'Europa Contemporània*. Patronato Municipal Josep Lladonosa, June 1996, Lleida, pp. 491–513.

21 Sansa, R. (1994) Il bosco fra difesa degli usi consuetudinari e conflitti di mercato, *Storia Urbana* 18 (69), 133–151.

18 Biodiversity Conservation in Central Himalaya: a People's Approach

Ajay S. Rawat

2B Sleepy Hollow, Naini Tal, UP 263001, India

Kumaon Himalaya or the Central Himalayan region is one of the most botanically diverse areas of the Indian subcontinent. The dramatic change in elevation has resulted in a unique biomass distributed over a variety of topographical and climatic zones. The situation has acted both as a bridge, facilitating the influx of many taxa, and as a barrier, promoting endemism in many areas.

This region is the catchment area of the Indo-Gangetic plain, where more than one-third of the Indian population resides. Continuous population growth, faulty planning and the ever-increasing biotic pressure have altered the landscape, with little or no concern for long-term environmental consequences and threats posed to biodiversity. Furthermore, the weak economic structure and minimal employment opportunities have added to the problem. Traditionally, a symbiotic relationship was maintained between the forests and the people in the hills. But, with the introduction of the colonial forest policy, the relationship waned. Even independence did not change the situation and the hangover of the British policy still continued. With colonial laws came a colonial mind-set. After establishing its own pre-eminence, the Forest Department allowed the claims of other vested groups to supersede those of the village people. In order to save the forests and the biodiversity of this fragile ecoregion, foresters should adopt a people's approach to make forestry a programme for the people, by the people and of the people.

(eds M. Agnoletti and S. Anderson)

1 Introduction

Himalayan biodiversity is a subject which is receiving global attention. It attracts a diversity of views and these views are as important as the subject itself. The Himalayan mountains, which have now come under a strong programme of development, support a variety of fragile ecosystems, whose existence is being endangered. Thus, the subsequent loss of biodiversity and its implications have become a matter of great concern.

Central Himalaya or the Uttar Pradesh (UP) hills are an integral part of the Himalaya. Ecologically, the area is divided into two geographical regions or the two Commissioneries, Kumaon and Garhwal. The Central Himalaya comprise three ranges, the Greater or the Inner Himalaya, the Middle or the Lesser Himalaya and the Outer or the Sub-Himalayan Tarai, all of which run almost parallel from east to west, with altitudinal variations ranging from a few hundred to over 8000 m. Here, the three factors latitude, longitude and altitude have moulded the characteristics, as well as the multiplicity of habitats, different microclimates and varied environmental niches of plants, animals and human beings.

Continuous population growth, faulty planning and ever-increasing demand on natural resources have altered the landscape, with little or no concern for long-term environmental consequences and threats posed to biodiversity. The weak economic structure, together with minimal employment opportunities have added to the problem. Conservation of what is left is probably the only option for overcoming this grave situation. But to draw a line between conservation and development is not easy. Development as it is understood in the common terminology is different from the development and conservation of biodiversity. In the Central Himalaya the area is being ravaged of its biodiversity in the name of development. To check this ecological damage, extraneous conditions, of whatever nature and magnitude they are, that are detrimental for biodiversity should be stopped immediately. But without the political will in this region, which is vying for statehood, it may not be possible.

The situation has become more alarming because of the unplanned growth of tourism, where rules are being tailored for the convenience of builders. Also the impact of plastic and non-biodegradable material brought by the tourist is wreaking havoc on the fragile ecosystem of the hilly areas. The visitors, with no respect for the environment, throw plastics, polythene bags, wrappers and other non-biodegradable material on the slopes of the hills, on the mountain, along trekking and pilgrim routes, in the lakes and in the rivers, leaving indiscriminate evidence of deliberately created garbage. The plastic collects on the slopes of the hills to be covered by autumn leaves of the season. This exercise is repeated year after year, with plastic being sandwiched between leaves and preventing rainwater from seeping and percolating and with the disappearance of the little streams and water resources on which the hill people rely. This deposit of plastic materials kills the green life on the slopes of mountains, pollutes water bodies and has a negative impact on aquatic life.

The unplanned growth of tourism has also affected the *bugiyals* or pasture-lands at high altitudes. The administration is attempting to establish hotels and motels on the *bugiyals* of Uttarkhand and has ignored factors that affect the ecological balance of mountain biodiversity. The *bugiyals* belong to the people. They are an ecosystem in themselves; it is not for humans to erode the sanctity of this area by putting up tourist houses and hotels.

Four representative areas were selected for the study of biodiversity conservation, two in Kumaon, namely Kilbury and Sitabani, and two in Garhwal Himalaya, namely Akaskamini catchment and the upper Yamuna valley.

2 Remedial measures

There is an increasing concern regarding the deteriorating environment in UP hills and the threat posed to biodiversity. But the planners do not seem to realize that the environment cannot be conserved with poverty and without meeting the day-to-day needs of the people. Poverty alleviation and developmental programmes that have been launched in the hills have not had an adequate impact. Professor Shankar Lai Sah, an agroeconomist, contends that the development approach has been fragmentary. The various development departments lack coordination, whereas multidisciplinary and coordinated efforts are needed to re-establish the linkages of land use and sod and water conservation. The gains of development are pocketed by the political village élite, who have become contractors and commission agents. Government development has largely been a matter of *batwara* (distribution), which has been, according to A.D. Moddie, thoroughly exploited by local politicians, down to the village, and which is now crippled by corruption at every door. Although there has been much discussion on the dangerous depletion of resources (mostly it is the people who are blamed directly for the dwindling forests), the environmentalists do not seem to understand that, in the wake of such lopsided development, the priority of the villagers in the UP hills is not forests and biodiversity but their own survival. The villagers' priorities today are sustenance, water, health, education, sanitation, fodder and fuel.

Damage here presents a greater risk than usually perceived because these lofty peaks are the source of the great Ganga system. It is the catchment area of the Indo-Gangetic plain, where more than one-third of the Indian population resides. It is important to preserve the forests and biodiversity of the UP hills, but not at the cost of the people. With the increasing political awareness created by the Uttarakhand Movement, the people have started to realize the ecological importance of this area. After years of callous neglect, the cauldron of the UP hills is boiling over with public anger and the local people are threatening to destroy the forests themselves. At the cost of our sustenance, why should we protect our forests for the benefit of those who have been thwarting our ambitions?

The attitude of the people towards the forests is apathetic and towards the Forest Department they have had a deep-rooted antagonism since the turn of the last century. The colonial forest policy, which promoted commercial forestry, deprived the people of their traditional forest rights. Even after independence, the situation did not improve. After establishing its own pre-eminence, the Forest Department in the hills allowed the claims of other interest groups to supersede those of the hill people. Several Uttarakhand Sangharsh Vahini activists, who at one time supported the Chipko Movement, felled thousands of trees throughout Garhwal and Kumaon during their 1988/1989 Ped Kato Andolan, or Felling the Trees Movement, which was launched to counter delays in environmental clearance for roads, electricity lines and water pipeline projects. Bipin Tripathi, a protagonist of the Chipko Movement in Kumaon, who led the 1978 struggle to save the forests of Chanchri Dhar near Dwarahat in Almora district from being felled by a paper-mill, has now diametrically changed his stance. Chanchri Dhar is the watershed of several streams supplying water to 18 villages. In January 1978, about 300 villagers camped there for 39 days and succeeded in saving the Chanchri Dhar forests. This was a people's movement and its leader Bipin Tripathi is now a changed man because he feels that several developmental schemes in the hills have been held up for environmental reasons. However, under the guise of tourism, rules are being brazenly flouted for the builders' mafia. As a protest against this dual policy, Bipin Tripathi and his supporters cut trees in 111 places where the government was using the Forest Conservation Act to hold up developmental projects. The hill people want trees but they want development too, reiterate the activists of Uttarakhand Movement. Their argument is over whether the trees are for the people or the other way round.

It is indeed an irony that the very region that gave birth to the Chipko Movement now has activists promoting the felling of trees. Even in Reni, the birthplace of Chipko, the womenfolk hate to hear the word environment. This is because, after the Chipko Movement, the conservationist element received greater emphasis, which widened the gulf between the local reality and the perception of the conservationists. In mindless reaction to environmental and political pressure, the government banned green felling above 1000 m.

Both the movements, Chipko and its antithesis, the Ped Kato Andolan, have one thing in common, the right of local communities to decide how they should manage and use their resource base. Meanwhile, the government has used the growing environmental concern to centralize environmental management without any concern for devolution of environmental rights and obligations. Thus, if the forests and biodiversity of the UP hills are to be preserved, the government will have to win the support of the people and manage the forests with their cooperation. It is the community that determinesthe success of natural resource management projects by virtue of their need, numbers and physical presence. The Forest Grievances Committee, which was constituted as a sequel to forest movements in the hills after 1916, concluded in its 1926 report that, 'Unless local villagers supported forest policies actively, no sustained management and regeneration of forests would be possible.'

In Uttarakhand more than 80% of the people live in rural areas and depend on subsistence agriculture. Basically, hill agriculture is biomass-based, the biomass being mostly obtained from forests. In Uttarakhand a village is an ecosystem which comprises cultivated land, clusters of houses, forests, grazing and pasture-land. From the mountain perspective, the role of forests in a hill farming system is altogether different from that in the plains. Forests – Civil Soyam/Panchyat/Reserve – supply fuel wood, fodder, compost, timber, staking and fencing material as well as food to the system. The problem with development strategies in the hills is that, when one talks about agriculture, it refers only to crop production and overlooks its linkages with forests. Forests in the hills are as integral a part of the farming system as arable land and livestock. Crop production, horticulture, animal husbandry and forestry constitute the four main closely and inseparably integrated components in hill farming, as shown below.

Thus, in this development approach, it is not agriculture versus forestry but agriculture and forestry.

The farming system in the hills has a number of features, such as inaccessibility, fragility and diversity, a great variety of crops, including perennial fruit, as well as fodder and different species of livestock. Thus, to achieve sustainable development and to maintain diversity in the forests of Uttarakhand, forest use and development will have to be treated as an overall resource use pattern and people's sustenance.

The dilemma of deforestation faced by planners today results from the fact that they did not react when the danger signals first appeared. Even after the authorities recognized the problems posed by deforestation and the subsequent loss of biodiversity, there was no unified land management strategy that sought to reconcile agricultural and forest-related uses. According to Professors J.S. Singh and S.P. Singh, each kilocalorie of agronomic yield requires the expenditure of about 7 kcal from the adjacent forests in the form of fodder, fuel and leaf manure. Even with this forest support system, agronomic production in the hills is sufficient for only about 50% of the population's need. The remaining half must come from the plains. In view of various problems, including the ruggedness of the terrain, neither the agricultural area nor its yield can be increased. Landholdings in the Lesser Himalayan region are small, averaging 0.8 ha per household. The present situation therefore calls for a change. The crop system in the Lesser Himalaya should be gradually replaced by tree farming and, to encourage the villagers as well as to prevent the impending ecological doom in the Indo-Gangetic plain, food grains should be provided for the village people at an extremely subsidized cost.

Education and promotion of environmental issues at the grass-roots level are the most effective means of overcoming peasant resistance to changing long-established customs. Moreover, there is a special need to involve women, because it is they who have to attend to domestic chores and are primarily responsible for agriculture, fuel and fodder collection and maintenance of livestock. Therefore programmes to organize women, mobilize them and improve their educational facilities should be given highest priority. Together with this, the government through NGOs should promote capacity building in the villages.

Local village people throughout the region depend on forests for meeting their fuel needs, and this leads to deforestation. It is recommended that alternative sources of energy, such as electricity, cooking gas, kerosene oil and pressure cookers, be supplied to them at subsidized prices to halt the destruction of forests, as well as to encourage tree farming. Also, fuel-efficient stoves and biogas plants should be popularized. Rajendra Singh Mehra, who is working with the villagers in Naukuchiyal Tal, has successfully installed 275 biogas plants, which are operational throughout the year. In his view, the biogas plants of the Deen Bandhu model have been extremely successful in the hills between the altitudes of 712 m and 1139 m. The size of the plant should be 2 m^3. In those areas where biogas plants were installed by the government agencies, they failed because they introduced those models which are suited for the plains. Then, owing to their size – 6 m^3 – the consumption of dung was greater.

Non-conventional energy sources, such as mini- and micro-hydropower generation and wind-power generation, should also be tapped. There is a vast opportunity for wind-power generation in the hills. The full potential of windy sites in the Himalayan region has not been estimated and an intensive survey of such locations is urgently needed. Finally, the sunny areas of the Himalayan region are suited to solar cookers, which could be provided free or at subsidized prices.

Tree farming certainly involves less sod disturbance than crop cultivation, which requires frequent ploughing. Further, the amount of inputs per unit of production would be less in tree farming. Deeper tree roots promote water absorption and check surface run-off, helping to prolong the life of springs, the major source of water for the villagers.

In Uttarakhand, with the depletion of vegetal cover, many villagers are facing a water crisis. In some districts, such as Pauri and Almora, during summers the scarcity of water is so acute that it is supplied to the villagers by tankers. Historically, the most important source of water for the villagers has been natural springs. But, with their drying up in hundreds of villages in Uttarakhand, carrying water as head loads from distant places (like fuel and fodder) has become a full-time activity. According to official statistics in Almora district, 2986 of 3003 villages had been covered under drinking-water supply schemes up to March 1995. However, a majority of these schemes are not fully operative and the situation has not improved. A.D. Moddie is of the view that in the mid-1970s the people of Uttarakhand began to articulate more clearly the growing threat of the phenomenon of drying and dying hill springs. Subsequently Dr D.D. Pant, the then Vice Chancellor of Kumaon University, undertook studies in the Bhim Tal subcatchment in Naini Tal district. The findings indicated an alarming fact: that 40–60% of hill springs were either drying or dead.

It has been surmised that the cause of this has been increasing human interventions over decades, including deforestation, extension of cultivation and grazing, and construction work in the mini-catchment areas of hill springs. But the microclimatic changes induced by reduced rainfall and vegetation decline are the most alarming. The average annual rainfall in Bhim Tal between 1971 and 1980 declined by 34% as compared with the previous 13 years. During the same period, 9.3% of the forest cover was lost due to various reasons. Similarly, in the Paharpani area the drop of 7.6% was most alarming between 1976–1980 and 1981–1986, and the extent of forest reduction has been 13.3% since 1978. In the Ballia subcatchment, rainfall declined by 9.5% in the decade 1974–1985, while the forest cover dropped to 17.4%.

Most of the forests in Uttarakhand are situated in the sensitive catchment areas of some rivers. These catchments are not only rich in biodiversity but they are intertwined with long-range ecological security and the perpetuity of glaciers and river systems. In the UP hills there are eight catchments, 26 watersheds, 116 subwatersheds and 1120 microwatersheds.

Afforestation and plantation programmes should be initiated in the subwatershed areas and the smallest unit – the microwatershed – should be prioritized. The small areas would be easy to manage and even difficult areas which have been overlooked hitherto would be tackled. Moreover, it would be easy to monitor the work done. The plantation should be established in the minicatchment areas of the hill springs and it should be conserved as far as possible as a sanctuary, with minimum human and livestock intervention. Deep-rooted species, such as oak and *utis*, should be planted to facilitate better groundwater seepage, which can enhance the discharge rate.

In Uttarakhand, the oak forests are being ravaged owing to apple cultivation and biotic pressure. Oaks supply leaf fodder for cattle, wood for fuel and timber for farm implements. Oak forest soil is rich in humus and is often used as a supplementary manure. Agricultural activities frequently extend into oak forests and, with increasing demands for fodder and firewood, the trees are repeatedly lopped and seed output is reduced. Pressure from seed predators, such as flying squirrel and langur, has reduced further an already diminished seed crop and the scanty young generation is grazed by livestock. As a result, these forests are fast disappearing.

At two of the project sites, Kilbury in Kumaon and Akash Kamini in Garhwal, it was found that the young seedlings of different species of oak were not being established. Only mother trees had been left, some of which were 200 years old. Once these mother trees die, in the absence of younger saplings, oaks will disappear from Uttarakhand. It seems that the current forms of disturbances (such as regular lopping of trees) lead to a reduction in seed output, photoinhibition in leaves and soil erosion.

The canopy gaps are formed very slowly. Thus controlled felling should be permitted so that the canopy gaps can be formed and then the site allowed enough time to recover. Conservation should be focused on seedling dynamics. Instead of saving the trees, the forests should be saved. Instead of a tree as a management unit, it should be replaced in totality by a forest ecosystem, where different processes occur.

Further deterioration of oak forests occurs by clearing for apple orchards. The temperate climate of Kumaon and Garhwal offered congenial conditions for apple cultivation between 1600 and 2700 m and this is the oak forest zone. The total area of apple orchards in Kumaon and Garhwal is 48,700 ha. According to the Department of Horticulture, this will extend to 67,400 by the year 2010.

Problems of soil erosion and loss of generic diversity, water scarcity and general environmental deterioration result from the destruction of natural oak forests. A case-study of villages in and around the Chamba-Mussorie fruit belt of Tehri Garhwal by the G.B. Pant Agricultural University indicated that the destruction of oak forests to convert land for apple orchards created an acute shortage of fuel, fodder, water and other necessities of life for the villagers. Crop production was reduced due to obstruction of the biomass flow from forests to crop-land, ultimately forcing outmigration of large numbers of people.

In the apple belt of Kumaon a new trend has been observed. The owners of large apple orchards have started selling their land to builders for group housing complexes. This will add to environmental degradation. It is essential that apple cultivation should not be promoted at the cost of oaks. In the existing orchards the production should be increased and the old orchards should be revived. Also there should be a ban on the sale of apple orchards for group housing activities.

A new threat posed to the forests and biodiversity of Uttarakhand is the growth in tourism. Builders are purchasing large tracts of land in this region and tourists have a significant effect on the environment. Tourists litter the scenic spots and trekking and pilgrim routes with plastic bags and other non-biodegradable material. The plastic disposed on the slopes of the hills is covered by the autumn leaves of the season. This process is repeated every year, with plastic being sandwiched between leaves preventing rainwater from seeping and percolating and thus causing an ecological imbalance. This deposit of plastic materials also kills the green life on the slopes of mountains.

The tendency nowadays is to seek redress in courts. But in Uttar Pradesh this has not been effective. In 1994, the Garhwal Mandal Vikas Nigam built prefabricated lodging houses as hotels for tourists on the slopes of the *bugiyal* near the ancient temple of Tungnath. On 26 October 1996, the High Court directed the Nigam to withdraw occupation of its campsite on the slope of the *bugiyal*. A decision is awaited by the Supreme Court.

The administration cannot shut its eyes and passively permit the builders and other despoilers of forests to violate the natural cycle. The Supreme Court has reiterated time and again that the local government has responsibility for the environment and the ecology of the region. But there is a wide gap between what is being said and what is being done. It seems that the administration has become a party to upsetting the ecological balance and the biodiversity of this region. Either rules are being framed to benefit the builders or, where they are not, the builders have the temerity to flout them brazenly. Mostly it is done with the connivance of the administration under political pressure. If an honest officer does not collude with the builders, then the politicians are there to oblige this lobby.

It is suggested that thorough investigation on meadow management and the impact of grazing on meadow diversity should be made. Most of the meadows are under great pressure, due to fuel wood collection by migratory graziers and tourists. Also extraction of medicinal herbs in these areas should be stopped. Instead, villagers should be encouraged to grow such species in suitable localities to achieve the desired balance between production and protection. Additional areas, such as wastelands, should be extended for cultivation of such species. But cultivation of herbs should not be permitted in close proximity to forests and meadows. It has been observed that, wherever farmers, especially some NGOs, were cultivating medicinal plants near the fringe areas of forests, the production became forest-based. An absolute ban should be enforced on illegal extraction of medicinal plants and herbs.

Similarly the timberline in Central Himalaya, which stretches between altitudes 3000 and 5000 m, is a unique high-altitude ecosystem which supports a high biological diversity. The timberline has a special plant and wildlife habitat. Its importance is greatly increased by the transitional position of this timberline within the latitudinal pattern of timberline plant distribution, as well as the Himalayan plant distribution. Unfortunately, there is very little information on this ecosystem and the threats its biodiversity is facing. It is unusually sensitive to even small disturbances and therefore well-researched data should be collected and thorough investigations should be made on this unique habitat. Professor Y.S. Pangtey contends that sufficient data are not available about the biodiversity existing in and around the timberline but his observations are that some common problems, such as indiscriminate tree felling, increased tourism, the establishment of cattle camps and some ill-planned developmental activities, have contributed to certain serious environmental problems, such as habitat degradation, habitat fragmentation, loss of biodiversity and loss of tree generation. The mountain environment becomes more stressful with its increasing altitude; the disturbances caused by outside agencies at this altitude enhance the stress and thus the vegetation is damaged. These disturbances lead to irreversible consequences, making the whole timberline zone in the Central Himalaya a special concern for conservation efforts.

The Central Himalayan region abounds with lakes. These lakes are the key sources of water-supply, both drinking and irrigation. In Naini Tal district there is a constellation of lakes. Some of these lakes have been abused by the inroads of tourism. Building activity in and around the catchment areas and disposal of domestic waste as well as sewage have accelerated the rate of eutrophication and affected its water quality severely.

In the last decade forest fires have caused tremendous damage to the forests and plant diversity of this region. The direct impact of forest fires is on the water springs, which have started drying up. The water retention capacity of the soil, as well as its percolation capacity, has been reduced and the annual growth of the forests stops for 1–3 years. Fire destroys the nutrients and has added to soil erosion, as well as aridity. Because of forest fires, chir pine is being propagated, unpalatable perennial grasses are coming up instead of annual grasses and nests of birds and their eggs, as well as animal life, including reptiles, are being destroyed. The accumulation of chir pine needles during the summer also provides fuel for additional fires.

These facts were reported by the villagers in different workshops. Their contention was that, though the villagers are normally blamed for these forest fires, actually it is the timber and resin mafia who set fire to the forest land and during the fires they have a field-day, as they use the fires to their advantage.

Despite the harsh reality, the Forest Department continues to underplay the destruction. Thus the statistics they provide cannot be relied upon. From data provided by the Indian Institute of Remote Sensing, Dehradun, for the year 1995, one can gauge the damage forest fires do to the ecosystem. But unfortunately these data are up to 27 May and forest fires ravaged Kumaon and Garhwal up to 20 June. The total area affected by forest fire in Tehri district has been estimated as 910.01 km^2 or 20.58% of the geographical area. This area includes forest burnt area as 168.88 km^2 or 3.82%, partially burnt forest area as 473.69 km^2 or 10.71% and the partially burnt fallow land/grassland/scrub land as 267.52 km^2 or 6.05% of the total geographical area. The overall accuracy of classification and subsequent estimation of these figures has been assessed as 88%.

The villagers admitted that fires are also caused by their traditional burning of pine needles and of pastures during summer to obtain freshly sprouting grass during the rainy season. Burning in pine forests by herders and villagers prevents the accumulation of slippery litter and makes grazing safer. They also agreed that the forest fires are caused by burning of agricultural residue, stubs and dry stems after the harvesting.

With the Forest Department poorly equipped to handle fires and the villagers indifferent to the blaze, the fires continue to consume more and more forests every year. Previously entire villages would turn up to fight fires, but today nobody shows up. Unfortunately, there is a growing divide between the Forest Department and the villagers, the bone of contention being the Forest Conservation Act. The villagers, who are denied their traditional rights to fuel and fodder, see this as a chance to spite the forest officials, who treat them as pests.

Traditionally, a symbiotic relationship was maintained between the forests and the people in the hills, but now this relationship is waning. The whole historical process in the wake of the development of forestry has alienated the people from the forests as well as from the Forest Department. The situation did not improve even after independence. The Forest Department should adopt a people's approach and ensure the villagers' role in forestry. To prevent recurrence of forest fires, the establishment of a rapport between the Forest Department and the villagers is essential. This can be done by taking a sympathetic view of the rights and privileges of the villagers in the adjoining forests. This is a conflict that needs to be resolved. With increasing biotic pressure, such tussles have become an everyday affair. The administration should also focus on village workshops and mass campaigns regarding environmental issues, especially in areas where biodiversity is threatened.

Worse still, the fire lines in the forests have disappeared beneath the undergrowth. They were natural protectors against fire but with paucity of funds the lines could not be maintained. Similarly the earlier system of appointing one fire-watcher for each beat to assist the guard has been scrapped for lack of funds. Previously, the Forest Department would carry out controlled fire sessions in February to make sure that by summer there were no excessive pine needles on the ground. But today the funds given for this are so abysmally low that not enough of this is done. For fire-fighting operations up to 1995 the Forest Department was allotted only Rs 1 ha^{-1}.

There is thus an urgent need to improve the Forest Department communications on fire-fighting. Additionally a ban should be put on the burning of forests for grass by the villagers. More importantly, however, the conflict needs to be resolved and proper funds need to be given to the department for forest fires. Finally, satellite remote-sensing data hold great promise to provide timely information and can be reliably used to identify forest fire-affected areas. This information represents an important input for planning forest fire management strategy.

Non-traditional forest plants (NTFPs) are a significant source of subsistence products, employment and household income in areas near forests. Tribal populations depend heavily on NTFPs for income and subsistence. One reason why NTFPs are so important for tribal people and forest dwellers is that they have been continuously pushed towards more marginal areas, especially in the Tarai region of Kumaon. In Uttarakhand there is a need for an institution to promote NTFPs, with the following objectives:

1. To develop a more practical and effective methodology for making an assessment of the availability of important NTFPs.
2. To promote cooperative marketing of NTFPs collected by the villagers.
3. To increase their production by way of growing in places of their occurrence and cultivating in some suitable places.
4. To collect market information for the advantage of growers, collectors, processors and traders.
5. To ensure higher earnings and generate employment opportunities for villagers by taking up projects based upon NTFPs.
6. To improve collection, storage, processing, transportation and marketing.
7. To fix minimum procurement prices for NTFPs.
8. To undertake extension and training programmes among tribal and non-tribal communities. Since women play an important role in the village economy, their participation must be ensured.
9. To ban the illegal and unscientific extraction of NTFPs.

The oil growers in India are facing a serious problem owing to cultivation in unirrigated lands under rain-fed conditions. An important source for the production of oils may be found in the oil-seeds of forest origin. In the Kumaon and Garhwal hills there are more than 50 species of trees and plants which produce oil-seeds, but a negligible percentage of their potential is being exploited. There is a high concentration of *chullu* or wild apricot trees in several areas of Kumaon and Garhwal. A detailed survey of *chullu* was conducted in the villages of Purola and Naugaon of the upper Yamuna valley and it was observed that, out of 257 villages which were surveyed, there is a good concentration of *chullu* trees in 200 villages. *Chullu* fruit are eaten by the villagers and for making oil and chutney. Even then, only 25% is consumed and the rest goes to waste. *Chullu* oil is used as a substitute for almond oil and costs much less. If we calculate the total oil potential of Purola and Naugaon, it comes to almost 200 t annually and the fruit potential to 8000 t. If this resource could be exploited to its full potential and the infrastructure for its marketing could be developed, it would improve the economy of the people. Further, it would be propagated by the villagers, which ultimately would increase the tree cover.

The Central Himalaya with its foothills, the Tarai and Bhabhar, has been nourishing the agriculture of the Indo-Gangetic plain. Increasing demand for renewable natural resources, both for sustaining an increasing human population and for supporting their gradually changing lifestyle, denuded the vital ecosystems in these critically important hill slopes. One after another, previously inaccessible areas were made accessible and forests were wantonly destroyed.

Due to demographic reasons most of the Tarai belt was gradually brought under the plough to increase agricultural production. The loss of natural forest cover on the hills had already been instrumental in increasing the frequency of floods. And now the reclamation in Tarai and Bhabhar has increased flash floods.

Organized encroachment, with tacit political support, on forest areas has added to the problem of deforestation and conservation of biodiversity. Encroachment in Tarai and Bhabhar started in 1953 in the Damuwadhunga forest block. According to the latest data available, for 1987, by this year 8005 ha of valuable forests had been encroached upon and destroyed.

In a bid to repair the damage done, the Forest Department has introduced programmes of social forestry in sub-Himalayan Tarai and is doing commendable work. But, instead of indigenous species, fast-growing species, such as eucalyptus, are being propagated. In Bajpur the farmers have reported that, owing to eucalyptus, the water regimen has been affected. During the field study in the Bajpur development unit it was observed that the water-table in several villages had gone down. Previously water could be obtained at 10 m for sinking a tube well. Now the villagers have to dig down to almost 70 m. This drastic change cannot be attributed to eucalyptus alone. Agricultural activity, which has reduced forest, is also responsible. After colonization, the Tarai region lost its water retention capacity to a great extent. Before reclamation this area acted as a natural sponge, absorbing excessive water during the rains and releasing it during the lean period.

Previously, no effort has been made by the Forest Department for extension activities here to increase people's awareness and enthusiasm to adopt social forestry on a large scale. One of the immediate needs in the social forestry programme is to develop a good extension effort, aimed at convincing people of the importance of forests and of their duty to protect, conserve and develop them. Prior to losing its rich forest cover, Tarai was the basis of the fertility of the Gangetic plain and ensured the perenniality of its rivers. It could even now turn out to be the basis of fertility, provided, of course, that its human as well as natural resources are utilized appropriately. With social and agroforestry, along with the farmers taking to horticulture in a big way, the vanishing green cover of the Tarai can be retrieved and replenished considerably.

While discussing the forest issues, whether in workshops or seminars, it is customary to start with the Forest Department in the witness box for all the ills that are happening. There are suggestions galore and some are very radical. In the Central Himalayan region there are numerous environmentalists and environmental activists who believe that the Forest Department should be discontinued and the forests should be given to the people. But these radicals do not seem to realize the financial and other constraints under which the officials of the Forest Department are working.

During the last decade several overlapping environmental projects have been launched in the hills. There is also uncertainty over the release of funds and enough scope for bungling. The procedure for releasing funds is very intriguing. The Ministry of the Environment in New Delhi releases the funds to the State of Uttar Pradesh and, after passing through various channels, such as the Uttarakhand Department, the Chief Conservator of Hills and the concerned Conservator, it reaches the Divisional Forest Officer (DFO). Subsequently the DFO releases the fund to the Range Officer through the treasury. Finally the Range Officer releases money to the Forester and the Forest Guards. The whole process of channelling funds is extremely cumbersome and circuitous.

There is another impediment in this regard. Chaos is created when the money is to be released from the Conservator to the DFO, because at his level he cannot withdraw the money directly unless there is a direction from the Finance Controller in Lucknow to the concerned treasuries, who in turn issue instructions to the concerned banks.

This leads to unnecessary delays in the releasing of funds and offers ample scope for bungling of money allocation and commissioning, which in turn has a direct bearing on the quality of work done. During the current year the funds to be released for the environmental projects in April 1995 were delayed until March 1996. Subsequently the payments were made to the labourers in one instalment. This again provided opportunities for the middlemen to take advantage of the predicament of the labourers receiving late payment, as well as for bungling of muster rolls. Ultimately it is the forest which has to bear the brunt of this financial laxity and the people are omitted from the objectives of the environmental projects.

The people in the hills do not have faith in the Forest Department. They have been alienated since the time the British introduced forestry in the Central Himalayan region. When the British occupied Kumaon, the colonial administrators encouraged an increase in population in this sparsely settled region and developed a relatively self-sufficient agricultural economy, not only in the lowlands but also in the uplands. Expansion of agriculture by the clearing of forests was promoted, which resulted in the reckless felling of trees.

In 1893, the British introduced organized forestry in the hills and this disturbed the former relationship between the people and the forests. With the establishment of forest reserves, the authorities increased restrictions on the felling of trees and grazing of animals, so that the people felt that their rights were being encroached upon.

Furthermore, the needs of the villagers were not considered before forest policies were introduced. No compensation was awarded to claimants whose freedom to exploit forests was restricted. The paradox that resulted from the British policy is evident. The extension of agriculture and consequent growth of population were encouraged, but access to most of the forests was restricted to ensure commercial timber production, especially after 1920. This resulted in a drastic reduction of the forest support base for agriculture on a unit area basis.

When resentment reached a critical level, people began to burn forests. In July 1921 alone, about 113,400 ha of reserved forest in Kumaon Division were burned. Subsequently the government appointed a committee, which produced a report popularly known as the Kumaon Grievances Committee's Report. One of the main recommendations, issued in 1926, was for isolated forests to be excluded from forest management and ultimately made *panchayati* or community-owned. As a result, forest panchayats were formed, keeping pasture lands within easy reach of communities in the hills. In spite of these concessions the situation did not improve: hence the feeling of animosity generated by the British forest policy, which even independence could not entirely remove. Even today there is a feeling of mistrust towards the Forest Department.

In the prevailing situation, joint forest management can bring about a congenial relationship between people and the Forest Department. The tradition of community forestry or Van Panchayats already exists. In Uttarakhand there are 4991 Van Panchayats today; if these Van Panchayats are strengthened and bottlenecks in their functioning are removed, joint forest management will become a reality. It will remove the scars left by the British forest policy and forestry will literally become a people's programme in the UP hills.

3 Sources of information

The information I have gathered in writing this research paper is through intensive field study, village workshops and interviews with the field staff of the Forest Department, forest officers and environmentalists. Details of the methodology are given in Table 18.1.

Table 18.1. Sources of information.

A. Village workshops in the study area

Sl no.	Study area	Number of village workshops	Venues
1	Sitabani forests	4	Chopra, Leti, Rampur, Kotabagh
2	Kilbury forests	5	Panguth, Budla Kot, Dariya, Suryani Pali, Bagar
3	Upper Yamuna valley	8	Damta, Barkot, Naugaon, Khand, Sylab, Basrale, Kotela, Kandari
4	Akashkamini catchment	7	Nail, Vishal, Raduwa, Jaurasi Kande, Banthal, Uda Manda, Makku
Total number of workshops		24	

Continued overleaf

Table 18.1. *Continued.*

B. Interviews with field staff of the Forest Department and village headmen as well as villagers. In total 250 people were interviewed personally in 18 months throughout the Central Himalaya

Sl no.	Name of the person interviewed	Occupation
1	Dr R.S. Tolia	Director, UP Academy of Administration, Naini Tal
2	Mr T.G. Joseph	Principal Secretary, Forest and Environment, UP Government
3	Mr Rakesh Sharma	Director, Joint Forest Management Project, UPAA, Naini Tal
4	Mr M.P. Ghildiyal	Chief Conservator of Forest, Uttrakhand
5	Mr I.D. Pande	Chief Conservator of Forest, Kumaon
6	Mr S.C. Varshney	Conservator of Forest, Western Circle
7	Mr J.P. Kandpal	Director, Forest Training Institute, Haldwani
8	Mr Prakash Bhatnagar	Conservator, Forest Working Plan, Naiifi Tal
9	Mr M.P. Mishra	Divisional Forest Officer, Soil Conservation, Naina Tal
10	Mrs Indu Sharma	Divisional Forest Officer, Naini Tal
11	Prof. S.P. Singh	Head, Botany Department, Kumaon University, Naini Tal
12	Dr Y.S. Pangtey	Professor of Botany, Kumaon University, Naini Tal
13	Mr D.P. Joshi	Chairman, Central Himalayan Environment Association, Naini Tal
14	Mr Mahendra Kunwar	Chairman, Himalayan Action Research Centre, Dehradun
15	Mr A.D. Moddie	Environmentalist and retired bureaucrat and agroeconomist
16	Prof. S.L. Sah	Environmentalist
17	Mr R.S. Bhandari	Environmental activist
18	Mrs Manjula Tolia	Member, Women's Commission, gender expert
19	Mrs Jaya Gupta	Gender expert
20	Mrs Sushma Rawat	Gender expert
21	Prof. Shekhar Pathak	Social and environmental activist

Acknowledgements

Mr M.P. Mishra, Expert in Social Forestry and Wildlife Management, Gorakhpur, India.

People's Struggle Against the Institution of Royal Forest Reserves in England in the 11th–14th Centuries

19

Julia J. Serovayskay

University of Karaganda, University Str. 28, 470012 Karaganda 12, Kazakhstan

The main aim of this study is to reveal the principal forms, character and tendencies of the people's struggle against the institution of royal reserves in England, in the 11th–14th centuries. In concept, the 'people's struggle' included all the forms of social protest and the whole arsenal of methods of fighting of peasants and artisans who lived in the bounds of the royal forests.

The principal and most valuable source of information for a reliable reconstruction of both individual instances and the general picture of the popular masses' struggle against the system of royal natural reserves in England are Forest Eyre Rolls and Forest Inquisitions. A complex study of these and certain other types of sources, such as documents of special 'forest' laws, secular and monastery chronicles of the 11th–14th centuries and folklore materials, provide sufficient data to solve the problem in question. The topical character of the problem is determined by the fact that it has never been studied either in prerevolutionary Russian and Soviet historiography or in the works of foreign authors.

The sources analysed indicate that the struggle of peasants and artisans against the system and regulations of royal forest reserves in England during the 11th–14th centuries was, in its social character, an antifeudal struggle directed against one of the institutions of feudal property and the forms of exploitation inherent in it. The struggle quite distinctly involved such forms as: (i) individual protests of local peasants; (ii) collective (communal) peasant uprisings, also limited locally; and (iii) actions of peasant highwaymen bands directed against judicial and administrative

(eds M. Agnoletti and S. Anderson)

authorities of the reserves. All these various forms of the struggle had one thing in common: they all represented a number of uncoordinated actions, limited locally and never achieving the highest forms of class struggle. Nevertheless, it does not diminish the historical role of the struggle, as it formed a significant part of the general direction of the class struggle of the English peasants and objectively facilitated the coming of the uprising of 1381.

The royal forests (*foresta*) in medieval England were extensive territories, which were under the control of the forest jurisdiction and were excluded from the common law. The rise of the royal forests in medieval England as a legal institution was connected with the Norman conquest of England in the year 1066. Subsequently developed and formed at the very end of the 11th century and during the 12th century, they represented a special and peculiar form of royal and feudal property of the English Crown, the basis of which was the monopoly of the king on the woods and their resources. From the point of view of property relations, the royal forests were a rather complicated phenomenon, because they consisted of the sovereign lands of the king (*boscus dominicus regis*) and woods and other lands of the clergy and secular feudals (*boscus baro, boscus priori, boscus mili*, etc.).

The royal forests were a favourite and zealously guarded prerogative of the English Crown. The territory of the royal forests was under special forest laws, which were supported by the considerable staff of officials in the graduated system of forest fines and punishment. The forest laws secured the preferential rights of the king on the incomes from the use of the woods and privileges of the other owners of the forests. In the Middle Ages, the royal forests became the arena of the struggle of different feudal groupings against the king.

As a system of exploitation, the royal forests were detrimental to the indigenous producers, peasants and artisans who lived within the limits of the forests. They were deprived of rights to use the wood, wasteland, pastures, waters, etc., from which they derived their economic activity. For this reason, the working population within the regime of the royal forests was deeply discontented.

The most valuable and reliable source of information from which to reconstruct the people's struggle against the system of royal forest reserves in England are Forest Eyre Rolls and Forest Inquisitions (Turner, 1901).[1] They reflect the legal practice concerning affairs of the royal forests, the policy of the Crown and its special forest administration as regards the area of royal forests.

A study of these records and other sources, such as documents of special 'forest' laws (Charter of the Forest 1184, Forest Charter of 1217 and other laws), assessments (assizes), statutes, charters of the kings Henry I, Stephen, Henry II, John, Henry III, Edward I[2] and materials of the constitutional history of England,[3] secular and monastery chronicles of the 11th–14th centuries and folklore, provides sufficient data to adequately describe the struggle.[4] Previously, some aspects of the struggle were investigated in researchers of such authorities as Petit-Dutaillis (1913),[5] Turner (1901),[6] Holt (1961, 1965),[7] Hilton (1949, 1958, 1967),[8] Hart (1966)[9] and others.

One of the most widespread forms of expressing protest against the regime of the royal forests were offences constantly committed in their area. Such qualification of delinquencies in the forests seems quite justifiable, because actions of delinquents were directed against standards of the 'forest' laws, representing the legislation of the English feudal state. The severe 'forest' laws served as a permanent occasion for displeasure of the whole working population and evoked social conflicts. The legal records of assizes concerning royal forests are the original chronicle of evidence about the struggle of the peasants with the regime of *foresta.*

In pleas of the forest, the range of 'forest' offences was signified by all kinds of assizes for offenders against the assets of the royal forests (game, wildfowl, ground litter or timber, pasture and fattening cattle without pay (*herbagium* and *pannagium*)), illegal assarts, duty-free passage with loads on paths and roads, collection of fuel, acorns, nuts, honey of the wild bees, burning of charcoal, use of minerals and, finally, hostile actions against the legal and administrative apparatus of the forests.

Conscious and intentional actions of peasant-type trespassers of forest laws were analogical to those operations of peasants which infringed upon the standards of ordinary and common law. This practice was reflected in records of manorial and other seigniorial courts.[10] The analysis of the learned select pleas of the forest Eyre Rolls for a number of counties (Northamptonshire, Rutland, Shropshire, Huntingdonshire, Somerset, Surrey, Nottinghamshire) of separate years during the period from 1209 up to 1334 permits us to reveal some tendencies of the social structure of trespassers. Among them were secular and clerical feudals (48 persons – 9.7%; 40 persons – 8.1%), representatives of legal and administrative staff of the forests themselves (73 persons – 14.8%), ministerials and serfs (41 persons – 8.3%), peasants and artisans (127 persons – 25.8%) and also some persons of unidentified social status (164 persons – 33.3%) (Table 19.1).

The records of the Forest Eyre Rolls indicated that, in addition to individual trespassers, peasant communities were also brought to trial. In 127 cases, 92 peasant communities figured in lawsuits. According to J.C. Russell's (1948) figures, these lawsuits included a minimum of 4692 and a maximum of 9200 people of peasant origin.[11] This indicates that, among all the offenders of the forest laws, the peasant strata make up the greater part.

The study of the character of offences committed by peasant communities on the territories of the royal forests shows the common opposition of peasantry in regard to their regime. This opposition was expressed in different ways by the demonstrative refusal of peasant communities to execute duties,

Table 19.1. Social structure of trespassers in royal forests (from materials of Forest Eyre Rolls and Forest Inquisitions).*

County	The year of the Forest Eyre	Common number of fixed deliquencies	Trespassers – the whole townships (peasant communities)	Trespassers – individuals: Representatives of the legal and administrative apparatus of the forest		Secular feudals		Clerical feudals		Ministerials and serfs		Peasants and artisans		Persons of unknown social status		
				Number	%	Number	%	Number	%	Number	%	Number	%	Number	%	Total
Northamptonshire	1209	38	8	2	3.8	8	15.4	4	7.7	3	5.8	18	34.6	17	32.7	52
Northamptonshire	1255	35	34	7	5.1	11	8.1	11	8.1	10	7.3	54	39.7	43	31.7	136
Northamptonshire	1272	5	0	7	30.4	5	21.8	0	0	8	34.7	0	0	3	13.1	23
Rutland	1209	18	4	12	48.0	4	16	1	4.0	1	4.0	2	8.0	5	20.0	25
Rutland	1269	16	4	3	21.4	2	14.3	3	21.4	0	0	2	14.3	4	28.6	14
Shropshire	1209	22	7	2	5.3	5	13.1	0	0	4	10.5	1	2.6	26	68.3	38
Huntingdonshire	1255	35	57	36	27.6	9	6.9	19	14.6	12	9.2	23	17.7	31	23.8	130
Somerset	1257	3	12	2	50	0	0	1	25.0	1	25.0	0	0	0	0	4
Somerset	1270	4	0	0	0	2	20.0	0	0	0	0	3	30.0	5	50	10
Surrey	1270	25	0	0	0	2	5.0	1	2.5	2	5.0	9	22.5	26	65.0	40
Nottinghamshire	1287	2	0	2	40.0	0	0	0	0	0	0	3	60.0	0	0	5
Nottinghamshire	1334	14	0	0	0	0	0	0	0	0	0	12	75.0	4	25.0	16
Total		217	126	73	14.8	48	9.7	40	8.1	41	8.3	127	25.8	164	33.3	493

*Source: Turner, G.J. (ed.) (1902) *Select Pleas of the Forest*, Vol. 13, Selden Society, London.

charges against them, in forest inquisitions, including: (i) non-payment; (ii) refusal to catch trespassers; (iii) refusal to participate in inquisitions on forest affairs; (iv) refusals to testify in courts before justices on forest law; (v) non-attendance of legal inquisitions (local and common) by communities, completely or partially; (vi) refusal to witness evidence about offenders of the forest laws; (vii) oral declarations and protests against prosecutions to inquisitions of all ranks.

Additionally, peasant communities not only committed all sorts of trespassing in the royal forests themselves but also helped offenders from their townships to escape, protected outsider trespassers, received them in communities during seasonal harvests as reapers and participated in other modes of concealment of offenders (Table 19.2). These actions by peasant communities can be characterized as organized sabotage, which took place in spite of threats and other repression on the part of the administration.

According to the records, 103 cases were taken to trial for non-attendance at legal inquisitions by communities, as a whole or in part. This represented 81.2% of all offences. Moreover, some peasant communities were taken to court for evasion of their duties several times. Peasants of the townships of

Table 19.2. Peasant communities taken to court for delinquencies in the royal forest.*

Character of the delinquency	Number of communities	Per cent of the total number of delinquencies
Unprovided guarantor	6	4.7
Escape of one member of a township who committed a delinquency	2	0.8
Protection for trespassers (receiving them in the village as reapers, concealment of offenders, etc.)	4	3.1
Suspicion of the whole community of participation in trespass by of one of its members as accomplices in trespass	4	3.1
Refusal to participate in catching trespassers	2	1.6
Refusal to participate in inquisitions on forest affairs	2	1.6
Refusal from previous evidences and depositions given in inquisitions before justices on forest law	5	3.9
Non-attendance of legal inquisitions by whole community, completely or partially	103	81.2
Total	128	100

*Source: Turner, G.J. (ed.) (1902) *Select Pleas of the Forest,* Vol. 1, Selden Society, London. The unit adopted is every separate case of bringing peasant communities to trial. For example, the community of Brampton (Huntingdonshire) was brought to trial seven times. All these and analogical cases were taken into consideration in our calculations.

Torn, Underwood and Cotinham (Rockingham Forest, Northamptonshire, 1255), were charged outright because they 'had no desire to come' on the summons of judges.[12] On occasions where peasant communities did appear in legal inquisitions, they gave evidence lacking in useful information. In 20 cases, reflected in the learned Forest Eyre Rolls, communities usually declared that 'they didn't know anything about this deed and nobody was suspect'.[13] Very often peasants would declare that they suspected trespass by lords and representatives of clergy and would provide actual names and titles.

Thus, villagers of the township of Deresburg (Gloucestershire) bore witness that Robert Clare, the Count of Gloucester, had trespassed in the royal forest (1255).[14] The townships of Woodford, Islip, Truvel, Slipton, Adington and Irtminburg (Northamptonshire, 1255) in their statements accused Count Ferers of shooting deer in the royal forests.[15] In only five cases where villagers testified was the suspect in question of peasant origin.[16]

Sometimes the population of the royal forests resorted to more active operations. For example, the inhabitants of the township of Saint Brivell (Gloucestershire, 1282) accomplished systematic devastations by cutting timber in the Forest of Dean. They declared in court that they had the full right to take all necessary things there in unlimited quantity, according to their desires, and that they always used this right.[17] Even at the end of the 13th century, when the royal forests already had a long history, the peasants still remembered their previous free access to woods and forests and tried to restore their rights of use.

In the struggle with the regime of the royal forests, the peasant townships repeatedly came out in courts with open collective accusations, complaints and petitions, directed against more hateful forest officers. In the year 1255 in Northamptonshire, the townships of Gedington, Brigstone, Stansi and Okli together accused the chief forester of the county, Gugo Goldinham, saying that he, in the company of others, who they also named, caused damage, pastured and fattened cattle and constantly hunted in the forest. Their complaint was examined in the forest inquisition in the presence of forest staff, knights and freeholders. But the judges' final determination was that the townships falsely accused the chief forester and other persons through personal hatred and the peasants of all those villages were fined.[18]

Not only did the peasant townships render resistance to the regime of the royal forests, but so did peasant individuals and separate groups of villagers. The records of the forest inquisitions are filled with many cases of bringing to trial peasant individuals who were fined for lying, false complaint and false charges, stupid acts and foolish speeches in courts before forest judges and the juries of travelling judges.[19] Sometimes the inhabitants of forest areas resorted to armed collisions with forest officers.

One such clash took place in 1298 in Kingswood Forest (Essex), when a forester attempted to detain a group of trespassers consisting of 16 persons on foot and two horsemen. They resisted, beat and wounded him with nets, and took away his sword and horn.[20] In 1246 in Northamptonshire, a group of well-armed trespassers (five persons) attacked foresters and heavily wounded one of them.[21] A similar armed clash took place in the same county in 1257 (Forest of Rockingham).[22] The legal records of Northamptonshire included an inquiry into the murder of forester Stephan Multon by trespassers in 1248.[23] In 1251, a group of 12 armed trespassers shouted at foresters in Worberg Forest (Huntingdonshire). One person of this group was identified as a peasant named Gervaisy Dean.[24] These and similar classes and skirmishes by trespassers, armed with bows and arrows, with the forest administration apparently resulted in local armed conflicts, in which the peasant population took part.

The whole situation in the areas of the royal forests at the end of the 13th century and the peasant moods connected with it produced a social soil out of which grew the ballads of Robin Hood – legendary hero, 'kind robber', fighter against the arbitrariness of sheriffs, bailiffs and judges. The image of Robin Hood repeatedly became the object of controversy, even among English historians.[25] According to R.H. Hilton, the activity of Robin Hood carried social significance. M. Keen considered that his actions were closely connected with the situation provoked by the regime of the royal forests. Analogical episodes, colourfully described in ballads, can be found in the legal records of the forest inquisitions. The sentiments of the ballads of Robin Hood were realized in context with the materials about the history of Royal Forests. Although the writing of ballads was begun during the 15th century, they lived in oral tradition and in people's memory in the 14th and may be reflected even earlier in the periods of the 12th and 13th centuries, when the institution of the royal forests had reached its apogee.[26]

Local transgressions of peasants against sheriffs, bailiffs, foresters and judges, similar to the actions of Robin Hood and his 'woodland brothers', represent the trend of the people's struggle, which was called by E.J. Hobsboum 'social banditry'[27] and was a transitional form between local conflicts and large-scale revolt, such as the uprising of 1381. There was a strong connection between the latter and peasant movements against the regime of the royal forests. The participants in the revolt demanded that all parks, warrens, woods and forests should be made common, so that every man, poor or rich, could freely use his right of chase and catching fish. In accordance with these demands in the period of uprising, the inhabitants of St Alban's, Cambridge and other places entered into possession of woods and water resources and hunted, caught fish, felled timber and chopped firewood there. In other words, they acted in the same way as the inhabitants of forest areas, but in a more organized and united manner.

The sources indicate that the struggle of peasants and artisans against the system and regulations of royal forest reserves in England in the 11th–14th centuries was, in its social character, an antifeudal struggle directed against one of the institutions of feudal property and the forms of exploitation inherent to it. The struggle quite distinctly involved such forms as: (i) individual, locally limited, protests of peasants; (ii) collective (communal) peasant uprisings, also limited locally; and (iii) actions of peasant highwayman bands directed against the judicial and administrative authorities of the reserves. All these various forms of the struggle had one thing in common: all of them represented a number of uncoordinated actions, limited locally and never achieving the highest forms of class struggle. Nevertheless, this does not diminish the historical role of the struggle, as it formed a significant part of the general direction of the class struggle of the English peasants and objectively facilitated the coming of the uprising of 1381.

Notes

1 Turner, G.J. (ed.) (1901) *Select Pleas of the Forest*, Vol. 13, Selden Society, London; Boulton, H.E. (1965) *Sherwood Forest Book*, Thoroton Society Record Series, Vol. 23, Nottingham.

2 Petrushevsky, D.M. (1938) *Monuments of the History of England XI–XIII Centuries*, Moscow.

3 Stubbs, W. (1870) *Select Charters and other illustrations of English Constitutional History from the Earliest Times to Reign Edward I*, Oxford.

4 Thorpe, B. (ed.) (1861) *The Anglo-Saxon Chronicle*, Vols 1–2, Rolls Series, 23, London; Stubbs, W. (ed.) (1887) *Historiae novellae, libri tres*, Vol. 1, *Willelmi Malmesbiriensis. De gestis regum Anglorum libri quinque*, Rolls Series, 90, London; Chibnall, M. (ed.) (1969) *Order icus Vitalies, Historia Ecclesiastica*, Vol. 2, Books 3–4, Oxford; Thorpe, B. (ed.) (1849) Florentii Wigorniensis. Chronicon ex chronicis, Vols 1–2, London; Child, F.J. (ed.) (1957) *The English and Scottish Popular Ballads*, Vol. 3, New York.

5 Petit-Dutaillis, C. (1913) *Les Origines Franco-normandes de la 'forêt' Anglaise*, Mélanges d'histoire offerts à Charles M. Bemont, Paris.

6 Turner, G.J. (1901) *Introduction to Select Pleas of the Forest*, Vol. 13, Selden Society, London.

7 Holt, J.C. (1961) *The Northerners. A Study in the Reign of King John*, Oxford; *idem* (1965) *The Making of Magna Carta*, Cambridge.

8 Hilton, R.H. (1949) Peasant movement before 1381. *Economic History Review* 2, 15–27; *idem* (1958) The origins of Robin Hood. *Past and Present*, pp. 30–45; *idem* (1967) *A Medieval Society. The West Midlands at the End of the 13th Century*, Weidenfeld and Nicholson, London.

9 Hart, C.E. (1966) *Royal Forest, A History of Dean's Woods as Producers of Timber*, Clarendon Press, Oxford.

10 Maitland, F. (ed.) (1889) *Select Pleas in Manorial and Other Seignorial Courts*, Vol. 1, Selden Society, London, pp. 7–11, 21–22, 27, 30–31, 45–47.

11 Russell, J.C. (1948) *British Medieval Population*, Albuquerque, p. 307.

12 Turner (1901) op. cit., pp. 33–36.

13 Ibid., pp. 69–116.

14 Ibid., p. 77.

15 Ibid., p. 105.

16 Ibid., pp. 64, 88, 90, 106, 109.

17 Maclean, J. (ed.) (1889) Perambulation of the Forest of Dene. *Transactions of the Bristol and Gloucester Archaeological Society* 14, 356–369.
18 Turner, G.J. (1901) op. cit., p. 103.
19 Ibid., pp. 4–9.
20 Ibid., pp. 80–81.
21 Ibid.
22 Ibid., pp. 99–102.
23 Ibid., pp. 79–91.
24 Hilton (1958) op. cit., pp. 31–32.
25 Holt, J.C. (1960) The Origins and Audience of the ballads of Robin Hood. *Past and Present* 1899, pp. 89–104; *idem* (1961) Robin Hood. Some comments: Communication, *Past and Present* 19, 16–19; Keen, M. (1961) Robin Hood – peasant or gentleman? *Past and Present* 19, 7–16; Aston, T.H. (1962) Robin Hood: Communication, *Past and Present* 20, 1–10.
26 Wilson, R.M. (1972) *The Lost Literature of Medieval England*, London, pp. 128–130.
27 Hobsboum, E.J. (1974) in Landsberger, H. (ed.) *Social Banditry. Rural Protest. Peasant Movements and Social Change*, London, New York, pp. 142–155.

In the Wood of Neglect 20

John Dargavel

Centre for Resource and Environmental Studies and Department of Forestry, Australian National University, Canberra, ACT 0200, Australia, and Australian Forest History Society, 20 Laidley Place, Florey, ACT 2615, Australia

The great wood of neglect includes all the forests and woodlands outside the formally planned forest cared for by foresters or park rangers. The wood of neglect has a forest but not a forestry history and in many countries forest historians have given it less attention than the well-documented forest of care.

The case of Australia is considered in this chapter. As in other lands of new settlement, the colonists dispossessed the indigenous people and transformed the landscape to develop agriculture, mining and settlement. Eventually some state forests and national parks were reserved from deforestation and state forestry services were created. Forestry became identified with the management of state forests and industrial plantations.

The forests and woodlands on farm and grazing land, which make up the wood of neglect, have been largely ignored by foresters and forest historians, even though their area (125 Mha) is larger than that of the forest of care (31 Mha). The effects of grazing, the condition of patches of forest and isolated trees on farms and the natural regeneration of some areas are discussed. Although government and non-government organizations have encouraged tree planting, with some success, there have been few attempts to halt deforestation or avert forest degradation on private lands. It is suggested that forest historians need to turn some of their attention to the wood of neglect.

(eds M. Agnoletti and S. Anderson)

1 The wood of neglect and the forest of care

The great wood of neglect lies beyond the forest of care. Its history is less known, few foresters inspect its groves, patiently measure its trees, make plans for its management or leave records in the government archives. Yet it is also a place of human intent, of use and abuse, with a history to be deduced from its very trees, learnt from people still alive and whatever scraps of information can be found from its distant past. It has a forest, but not a forestry, history and, in places, a conservation history of concern over its neglect and of some remedial measures.

The contrast between the wood of neglect and the formally planned forest of care can be seen on a world scale in the contrast between the degraded tropical forests of many developing countries and the managed temperate forests of the developed, industrialized world. It lies at the heart of the world forestry debate, in the call made at the United Nations Conference on Environment and Development at Rio de Janeiro in 1992 to put the world's forests on a basis of sustainable management and in the many policies and projects which have followed since. What has been neglected is to be cared for, or so it is said.

There is an important difference between the forest history of many European countries and that of colonized lands. In most of Europe, the pattern of forest use and agriculture evolved together over a very long period to create much of the present humanized landscape. The ideas of organized forestry, which evolved in Germany and France to counter overuse, created the European forest of professional care that we know today. However, in most colonized lands a sudden, often destructive break – rather than a gradual evolution – was imposed to advance the immediate interests of the new settlers. They created the present wood of neglect by destroying whatever patterns of forest use and care the precolonial societies had evolved. This essential disjuncture has rarely, if ever yet, been fully resolved, either socially or in terms of forest use. Notably, it has been one of the major roots of modern environmentalism, as Grove (1995) has revealed.

The contrast exists in various, complex ways within many countries. They are explored in this chapter for the case of a 'land of new settlement', Australia. Lands of new settlement or lands of 'demographic takeover', as they are sometimes called, such as Argentina, Canada, New Zealand and Australia, reveal the contrast sharply because they have features both of developing countries with a colonial past and of developed countries with urban and industrial bases. The chapter starts by looking at the use of Australian forests by the indigenous population and then examines the impact of British colonization. It then sketches the imperial model of forestry, which was introduced from the 1870s and which created, by default, the other – the wood of neglect. The following sections discuss the almost unknown history of this wood, some early remedial measures and, finally, more recent measures to counter land degradation.

2 'Caring for country'

Australia has been peopled for some 50,000 years, probably much longer, during which time both ecological and human communities adapted together to the slow changes in climate, the rise and fall of seas, the continual reshaping of the landscape and the periodic droughts which, as we now know, are correlated with the El Niño/southern oscillation climatic cycle. The whole continent was occupied, although the least hospitable parts of the coldest, wettest forests were scarcely used during the current climatic period. The features of the landscape, its hills and rivers, rocks and waterholes, plants, birds, animals and humans are seen as the creations of the beings of the 'dream-time'. All are intimately and spiritually connected in ways which compel human responsibilities for managing the land by traditional regimes, which cover the use of fire, restriction of access to sacred places, hunting and gathering rules for different species and seasons and the conduct of songs and ceremonies. Collectively, these responsibilities are often referred to as 'caring for country' (Young *et al.*, 1991; Horton, 1994). Although indigenous use and management may have altered the ecology to some uncertain degree (Flannery, 1994), neglect was to be a European creation.

3 Colonization and settlement

The British invasion in 1788 was sudden and brutal for both people and forests. The invaders arrogated all the land to the Crown so that it could be parcelled out to the new settlers. Once they explored past the coastal strip and east of the great dividing range, they found that they could ride their horses through a great belt of country with grass and openly spaced trees, 'like an English park', ideal for sheep and cattle. They did not, or perhaps chose not to, know that the openness of these savannah woodlands they so admired was partly a product of the millennia of 'fire-stick farming' (Jones, 1969; Pyne, 1991) by the people they called 'Aborigines'.

The new settler population expanded and thrived in the new land. Wool and gold, wheat, meat, fruit and even some timber proved profitable exports. By the end of the 19th century there were 3.7 million people, with one of the highest standards of living and health in the world. In contrast, the Aboriginal and Torres Strait Islander population had fallen from an estimated 314,000 people – although there may well have been more than double that number – to an estimated 94,000, due to murder, introduced diseases, dispossession and despair. The settler population continued to expand and their separate colonies federated in 1901 to create the Commonwealth of Australia with a national parliament. However, it was not until 1967 that the original people were treated as citizens and not until 1992 that their right to a 'native title' over some of the land was acknowledged. By now there are 17.8 million people in Australia of whom 353,000 are recorded as Aborigines or Torres Strait Islanders.

For its size, Australia is a poorly wooded country, as only about 20% of the continent is forest and woodland. It has a 'dead heart' of arid land, extremely variable rainfall with periodic severe droughts, mostly poor leached soils (they are geologically much older than those of Europe), few conifers and relatively small areas of moist tropical rain forest. However, the 700 species of eucalypts and 1000 species of acacias are adapted to these conditions and plantations of exotic conifers, notably *Pinus radiata*, have proved successful.

The new settler population transformed the landscape. Forests were cleared to make the farms, woodlands were cleared and thinned to grow more grass and areas around the gold fields were denuded of their trees. As a result, about 26 Mha of the better forest land (with trees over 10 m tall) were deforested and, of the 43 Mha left, much was too remote, too steep or on such poor soil that it was not suited to agriculture. Much of the 112 Mha of woodlands remaining are used for grazing. Little of the precolonial landscape escaped agriculture, mining, logging, settlement and other modifications, but what did is now prized for what the settler society calls 'wilderness' or 'old growth' and which the indigenous people call their 'country' and knew as their home.

4 Demarcating the forest of care

The history of Australian forestry, from the 'case for forest conservancy' argued in the 1860s to the 'cause of forestry' advanced in the 1920s, has been described elsewhere (Carron, 1985; Dargavel, 1995). All that needs to be noted here is the attributes of the imperial model of forestry which was adopted and its implicit definition of 'the other', the wood of neglect.

The imperial model emerged in India during the 19th century from several strands: an adaptation of the structure of earlier princely forests preserved for hunting, concern for the environmental influences of forests on health, rising demands for railway timbers, demands for government revenue and the concepts of the deliberate, science-based, professional forestry practice which had been developed in Germany (Tucker, 1983; Grove, 1988, 1995; Lowood, 1990). The model consisted of the survey, or 'demarcation', of the best forests, their legal reservation as 'state forests', the dispossession of the original inhabitants or diminution of their traditional rights, and management by a cadre of professionally trained foresters, recruited at first largely from Germany, arrayed in state forest services. The state forests, once reserved, were to be managed primarily for the sustained production of timber according to formal plans (D'Arcy, 1898). As the planning system developed, the provision of some 'minor products' needed by local people gained some recognition and catchment areas were designated as 'protection forests'.

The model, even to the detailed headings of its working plans, spread throughout the British Empire, greatly influenced by Wilhelm, later Sir William, Schlich (1840–1925), who set up the system for working plan control in India and rose to become Inspector-General of Forests. He then moved to Britain to set up a forestry school, which became part of Oxford University, to train British foresters for the Indian and colonial forest services as they emerged. His comprehensive, five-volume, *Manual of Forestry* (1889–1906) became the Empire's textbook. The corpus of professional knowledge was expanded by French experience and particularly by the development of the pine forests and plantations on the sandy heaths of Les Landes region of south-west France.

In Australia, forestry ran against the settlers' dominant ethos of clearing the land for agriculture; 'the invaders hated trees', as Hancock (1930) has aptly put it. Change came from a combination of local concern about the loss of timber resources through waste and fire, sawmillers' needs for resource security and the urgings of visiting imperial foresters: Vincent in 1887, the Inspector-General from India, Ribbentrop, in 1895 and Hutchins in 1916 (Dargavel, 1984; Carron, 1985). No attempt was made to halt deforestation on private lands and the far easier route of declaring state forests from unalienated Crown lands was followed. Progress was slow, so that the national target to have 10 Mha of state forest, which was set in 1920, was not reached until 1965. In every state, the professionally trained foresters who staffed the forest services concentrated their attention almost exclusively on the care and management of these state forests and on the establishment of industrial plantations to supplement their resources. They had neither time nor remit to extend their care to the wood of neglect elsewhere.

The environmental movement, which developed very strongly in Australia during the 1970s and 1980s, found the quality of care and management of the state forests wanting and sought to have them transformed into national parks. As the state forests were in public ownership, they readily became the subject of fierce social and political contests in every state and between the state and federal governments. The point to be made is that the environmental organizations, as well the forestry, forest industry, unions and governments, focused virtually all their care and attention on the public land – the state forests and national parks. The wood of the environmentalists' neglect was also elsewhere.

5 The wood of neglect

The wood of neglect survived colonization by default. It was too steep to clear easily, too stony or too remote, its stands were too poor to be demarcated as state forests or it lacked some quality that might have made it a national park. It is more extensive and more varied than the forest of care, but, although its history is yet to be written, we know something about its extent and can infer from general observation something about the major influences on it. However, its use and management by thousands of individuals is far more varied than that in the forests of care managed according to the centralized regimes of the forest bureaucracies.

5.1 Extent

Australia's National Forest Inventory has estimated the areas of each type of forest and woodland in each type of land tenure. For the purposes of this chapter, the state forests, timber reserves and national parks and other nature conservation reserves have been grouped as comprising 'the forest of care', while the privately owned farm land and the extensive leasehold grazing land (with other Crown, private and leased land) have been grouped as comprising 'the wood of neglect' (Table 20.1).

5.2 Grazing

Australia, it was said, 'rode on the sheep's back' from the 1830s, as the flocks multiplied across the landscape and wool became a major export staple. Later, cattle too multiplied across the woodlands, especially in northern Australia, where they are ranched on a large scale. Grazing sheep and cattle relied on the native grasses between and underneath the trees at first, but on the more settled farming areas the pastures were gradually 'improved' by cultivation, fertilization and the introduction of exotic grasses and legumes.

The graziers burnt the forest and woodlands, not as the indigenous people had in patches and with care, but ferociously. Moreover, the hard-hooved sheep and cattle compacted the soil in a way that the gentler tread of kangaroos and emus never had. The combined effects inhibited tree regeneration and virtually prevented it, as rabbit plagues spread across most of the continent from the 1870s. Although rabbit populations were drastically reduced by the spread of myxomatosis in the 1950s, the progressive degradation by grazing and fire continued inexorably.

Ellis (1993) studied the tall open forest of gum-topped stringy bark, *Eucalyptus delegatensis*, and mountain white gum, *Eucalyptus dalrympleana*, on large grazing properties on the central plateau of Tasmania. He examined how ring-barking the large trees to kill them and get more grass had been encouraged by government policies, even though maintaining some shelter for stock during the cold winters is highly desirable. This, the use of fire and extensive felling for the Japanese woodchip export trade since the 1970s had drastically altered the structure of the forest. Ellis's work is one of the very few examples of work designed to develop appropriate silvicultural systems for the forest of neglect.

Table 20.1. Areas of forest and woodland, Australia.

Forest type	Forest of care (Mha)	Wood of neglect (Mha)	Total (Mha)
Rain forest	1.9	1.7	3.6
Tall open eucalypt forests	4.4	1.0	5.5
Medium open eucalypt forests	9.0	13.7	22.7
Low open eucalypt forests	0.1	0.3	0.4
Tall eucalypt woodland	0.0	1.0	1.1
Medium eucalypt woodland	7.6	61.2	68.8
Low eucalypt woodland	0.7	13.6	14.3
Eucalypt mallee forests and woodland	3.8	8.0	11.8
Callitris forests and woodland	0.4	0.5	0.9
Acacia forests and woodland	0.4	11.9	12.3
Other forests and woodland	2.6	12.0	14.6
Total forest and woodland	30.9	124.9	155.8
Plantations	1.0		1.0
Total forest estate	31.9	124.9	156.8

Source: Adapted from *Australia's First Approximation Report for the Montreal Process* (1997).

Although the grouping in Table 20.1 is a crude one, three important features stand out. First, the most obvious feature from the conventional forestry perspective is the relatively small area of the types suitable for commercial timber production: the rain forests, the tall (over 30 m) and part of the medium (11–30 m) open eucalypt forest. This is due climatically to the limited area of the continent with sufficient rain to grow tall and medium trees and historically to the fact that agriculture and settlement sought out land primarily in the same regions. The state forests were demarcated in the remaining areas of these types, with the addition of some areas of low, open eucalypt forest needed to make convenient boundaries, and some of the *Callitris* woodland needed to provide softwood timber.

Secondly, from the environmental perspective, these same forest types are also the most important for the preservation of biological diversity. The 1.9 Mha of tropical rain forest are particularly important as they contain a wealth of endemic species. Large areas of these and other rain forests have been placed in parks and conservation reserves. However, there is an almost equivalent area, mostly made up of many small patches on private land, which has been grouped in the wood of neglect column.

Thirdly, the wood of neglect is concentrated in the low forests and woodlands of little commercial value for timber production. Only for the *Callitris* type does the area under formal care match that of neglect, but that is largely because extensive areas of the original forest have been cleared for dry-land arable farming.

5.3 Trees and woods on arable and mixed farms

Individual old trees and small patches of forest, or 'wood lots', remain on many farms for a variety of reasons and accidents of history. Some are on land which was too steep or stony to clear conveniently or was somehow missed. Some were deliberately kept to provide shelter or shade for sheep or cattle and a few of the largest trees were commonly left to create the 'park-like' landscape of large trees with cattle grazing peacefully beneath, which is the idealized image of Australia.

The condition of these remnant patches and isolated trees is, however, deteriorating. Many of the large trees, which originated before colonization, are coming to the end of their lives and there are few young ones to replace them, due to grazing and cultivation. In some regions, death of the residual trees is widespread, due to complex interactions of fertilization, insect defoliation, soil compaction and changes to the water-table. The combined effects of excessive land-clearing and tree death have resulted in the serious dry-land salinity problems, discussed later.

One of the few historical studies of the wood of neglect has been made by Banks (1997) regarding the condition of the red gum–yellow box woodlands, *Eucalyptus melliodora–Eucalyptus blakelyi* alliance, which stretch across extensive areas of south-east Australia. No large intact areas remain, but the nature of the precolonial vegetation can be deduced from some residual stands along road reserves or in cemeteries. Most areas have been grazed for up to 175 years, which has caused profound changes by fragmenting and thinning out the woodlands. Banks's dendrochronology established a range of ages for the largest trees remaining in the Canberra region of 130–170 years, with one tree estimated by carbon dating to be about 400 years old.

The romantic image of a stable, peaceful pastoral Australia dominated by a scattering of such old trees has turned into one of widespread environmental degradation.

5.4 The trees strike back

Not all attempts to clear the land of its trees were successful, for either ecological or economic reasons. The brigalow and the Pillaga forest regions are examples of ecological resilience, while some dairy and other farms are examples of economic failure.

Brigalow

Nix (1994) has written a history of the vast open forest and woodland region of southern Queensland and northern New South Wales dominated by brigalow, *Acacia harpophylla*, a tree or bush of no commercial timber value, interspersed with patches of eucalypts. At the time of the British invasion there were over 30 tribal groups of Aboriginal people, amounting to some 15,000 people, in the region. Their burning regime kept the country as fairly open woodland with plentiful grass beneath, mixed with patches of impenetrable 'whipstick' growth of brigalow suckers. The new settlers took over the region from the late 1820s and, by 1890, the number of Aborigines had fallen to about 1500, due to dispossession, massacres and European diseases against which they had no inherited immunity. The settlers cleared areas of the brigalow by hand, but, although grazing was initially successful, their hard-hooved animals compacted the soil, the native grasses and waterholes diminished, periodic droughts (such as that in 1899–1903, which killed up to 90% of the sheep) and the infestation of prickly pear (an introduced cactus controlled biologically in the 1930s) drove many from the land. The brigalow returned as dense stands of 'whipstick' shoots on many of the areas which had formerly been partially cleared. The 'second assault' on the forest was made with bulldozers and 2,4,5-trichlorophenoxyethanoic acid (2,4,5-T) weedicides during the 1950s and 1960s. It was far more effective, so that now there are only remnant patches of brigalow across the agricultural landscape and some representative samples in national parks.

Pillaga

Rolls (1981) has written a history of the Pillaga forest region, which is dominated by white cypress pine, *Callitris glauca*, forest interspersed with patches of open eucalypt forest and belah, *Casuarina cristata*. Its history of Aboriginal land management, dispossession, murder, disease and population decline is miserably similar to that in the brigalow region immediately to its north. Although the history of grazing, settlement, drought and prickly pear infestation was similar, the forest responded differently. The change from the Aboriginal regime of frequent, low-intensity fires in patches across the landscape to the settler regime of intense, large fires resulted in regeneration of many thousands of young stems per ha in years when the rainfall was suitable. Whereas the settlers had first been able to ride their horses through the forests and graze stock on the native grasses, they later had to battle the dense growth of young trees by hand until bulldozers and herbicides became available. Unlike brigalow, cypress pine produces a commercially valuable timber, which supplied sawmills from the 1870s and led to some state forests being reserved from clearing for agriculture. Although much of the forest outside the state forests has now been cleared for agriculture, some blocks with modified forest are still managed as rough grazing land.

Abandoned land

Agriculture did not always succeed and some types of land were abandoned during periodic droughts or economic depressions. There are innumerable variations in the forest consequences, but little recorded history. Three can be mentioned here.

The first occurs very commonly and consists of poor, steep or stony land that had been partially cleared and grazed at one time but which proved not to be economically viable for agriculture. It has now reverted to a two-layered forest of scattered precolonial trees and postgrazing regrowth trees. It is generally too poor to have any economic prospects for either agriculture or forestry.

The second is less common and usually consists of somewhat better land that had been more extensively cleared, and sometimes cultivated or very severely burnt, but which has now reverted to even-aged pole stands with few, if any, precolonial trees remaining. These stands can appear promising for conversion to high forest through a thinning regime, but many prove to be very slow-growing and have high levels of internal defects in the wood due to past fires.

The third category occurs in specific regions and consists of fertile, often volcanic soils on steep slopes in high rainfall regions that were cleared of high-quality, wet sclerophyll forests and used successfully for dairying between the 1880s and the 1920s. Some of this land was abandoned from the 1930s in favour of improved and fertilized pasture on flatter land or when the creation of the European Common Market drastically reduced Australian exports of dairy products. Plagues of rabbits and the invasion of noxious exotic weeds, such as ragwort, *Senecio jacobaea*, and blackberries, *Rubus fructicosus*, added further difficulties. The land reverted to bushes rather than high forest, due to the difficulty of obtaining natural regeneration on these better sites. Forest is being re-established by planting in most of these areas (e.g. Noble, 1978).

6 Replanting the trees

The case for forest conservancy, which was argued in Australia from the 1860s, led not only to the imperial model of state forestry but also to establishing state pine plantations on Crown land unwanted for agriculture and to a general movement to encourage planting trees on farms. Several strands contributed to this.

Whereas the management of the native forests owed much to the Germans, tree planting, or 'arboriculture', as it was called, owed much to the Scots. Plantations, which had long been part of the British land improvement movement, were being established energetically on the great Scottish estates from the late 18th century. As Scottish doctors, scientists and clergymen occupied many positions throughout the Empire, their advocacy of forest conservation could readily encompass planting (see Grove 1988, 1995 and especially 1997). For example, Dalhousie, the Governor-General of India, traversed the plains of the Punjab, noted the 'almost total absence of forest trees' and declared a policy of encouraging villagers to plant trees, while ordering roadsides and government compounds to be planted as a public work (Dalhousie, 1868). Scottish estates supplied Australia with people trained in landscape gardening and horticulture, including plantation establishment, who took prominent positions in the early forest services, botanical gardens and plant nurseries. For example, John Ednie Brown, a Scottish 'arboriculturalist', became South Australia's first Conservator of Forests in 1878.

The economic prosperity in Australia from the 1860s to the 1890s and in the 1900s and 1910s enabled wealthy landowners to have the surrounds of their houses landscaped and planted with ornamental trees. European trees were planted for nostalgic reasons and specimen trees from many parts of the world were planted as part of the general 'acclimatization' movement of the period. Planting extended from the domains of the wealthy to the surrounds of most farms. In the towns and cities, public gardens and tree-lined avenues were established, again mostly with exotic trees. The botanic gardens in the state capitals and some provincial cities proved what planting could achieve, especially in Melbourne, where Guilfoyle's landscaping was outstanding (Pescott, 1974).

Plants for gardens and estates were supplied initially from the botanic gardens and from a nursery industry, which thrived from the mid-19th century. As the state forest services started to establish state plantations, they set up their own tree nurseries: Victoria in 1872 (Moulds and Burns, 1997), South Australia in 1876, New South Wales in 1912, Western Australia and Queensland about the same time and Tasmania in the 1920s (Carron, 1990). As the planting movement gained strength, most of these state nurseries started supplying plants not only to local governments for street and other amenity planting, but also to rural landowners and schools.

This extension of state activity owed much of its force to the first conference of the heads of the state forest services in 1911, which supported the foundation of the Australian Forest League. The League was formed in each state as a coalition of organizations that supported forest conservation. In New South Wales in 1933, for example, these included the Wildlife Preservation Society, the Council Tree Lovers League, the Wattle League (a group for schoolchildren), the Institute of Architects and the Timber Merchants Association; its Patron was the Governor and its Life President was a former Forest Commissioner, R. Dalrymple-Hay. The Council of its Victorian branch was similarly prestigious and well connected with the state's forest service, the timber industry and conservation groups. The objects of the League included supporting the state forests and 'fostering of tree-planting generally – as shelter for stock, crops, homesteads and for ornamental purposes'. Its branches produced newsletters – *The Gum Tree* in Victoria and *The Tree Lover* in New

South Wales – and encouraged schools to plant trees and make their own plantations. Australian branches of Men of the Trees was formed in the 1930s after a visit from the British tree advocate, Richard St Barbe Baker.

Avenues of trees were often planted as memorials to the servicemen and women who died in the 1914–1918 and 1939–1945 wars. Some towns named each tree for a person lost to that place, so that the length of the now pleasant avenues is a poignant reminder of just how heavy were the losses to some quite small places.

After Victoria's disastrous forest fires in 1939, a Save the Forests Campaign was launched in 1944, which became the Natural Resources Conservation League (Robin, 1991). It was concerned not only with forest fires, but equally with the protection of water catchments and tree planting. Its Council was similarly prestigious but was able to attract a wider range of government departments and organizations. It ran education and school campaigns, produced a magazine, *Victoria's Resources*, promoted tree planting and established its own nursery, which distributed trees across the state. By the 1990s, it was sending out over 1 million trees a year.

The establishment of pine plantations, first by the state forest services and from 1950 by large industrial companies, led to a series of policies and schemes to encourage farmers and other landowners to establish their own plantations. The response has been relatively small, as most farmers have not considered tree crops to be economically attractive to them. Leasing systems are being developed and some of the tax arrangements are being revised, but their overall effect is as yet unknown. However, over the years, the planting movement succeeded in having many millions of trees planted across rural Australia in windbreaks, in school plantations, along roads, around farmhouses and station homesteads, to halt the erosion of watercourses and in some farm wood lots. Although the various state forest and soil conservation services supported the movement by providing cheap trees and free advice, they avoided commentary or control of the deteriorating condition of the native forests on privately owned land. The areas of private forest and woodland were neglected even while trees were being planted on other parts of the farms.

7 Restoring the land

The consequences of neglect for soil, environment and wood supplies took a long time to appear, longer to be recognized and still longer to prompt state and community action.

Great areas of Australian soils are old and derive from marine sediments so that, when much of the deep-rooted tree vegetation was removed, salt from deep in the ground of the upper slopes gradually moved through the groundwater to salinate lower slopes, flat ground and valley bottoms – the best agricultural land. Some 1–1.5 Mha of the Murray–Darling River basin are likely to be affected by dry-land salinity, which seriously reduces agricultural production. There were similar problems on irrigated land and a serious rise in the salinity of rivers. By the 1980s, the importance of the woodland remnants scattered on farms was better recognized for the wildlife habitat it provided, and its importance as shelter for stock was reiterated as a further argument in the advocacy for its preservation (Breckwoldt, 1986). Private land provided about 30% of Australia's hardwood sawn logs, certainly until the 1960s, from areas being deforested for agriculture and from the present wood of neglect. It also provided pulpwood for Australian pulp mills and from the 1970s for export as woodchips to Japan. The resource statistics for the wood of neglect are neglected and there are no reliable forecasts of its decline as a future wood resource.

It was the combination of soil degradation, rising salinity in domestic water-supplies, widespread tree death and loss of wildlife habitat, rather than declining wood resources, which led to action. Although regulations were introduced to try and control deforestation in some states, the major response came from the community-based land-care movement.

The movement started from Farm Tree Groups, first formed in Victoria in 1980, which promoted planting and protection of remnants of native vegetation. They continued the philosophy of the planting movement but consisted of local, rural groups rather than centralized state organizations like the Forest League or the Resources Conservation Council. Their scope enlarged in 1986 to become an integrated approach to the rehabilitation of agricultural and other land and extended in 1988 to Western Australia, where salinization was particularly severe (Campbell, 1994). First state and then federal governments supported these groups, which sprang up all over Australia. Now there are 3000 groups, typically with 20–30 members, which determine their own priorities and activities, such as planting trees, controlling weeds and feral animals, altering tillage practices and preventing erosion. Governments provide regional coordinators, advice and partial grants for some of the work.

While the land-care movement is having a significant effect on farmers' attitudes and has undertaken innumerable small-scale activities with enthusiasm, it is not large enough to offset continued deforestation and the degradation of the wood of neglect.

8 A concern for forest history

This chapter has sketched some of the origins and categories of and influences on the wood of neglect, but the conclusion can only be that we know very little of its diverse history. Forest historians may be able to provide some deeper understanding of how it has been used and abused and of the consequences for its present condition. Such studies are a proper concern if the wood of neglect is to be cared for.

References

Banks, J.C.G. (1997) Tree ages and ageing in yellow box. In: Dargavel, J. (ed.) *Australia's Ever-changing Forests III: Proceedings of the Third National Conference on Australia's Forest History*. Centre for Resource and Environmental Studies, Australian National University, Canberra, pp. 35–47.

Breckwoldt, R. (1986) *The Last Stand: Managing Australia's Remnant Forests and Woodlands*. Australian Government Publishing Service, Canberra.

Campbell, A. (1994) *Landcare: Communities Shaping the Land and Future*. Allen and Unwin, Sydney.

Carron, L.T. (1985) *A History of Forestry in Australia*. ANU Press, Canberra.

Carron, L.T. (1990) A history of plantation policy in Australia. In: Dargavel, J. and Semple, N. (eds) *Prospects for Australian Forest Plantations*. Centre for Resource and Environmental Studies, Australian National University, Canberra, pp. 11–23.

Dalhousie, Marquis of (1868) Minute 20 February 1851. In: *Select Papers of the Agri-Horticultural Society of the Punjab*. Lahore.

D'Arcy, W.E. (1898) *Preparation of Forest Working Plans in India*, 3rd edn. Superintendent of Government Printing, Calcutta (1st edn 1891, 4th edn 1910).

Dargavel, J. (1984) Radical analysis of forest history: a Tasmanian case. In: Steen, H.K. (ed.) *History of Sustained-yield Forestry: a Symposium*. Forest History Society, Durham, North Carolina, pp. 297–305.

Dargavel, J. (1995) *Fashioning Australia's Forests*. Oxford University Press, Melbourne.

Ellis, R.C. (1993) Effect of settlement on the forests of the Central Plateau, Tasmania. In: Dargavel, J. and Feary, S. (eds) *Australia's Ever-changing Forests II: Proceedings of the Second National Conference on Australia's Forest History*. Centre for Resource and Environmental Studies, Australian National University, Canberra, pp. 55–65.

Flannery, T.F. (1994) *The Future Eaters: an Ecological History of the Australasian Lands and People*. Reed Books, Chatswood, New South Wales.

Grove, R.H. (1988) Conservation and colonialism: the evolution of environmental attitudes and conservation policies in St. Helena, Mauritius and in western India 1660–1854. In: Dargavel, J., Dixon, K. and Semple, N. (eds) *Changing Tropical Forests: Historical Perspectives on Today's Challenges in Asia, Australasia and Oceania*. Centre for Resource and Environmental Studies, Australian National University, Canberra, pp. 19–45.

Grove, R.H. (1995) *Green Imperialism: Colonial Expansion, Tropical Island Edens and the Origins of Environmentalism, 1600–1860*. Cambridge University Press, Cambridge.

Grove, R.H. (1997) Scotland in South Africa: John Croumbie Brown and the roots of settler environmentalism. In: Griffiths, T. and Robin, L. (eds) *Ecology and Empire: Environmental History of Settler Societies*. Melbourne University Press, Melbourne, pp. 139–153.

Hancock, W.K. (1930) *Australia*. E. Benn, London, p. 33.

Horton, D. (1994) *The Encyclopaedia of Aboriginal Australia*. Australian Institute of Aboriginal and Torres Strait Islander Studies, Canberra.

Jones, R. (1969) Fire-stick farming. *Australian Natural History* 16, 224–228.

Lowood, H.E. (1990) The calculating forester: quantification, cameral science, and the emergence of scientific forestry management in Germany. In: Frangsmyr, T., Heilbron, J.L. and Rider, R.E. (eds) *The Quantifying Spirit in the 18th Century*. University of California Press, Berkeley, pp. 315–342.

Moulds, F.R. and Burns, M.J. (1997) *From Little Seeds: a History of the Macedon State Nursery 1872–1995*. Department of Natural Resources and Environment, Melbourne.

Nix, H. (1994) The brigalow. In: Dovers, S. (ed.) *Australian Environmental History: Essays and Cases*. Oxford University Press, Melbourne, pp. 189–233.

Noble, R. (1978) *The Strzeleckis: a New Future for Heartbreak Hills*. Forests Commission of Victoria, Melbourne.

Pescott, R.T.M. (1974) *W.R. Guilfoyle, 1840–1912: the Master of Landscaping*. Oxford University Press, Melbourne.

Pyne, S.J. (1991) *The Burning Bush: a Fire History of Australia*. Henry Holt, New York.

Robin, L. (1991) *Building a Forest Conscience: an Historical Portrait of the Natural Resources Conservation League of Victoria (NRCL) 1944–1990*. The League, Springvale South, Victoria.

Rolls, E.A. (1981) *A Million Wild Acres: 200 Years of Man and an Australian Forest*. Nelson, Melbourne.

Schlich, W. (1889–1906) *Manual of Forestry*, 5 vols. Bradbury Agnew, London (several revised editions to 1925).

Tucker, R.P. (1983) The British colonial system in the forests of the Western Himalayas 1815–1914. In: Tucker, R.P. and Richards, J.F. (eds) *Global Deforestation and the Nineteenth-century World Economy*. Duke University Press, Durham, North Carolina.

Young, E.A., Ross, H., Johnson, J. and Kesteven, J. (1991) *Caring for Country: Aborigines and Land Management*. Australian National Parks and Wildlife Service, Canberra.

21

Law, Forest Resources and Management of Territory in the Late Middle Ages: Woodlands in Tuscan Municipal Statutes

Francesco Salvestrini

Department of History, University of Florence, Piazza S. Remigio, 1, 50122 Florence, Italy

The late Middle Ages cities' and villages' statutes of Tuscany devoted a large space to the set of rules concerning the economic exploitation of woodland areas. The legislation of the major urban centres, such as Florence, or of the most important towns among the smaller ones, such as Montepulciano and San Miniato al Tedesco, presented, at different levels, environmental planning interventions for the correct exploitation of the woodland space. The rural communities' codes, instead, prohibited only the great stripping of public and private woods, trying above all to prevent foreigners' access into them. About wood and its utilization, between the 13th and 16th centuries, the needs of these smaller communities, linked to a subsistence economy, clashed with those of the cities. In fact, the big *municipia* needed great quantities of firewood and timber, and aimed to impose a set of rules in order to break the villages' isolationism and to stop their exclusive utilization of the non-agricultural soil.

The town statutes, i.e. the fundamental laws of cities and smaller centres, represent one of the most important normative expressions produced by Italian municipal civilization between the late Middle Ages and the early modern period. The nature of their local legislation, supplying an update and modification of the over-generalized old Roman right, made them one of the major significant witnesses, not only for the institutional organization's history, but also for the social, economic and cultural evolution of the single communities that provided themselves with a town law. Most of the statutes concerning

(eds M. Agnoletti and S. Anderson)

territorial organizations composed in north–central Italy between the late 12th and the second half of the 16th century aimed to regulate almost every activity of the social context to which they referred; and they became, perhaps ironically, its most emblematic and immediate expression.

The management of natural resources, and particularly that of wood, occupies a large place in these legal books. This space is proportional to the economic importance that those environments had in the *ancien régime*'s communities. About this, we can say that mentions of the use of the forestal ecological system appear in almost every major code of cities, although in different contexts, as well as in the records of many castles, small villages and rural *villae*. This reality is particularly evident in the statutes of the late Middle Ages in Tuscany.

The main type of legal intervention concerning the non-agricultural management of the soil seems to consist of specific rules prohibiting or regulating the exploitation of the 'spontaneous' vegetation areas, which were legally considered as precious as the tilled lands for the community's status. So we can find articles that aim to limit the cutting down of woods; to regulate the access of woodcutters and grazing livestock to *silvae*; to create reserves and protected farms; and to create the fruitful specialization of single woodland areas (for example, the so-called *silvae palares*).[1] The presence of *rubriche* (i.e. single rules of statutes) concerning wood and pasture preservation rises numerically in the statutes of the region between the late 13th and early 15th centuries; then it is less frequent in the documents produced between 1450 and 1500, and becomes important again in the modern age. All this happens in parallel with the periods of increase in population (evident until the early 14th century) or population decrease (from the second half of the 14th century until the second half of the 15th), and periods of more or less intense pressure on the wood reserves and on the other untilled lands.

A significant fact to point out is that the municipal statutes, whose relevant spread in Tuscany starts more or less from the 1330s, contain, with rare exceptions, no rules encouraging or prohibiting the great breaking up of woodland surfaces, a phenomenon which had already occurred on many lands of the region. They present provisions aiming for a correct management of residual woods, which might have been damaged and impoverished by excessive exploitation by the rural population, who, in any case, being aware of their importance, hardly ever wanted their complete destruction.

Of course, the complexity of the major texts and, above all, the more compelling force of the city statutes in comparison with the simple regulations of communities involved different legal solutions that either integrated rural status or ignored it. There were at least three law intervention levels.

1. Generally speaking, the codes of small castles or *villae* prohibited only the grazing of livestock that did not belong to the community's inhabitants, or the dangerous manufacture of stoves (*fornelli*) in woods; and they limited the freedom to hunt wild animals, to pick wild fruits or to sell wood.

2. On the other hand, the codes of the major centres (the so-called 'half-cities'[2]), which had a certain political influence and the direct control of some rural villages, could present real territorial management projects. They programmed exploitation in sectors of the *silvae communes*, controlled the export of woodland products from the community's *districtus* (territory), imposed the planting of trees and care for arboreal products which were considered economically relevant (chestnut-trees, for example), and planned the protection or exploitation of woods in relation to the needs of important manufactures, such as glassworks or ironworks.

3. Finally, the main cities, virtually sovereign institutions, equipped since the early 15th century with wide territorial dominions and owning great state reserves, inserted in their lawbooks (characterized by extraurban profiles and very wide fields of competence) articles that aimed to regulate timber rafting, the influx of game to the city markets or the productive specialization of the municipal woods through the assigning of whole forests to the production of a particular work timber, destined for building or for the shipyard.

We shall now examine some concrete examples concerning different periods and geographical areas, linked, however, to the categories mentioned above. When examining Florence's statutes of the 14th century and the great codification of 1415, written when the city controlled more than one-third of Tuscany, we can observe that the authorities paid attention to the management of wood resources coming, above all, from the Apennine woods. The 14th century's codes of the Podestà and the People's Captain (*Capitano del Popolo*) maintain some characteristics that made them like the rules of smaller centres. In the 'Podestà's Statute' (1322), the legal intervention on the subject of resource control seems de facto non-existent.[3] However, the contemporary statute of the People's Captain, far more interesting on the theme mentioned, set limits to the purchase of chestnuts, whose supply was perhaps insufficient.[4] In a large article '*De deveto lignorum et eorum materia*',[5] the code prohibited the export from the district of Florence, '*precipue per flumen Arni*', of all work timber coming from Tuscany's Apennines and Pre-Apennines; thus, it prohibited the sale to the outside of the material for burning which arrived in the city and gave instructions about the ways of internal commerce for the whole product.

Tree-trunk rafting along the Arno River and its affluents towards Florence is the object of another chapter (*rubrica*) of the statute.[6] This one exempted the operation from the payment of tolls, indicated the methods of transportation (by '*foderi*', i.e. big rafts made of '*lignis abietum, arborum et ontanorum, faggiorum et tilliorum*') and fixed the optimum quantity to send for every shipping ('*duas salmas lignorum combustibilium ad minus*'). It also required that managers of the mills situated along the rivers eased the passage of big rafts towards the city. There are mentions of retailers' commerce in another article which regulated the sale of '*traina lignaminis*' (above all, firewood) on the famous Old Market's rich city square.[7] The great number of articles aimed at preventing fires in buildings, from which we get an image of 14th-century Florence that was largely made of wooden structures, also show how important wood was for building.[8]

A wider range of information is provided by the city code of 1415, which unites knowledge of previous wood requirements with a policy to preserve the arboreal resources produced in the lands that were politically subject to Florence. One of the articles of this statute punishes dependent farmers (tenants, sharecroppers) who damage '*arbores, seu nemora scoparum, seu lignaminus, e aliorum similium, seu arboretos, seu salcetos*' situated on the fields the landlords gave them.[9] Another one defines particularly the taxes '*lignorum pro ardendo*' that had to be paid at the city gates.[10] One more article fixes the quota of taxes imposed on work wood, underlining that the higher taxes were those requested for the entrance of planks of fir, chestnut-tree, walnut-tree and some fruit trees; it also shows the various transport methods, remembering that wood also arrived in Florence already cut into boards and small planks ('*assarelli*'), on average 10 ells long. So it pointed out that the material could be transported in tables ('*salme*') on donkeys and horses, or in wagon-loads ('*carrate*') (above all, lumber). Some further exact information underlines that the work wood came from woods specially cultivated in order to produce some types of material ('walnut to make hoops', 'wood to make crossbows', barrels, vats, small water tubs ('*tinelle*'), cask wood, 'worked box-tree's wood', 'pole to hunt', etc.).[11]

In this statute written in the 15th century, the prohibition on exporting wood from Florence's territory remained and was made more precise, in order to put that important raw material totally at the disposal of the urban population.[12] The articles aimed at preventing or putting out fires covered an even broader area.[13] A rule imposed which plants should be cut down and which not in case of disputes between neighbouring proprietors, both in the city and in the countryside, when a tree cast a shadow on other plants in the possession of a neighbour. In these cases it authorized the removal of oak branches and stems, but not the removal of plants producing '*fructus commestibiles ad usum hominis*', such as the chestnut-tree.[14] The last article we should notice is one that fixed taxes imposed on the heads of various types of game that were hunted in the countryside or outside the state and which were sent to the urban markets (wild pigs, roe deer, fallow deer, other deer, bears, horses, pheasants, ducks, pigeons, doves, 'large waterbirds', goshawks, tercels, partridges, thrushes, blackbirds, larks, quails, hares, stone martens, badgers, wolves and so on).[15]

With regard to the laws of some smaller centres that lacked the title of *civitates*, because they did not have an episcopal seat, but which had a population, productive activities, social structure and institutional organization like those of real cities – an urban typology well represented in Tuscany – the statutes of Montepulciano and San Miniato al Tedesco, both written in 1337, are interesting for their laws pertaining to wood (we could mention many others). The book concerning Montepulciano (a town situated in the south-east of

Tuscany) contains a whole section composed of 12 chapters (*rubriche*) and named '*capitula silvarum*'.[16] These articles are mainly centred on the repression of damage to wood reserves, but disciplinary intervention appears organized and motivated by a concrete interest in environmental planning. For instance, the rule '*De pena incidentis arbores infructiferas in silva Clanium*' fixed limited access to the municipal wood, through a special permit given by the authorities, perhaps to allow pasture only from time to time. It prohibited, in fact, damage to ash, hornbeam, olive and willow trees, and fixed a temporal limit to people's stay in the *silva*, a limit that was stricter for the district inhabitants and slightly more flexible for the inhabitants of the administrative centre.[17]

The second code, regarding the most important middle Valdarno's community, presents a smaller number of articles concerning wood. However, one of these, dedicated to safeguarding Camporena's municipal forest, which was situated in Valdelsa, near the borders of San Gimignano, stands out because of its complex and interesting structure. It established that timber cutting and pasture in the wood must take place following a scheme of rotation among coppice grounds, fallow grounds, pasturing zones and places reserved for fruit picking. In fact, the whole area in question was divided by the statute-makers (*statutarii*) into four parts by precise boundary marks. In one half of the fourth part, the southern one, the cutting down of trees took place in winter and spring, while in the other half it took place in the summer months. The following year this division was applied to another one of the four sections, and so on for 4 years, so the complete rotation permitted new growth of vegetation. Parts of the forest that were not cut down were alternately fallow and pasturelands. The same article, moreover, regulated the grafting of chestnut-trees, which were also widespread in this low hilly area; also it fixed periodical weeding of the soil, and regulated the access of hunters of the avifauna.[18] Other rules of the law code limited wood commerce, like Florence's statutes.[19]

As far as the rural communities' codes are concerned, if we observe some examples referring to the 14th to 16th centuries, situated in different geographic areas characterized by the presence of larger or smaller woodlands we notice that, without considering the obvious local peculiarities (villages of the mountain, hill, plain, coast, etc.), the fields of intervention were often the same. They mainly concerned the protection of the wood as a property, safeguarding a good, and the control of the woodland area because it was a productive resource for the whole community. In these two fields, management could be expressed in various ways: the institution of 'reserves' ('*bandite*'), the punctual marking out of the wood's boundary, the fixing of the sales price for wild fruits, the regulation of the wood's infrastructures ('*metato*', '*ricciaia*', etc.), the detailed provisions for plant cutting and grafting, and the strict repression of the so-called 'given damages' ('*danni dati*').

The protection of woodland as a property or good came to the municipal legislation from the classical laws and, above all, from Lombardic law. The determination of the so-called '*de furtis arborum*' crime dated in fact from the 8th century.[20] Even if the codes of the late Middle Ages provided for a greater attention to the economic data, the defence of woods as public and private goods continued to justify many compulsory interventions until the 15th and 16th centuries. Thinking of the care with which inhabitants of small centres tried to protect the possession of trees is enough. For instance, the statute of the Agliana and Montale league (1407, Pistoia countryside) stated that, in case of flooding of rivers and streams, the wood which was broken off and carried downstream by the waters had to remain there for 3 days, so that the owners could recognize their uprooted trees ('*ad hoc ut quilibet possit cognoscere et habere sua lignamina*') and pick them up.[21] Some codes of Chianti fixed heavy fines for persons stealing wood.[22] The 14th-century statute of Barga, which was a community in Garfagnana and very rich in woods, included whole articles dedicated to the marking out of boundaries of the wood areas.[23] The statute (1415) of Val di Greve, also in Chianti, allowed each landowner 'to put aside his properties, *selvi* . . . or pastures'.[24]

As far as the above-mentioned 'given damages' ('*danni dati*') are concerned, the codes included a great number of pecuniary fines and, sometimes, corporal punishment, for people who damaged on purpose or allowed their livestock to damage someone else's plants. An example of these kinds of interventions is provided by the articles entitled '*Della pena di chi mettessi fuoco in stoppie o boschi altrui*' or '*Della pena di chi dessi danno ne' beni altrui con bestie*' in the code of Barberino in Valdelsa (1570).[25]

Of course, if the protection of wood through statutes was strong when regarding private woods, it was even stronger when dealing with the state demesne. During the 15th century, the collective property of woods and pastures was still widespread in Tuscany, not only in mountain zones but also in hilly areas and near the main centres.[26] Very often these collective woods were transformed, through the statutes, from common woods to municipal woods.[27] The statute (1441) of Castagno, a community of the High Mugello situated in the Apennine Mountains, established precise limits to the cutting of plants in the woods managed by the community.[28] Generally, the statute writers were particularly careful when prescribing punishments for damage caused to particular woods – for example, chestnut woods or tree nurseries. The statutes of Castra and Conio (Pistoia Mountain, 1411) fixed a penalty '*del danno dato ne' querceti tagliati di fresco*' (for the damage done to the fresh-cut oak woods).[29] Special officials, generally called '*campai*', were frequently nominated: as the code of the Mountain of Florence (1371, Mugello) specifies, they had to take care of fields, woods and chestnut woods.[30]

Perhaps the most typical form of safeguard of the wooded areas made by the rural municipalities was represented by the statute rules that prohibited the pasture of livestock, cutting of plants, collection of dry branches or wild fruits, together with hunting and fishing inside the woods, by all persons who did not reside in the community's territory. These rules were justified by the fact that the wood resources in many cases were only sufficient for the local population.[31] However, this was also a typical legislative expression of the autarkic isolation, according to which each village tried to enclose its own productive resources. It is not by chance that the Florentine judges charged to approve the statutes of the subject commons often used to reduce these prohibitions, in order to avoid a situation where some centres had spare woods available while others had insufficient areas.

According to the same point of view, we should take notice of the rules aiming to protect single tree species, or those which regulated the commerce in tree fruits. For instance, the oak was mainly protected by the statutes of the hill communities, which used acorns as a major food for pigs. This is highlighted in the statutes of Montegonzi (Chianti region, 1402–1542) and Val di Cintoia (1397) or in those of Volterra.[32] On the other side, several statutes of the Apennine commons, but also those of the region's hill area, related to chestnuts and their fruits,[33] fixing in some cases (as in the statutes of Gaiole (1537) and Montegonzi (1484) in the Chianti region) the price, the destination for the sole community market and the collection times.[34] From the 16th-century code of Barberino, we come to know that the '*fronde d'arbori*' (branches of trees) also supplied a local commerce.[35]

A great part of the Tuscan rural codes dealt with the defence of fresh waters against pollution and excess fishing. We can mention, on this subject, the article of San Casciano's statute in Val di Pesa (1359), which specified a fine '*mictentium calcem in fluminibus aut aliquod archimium*'.[36] The codes related to the mountain and hill centres often regulated the production of coal.[37]

As for wood-growing techniques, the code of Poggibonsi (1382), central Valdelsa, imposed the protection of standards ('*matricine*') and of 'soft woods' ('*boschi teneri*'); while the above-mentioned charter of Montegonzi specified the separation between chestnut woods 'to produce flat staves or for the saw' (i.e. destined for cutting) and the woods of fruit-bearing chestnut-trees.[38] Finally, the statute of a coastal community, Castiglione della Pescaia (1418), established the plantation and care of 'domestic pine wood' and of 'palm wood' along the littoral.[39]

Of course, as already mentioned, the provisions contained in the city and rural statutes could be superimposed and be cancelled altogether or could determine clashing situations between the needs stated by the minor centres and the immediate needs of the urban commons. The former provisions were strongly linked to a survival economy, when they preferred the trees useful for pasture and food (such as chestnut and oaks) or plants that produced firewood. The latter provisions wished to have available great quantities of timber for fire or for work; they were also interested in plants such as firs, which were necessary for great building works, shipyards or road infrastructures. For this purpose they aimed to break the isolationist protectionism with which the small villages controlled the use of their own woods.

Without doubt, as time passed and with the affirmation of Florence as the most important city of the regional political area, the urban set of rules prevailed. The need to protect woods was addressed on a regional rather than a local level (e.g. the prohibition set in the 16th century by the Grand Duke Cosimo I on the cutting of mountain woods, in order to avoid floods and landslides to the low hills and plains).[40] However, the legislation of the Medicis on the environment, which has recently been studied and appreciated for its richness of interventions and far-sightedness,[41] was due to the experience of the municipal rulers in checking the territory, designing its economic use, and in order to manage with effectiveness also the resources of the woods, placing exact limits on its huge anthropization and imposing their dependence on minor centres and the cities.

Notes

1 *Silva palaris* was the name of woodland cultivated for the production of stakes especially destined to support vines (Di Bérenger, A. (1965) *Studii di Archeologia Forestale*, B. Coppini, Florence (1st edn, Treviso–Venice, 1859–63), pp. 143–144, 654).

2 The definition is given by Chittolini, G. (1990) 'Quasi-città'. Borghi e terre in area lombarda nel tardo Medioevo, *Società e Storia* 13(47), 3–26; repr. in: *idem* (1996) *Città, comunità e feudi negli stati dell'Italia centro-settentrionale (secoli XIV–XVI)*, Unicopli, Milan, pp. 85–104.

3 The code just protected the river ichthic fauna, and only mentioned timber and firewood imported into the city (*'De piscibus non tossicandis'*; *'Quod nullus de lignamine vel aliis rebus ingombret plateas, vias, seu alia loca publica'*. Caggese, R. (ed.) (1921) *Statuti della Repubblica fiorentina*, Vol. II, *Statuto del Podestà dell'anno 1325*, Ariani, Florence, Book II, ch. LXXXII, p. 240; Book IV, ch. LXVIIII, pp. 356–357).

4 *Statuti della Repubblica fiorentina*, Vol. I, *Statuto del Capitano del Popolo degli anni 1322–25*, Ariani, Florence, 1910, *'Quod nullus emat ultra duos starios castanearum'*, Book I, ch. XXIII, p. 32.

5 Ibid., Book V, ch. XX, pp. 234–235.

6 Ibid., ch. XXI, pp. 235–236.

7 Ibid., ch. XXII, p. 236.

8 Ibid., chs XXIII and XXIIII, p. 237.

9 *Statuta Populi et Communis Florentiae, Publica Auctoritate Collecta, Castigata et Praeposita, Anno Salutis MCCCCXV*, Friburgi [but Florence], Apud Michaelem Kluch 1778, Vol. II, Book IV, ch. XIX, p. 400.

10 Ibid., Vol. III, Book V, Part III, ch. XLV, p. 426.

11 Ibid., ch. LX, pp. 446–448.

12 Ibid., Vol. I, Book III, ch. CLXXXII, pp. 383–384.

13 For instance: *'De incendiis & vastiis'*, *'De elevando rumores contra incendiarios'*, *'Quod nullus teneat in domo ultra 20. fastellos scoparum, vel similium'* (dried grass, highly inflammable material); *'De modo, & forma tenendis circa extinguendum ignem in civitate Florentiae'* (Ibid., Vol. I, Book II, chs LXXV, LXXVII, pp. 171–173 and 176–177; Vol. II, Book IV, ch. LIV, p. 424; Vol. III, Book V, Part II, ch. II, pp. 2–11).

14 Ibid., Vol. II, Book IV, ch. LXVIII, pp. 437–438.

15 Ibid., Vol. III, Book V, Part III, ch. LVII, pp. 444–445.
16 Morandi, U. (ed.) (1966) *Statuto del comune di Montepulciano (1337)*, Le Monnier, Florence, Book III, chs CXX–CXXXI, pp. 242–249.
17 From April to November, townsmen had to leave the forest within the '*ora nona*' (ninth hour), *districtuales* within the '*ora terza*' (third hour) of the day (ibid., ch. CXXIII, pp. 244–245).
18 Salvestrini, F. (ed.) (1994) *Statuti del comune di San Miniato al Tedesco (1337)*, ETS, Pisa, Book IV, ch. 105 [110], pp. 394–397.
19 Ibid., Book II, ch. XXXII [XXXIII], pp. 152–158; Book IV, chs 54 [57] and 55 [58], pp. 346–348; '*De lignis vendendis*', ibid., ch. 64 [67], pp. 357–358; '*De lignis non portandis extra districtum Sancti Miniatis*', Book V, ch. 23 [24], p. 248; '*De kabella solvenda pro lignis exportandis*', ch. 62 [63], pp. 469–470.
20 Pratesi, F. (1985) *Gli ambienti naturali e l'equilibrio ecologico*, In: *Storia d'Italia, Annali*, 8, *Insediamenti e territorio*, Einaudi, Turin, pp. 73–74.
21 Archivio di Stato di Firenze, *Statuti delle comunità 'autonome' e 'soggette'* (= ASF, *Statuti*), 1, ff. 75v–76r.
22 See *Statuti di Barberino* (1570), *Lega di San Donato in Poggio* (1406–1414), *Staggia* (1422), *Lega di San Piero in Mercato* (1398), *Monteriggioni* (1380): ASF, *Statuti*, 65, Book II, ch. XV, f. 33r; 754, Book III, ch. 8, ff. 33v–35r; 865, f. 7r; *Statuto della lega di S. Piero in Mercato del MCCCXCVIII*, in: *Statuti della Valdelsa dei secoli XIII–XIV*, I, *Leghe di Gambassi, Chianti e S. Piero in Mercato*, ed. by Latini, A. (1914) *Corpus Statutorum Italicorum*, 7, Loescher, Rome, Book III, ch. XXXX, pp. 267–268; Prunai, G. (eds) (1961) *Statuti dei comuni di Monastero S. Eugenio (1352), Monteriggioni (1380) e Sovicille (1383)* [SI], Olschki, Florence, dist. IV, ch. VII, p. 92. See also Salvestrini, F. (1994) Il bosco negli statuti rurali del comprensorio chiantigiano (seconda metà del XIV–seconda metà del XVI secolo), *Il Chianti, Storia Arte Cultura Territorio* (Centro di Studi Storici Chiantigiani) 17, 79–106.
23 ASF, *Statuti*, 68, '*De signis silvarum de Barga*', chs LXXXII–LXXXXV, ff. XXVIv–XXVIIIv.
24 ASF, *Statuti*, 911, Book IV, ch. 117, f. 24r.
25 ASF, *Statuti*, 65, Book. IV, chs. V and X, ff. 40r-40v, 42r-43r.
26 See, for example, the presence of common rights on woodlands in *Statuti di Vinci* (Middle Arno Valley) and in *Statuti di Torri* (Florentine hills south of the Arno): Latini, A. (1922) Lo statuto del comune di Vinci [1418], *Bollettino dell'Accademia degli Euteleti – San Miniato (PI)*, 4(1), 32; Bicchierai, M. (ed.) (1995) Statuto et ordinato è . . ., *Torri in Val di Pesa, una comunità della campagna fiorentina nei suoi statuti quattrocenteschi*, CentroLibro, Scandicci (FI), *Introduzione*, pp. 32–37.
27 Cf. Imberciadori, I. (ed.) *Statuti del Comune di Montepescali (1427)* [SI], a cura Imberciadori, I. (1995) repr. Litografia Felici, Pisa (1st edn,. Siena, 1938), chs 88, 89, 92, 93, 94, pp. 81–83.
28 Di non '*tagliare o fare tagliare alcuno legname nell'alpe del comune . . . senza expressa licentia del consiglio*' comunale (ASF, *Statuti*, 162, ch. XXI, f. 8v).
29 ASF, *Statuti*, 205, ch. XLIIII°, f. non num.
30 ASF, *Statuti*, 447, ch. XIII, f. 9v.
31 Especially evident are, in this sense, *Statuti di Cascia* (Upper Valdarno, 1479 and 1490), ASF, *Statuti*, 149, ff. 2r–5r; and *Statuti di Castelfalfi* (Valdera), Salvestrini, F. (ed.) (1993) Statuti di Castelfalfi, 1546–1614, *Miscellanea Storica della Valdelsa* IC(1–2), 13–14 and 25–27. '*Della pena del forestieri che taglasse arbori nella selva et alpi del comune*'

(Zerboni Zoli, F. (ed.) (1985) *Statuti di San Godenzo* [1413, Mugello], All'Insegna del Giglio, Florence, ch. CVII, p. 68).

32 ASF, *Statuti*, 484, Book IV, ch. IIII, f. 22v; 240, Book IV, ch. 51, f. 17r; Fiumi, E. (ed.) (1951) *Statuti di Volterra (1210–1224)*, Olschki, Florence, ch. CLXXXVI, pp. 205–206.

33 See, for example, *Statuti di Barga, passim*; and *Statuti della Montagna Fiorentina*, rubb. XVI and XVII, f. 18r. See also Cherubini, G. (1992) Una comunità rurale della Montagna Casentinese ed il suo Statuto: Moggiona 1382, in *idem, Fra Tevere, Arno e Appennino. Valli, comunità, signori*, Tosca, Florence, p. 148.

34 ASF, *Statuti*, 911, f. 107r; 485, ch. 10, f. 5v.

35 ASF, *Statuti*, 65, Book II, ch. XV, f. 33r.

36 ASF, *Statuti*, 746, Book II, ch. XLII, ff. 39r–39v.

37 *Statuti di Cascia*, 1539, ASF, *Statuti*, 149, ff. 61v–62r.

38 ASF, *Statuti*, 65, Book IV, ch. X, f. 42v; 485, ch. 20, f. 10v.

39 ASF, *Statuti*, 199, f. 21r.

40 We refer to the ducal proclamation published in 1559, which forbade wood cutting within half a mile from the top of the Apennine Mountains ('*Alpe*'), and which remained in force until 1780 (*Legge Dell'Illustriss. & Eccellent. Signor Duca di Fiorenza & di Siena sopra el non poter tagliare, & lavorar l'alpe, nel dominio Fiorentino*, in: Cantini, L. (ed.) (1802) *Legislazione toscana raccolta e illustrata dal dottore*, Vol. III, *1556–1559*, Stamperia Albizziniana, Florence, pp. 328–330; *Bandi, e ordini del granducato di Toscana pubblicati in Firenze dal dì primo gennaio MDCCLXXX a tutto dicembre MDCCLXXXI*, codice decimo, Per Gaetano Cambiagi, Florence, 1782, LXVII, 24 October 1780). With regard to the pursuance of this law, see: ASF, *Magistrato de' Nove conservatori del Dominio e della Giurisdizione fiorentina*, 942 (years 1564–1574); ASF, *Auditore delle Riformagioni*, 6 (years 1560–1561).

41 Cascio Pratilli, G. and Zangheri, L. (1994) *La legislazione medicea sull'ambiente*, Vol. I, *I bandi (1485–1619)*; Vol. II, *I bandi (1621–1737)*, Olschki, Florence; Vol. III, *Indici*, ibid. (1995); Vol. IV, *Scritti per un commento*, ibid. (1998).

A Brief History of Forest Changes in the Sila Greca Mountains

22

M.A. Gallo[1] and F. Iovino[2]

[1]Liceo Scientifico Statale Decollatura (CZ), viale Stazione, 70, 88100 Gallina di Reggio Calabria, Italy; [2]Dipartimento di Agrochimica ed Agrobiologia, Università di Reggio Calabria, Piazza San Francesco di Sales, 89061 Gallina di Reggio Calabria, Italy

This chapter examines the relationship between social history and forest changes during the last two centuries in the Sila Greca mountains in Calabria (Italy). For many thousands of years human impact in this area has strongly influenced the landscape, and specifically soil use and forest cover.

The administrative boundaries were reconstructed in order to interpret the demographic changes that have had an impact on the forest area. The periods that have most strongly characterized the area's history are the following: (i) the late 18th century (1741–1807); (ii) the French occupation (1806–1814); (iii) the Bourbon restoration (1815–1859); (iv) the Kingdom of Italy (1861–1914); (v) the First World War and the immediate postwar period (1915–1921); (vi) from the Fascist era to the Second World War (1922–1945); and (vii) the post-Second World War period (from 1946).

The historical sources used are reliable and consist of pertinent bibliographical references and vast archival documents: land registers, the French cadastre, the acts of the Feudal Commission and the correspondence of the Regional Commissary for Civic Rights.

The documents that have been examined and their interpretation have led to the quantification of the effects of historical and political events on the forest cover of the Sila Greca area.

(eds M. Agnoletti and S. Anderson)

1 Introduction

The civil history of a people is related to the various contexts in which the life of the communities is expressed and manifested: from the social to the political organization, from the administrative to the religious institutions and from the culture to the economy. One of the little-known contexts, however, is the relationship between the civil history of a community and the forest in the area where that community lives.

Today, analysing and understanding the state of the Sila Greca forests requires a reconstruction, albeit brief, of the civil history of the area, at least over the past two centuries. This is because of the long periods necessary for the formation of forests and because of the particularly intense relationship that was developed with the neighbouring forests by the civic community that lived in close contact with them.

In this specific case the forest has played a role that goes beyond the usual economic sphere, with consequences for the surrounding area; by the first half of the 19th century, Regia Sila (and Sila Greca in particular) had already aroused the interest of the government and the Kingdom of Naples, and then it became a national issue in the parliament of the Kingdom of Italy.

2 The historical documentary sources

The interest of the national governments in the administration of the Sila forests, which dates from the Middle Ages and which was better defined during the modern era and intensified during the 19th century, produced a wealth of documentary sources, which are extremely useful in reconstructing the history of the Sila Greca forests. General interest in Sila during the past two centuries has led to the production of a vast bibliography; some publications, because of the authoritative documentation they contain, can properly be considered documentary sources. These include the writings of Zurlo (1852, 1862) and Barletta (1852, 1854), but it was mainly the national laws and local administrative measures that produced the documentation most suitable for outlining the general context in which the forest landscape of Sila Greca has gradually changed over the past 200 years and led to the current situation.

One description of the Sila Greca forests can already be found in the 18th-century cadastre of Naples,[1] but it is mainly from the beginning of the 19th century that the documentation becomes richer and more specific. We refer not only to the documents of the French *cadastre*[2] but mainly to the mass of documents accumulated at the Feudal Commission or the authorities that took over procedures for dividing the land into lots and in defining the age-old issue of civic rights of use.[3]

Thus, the history of the Sila forests is amply supported by a wealth of archive documents.

3 The administrative districts

First of all, we must consider the political–administrative organization of the area where Sila Greca is located. It is a generally homogeneous district in so far as the economy is concerned, but it is inhabited mainly by populations that share an interest in the exploitation of the forests.

This area, which extends over approximately 1282 km^2, is situated on the Ionic versant of central northern Calabria. With the exception of the coastal area, which can be defined as plains, the district consists of hills and mountains, characterized primarily by steep or very steep slopes. Currently it is divided into the municipalities of Bocchigliero, Calopezzati, Caloveto, Campana, Corigliano, Cropalati, Crosia, Longobucco, Mandatoriccio, Paludi, Pietrapaola, Rossano and Scala Coeli, all in the Province of Cosenza, and Umbriatico, which was recently shifted from the Province of Catanzaro to the new Province of Crotone. We shall also consider the town of Acri, due to its involvement in the domanial division into lots, but only as far as the area near the Sila Greca is concerned.

Bocchigliero and Longobucco are the physical centre of this section of Calabria. During the past two centuries, however, Rossano has been the political–administrative and religious centre. In fact, during the 19th century, the District of Rossano in the Province of Calabria Citra (Cosenza) encompassed most of the Sila Greca area.

With regard to administrative jurisdiction, we must point out that some towns – specifically, Acri, Bocchigliero, Campana, Corigliano, Longobucco, Rossano and Scala Coeli – did not undergo any changes following the legislation concerning the administrative districts issued during the French domination of the Kingdom of Naples.[4] Other towns, however, were involved in aggregations or separations which, from time to time, modified the administrative arrangement of the area. In 1820 the hamlet of Crosia was separated from the town of Calopezzati and was united with Caloveto;[5] in 1834 the territorial jurisdiction of the town of Umbriatico was reduced, following the separation of the hamlet of Pallagorio, which became an independent municipality;[6] in 1902 Crosia was separated from Caloveto and became an independent municipality;[7] in 1928 Caloveto Cropalati and Paludi became a single municipality and capital known as Cropalati;[8] Mandatoriccio and Pietrapaola did the same, as did Calopezzati and Crosia; in 1934 they broke away and established the towns of Caloveto, Paludi, Mandatoriccio and Pietrapaola;[9] and finally in 1937 Crosia separated from Calopezzati and became an independent municipality.[10]

Knowledge of the administrative districts was useful in this study because it permitted a better reading and understanding of the demographic statistics and provided a possibility for evaluating the modifications in the forest cover. The decrease in the population of Umbriatico from 1540 in 1815 to 1057 in 1901 is not a demographic drop, but is the result of the separation of a hamlet that became the independent municipality of Pallagorio, and this is only one example of the utility of defining and understanding the structure of the administrative districts.

4 Sila Greca in the historical heritage of the 18th century and the period of the French occupation

The political and administrative events that took place in southern Italy and Calabria during the course of the 19th century had considerable repercussions on the forests of Sila Greca. Early in the century the nobles were already usurping the Sila forests and the central government was powerless against them. The forests were also the object of old legal battles, with the citizens' interests – represented by the *amministrazioni universali* – pitted against those of the noble families. This situation, which was officially ascertained by Zurlo (1852, 1862) was dragged out endlessly in the courts.

During the French domination of the Kingdom of Naples (1806–1814), many laws were issued that had a profound impact on the organization of the state and the local governments, and which launched a project for economic and social reforms. Feudalism was suppressed in 1806.[11] The ecclesiastical mortmain suffered a similar fate, with the consequence that all property and in particular the country estates of the clergy were expropriated and became royal property. The result of these measures was that the nobles were able to keep all assets that were properly purchased and registered, but lost all their property of feudal origin.

In 1807 a Feudal Commission was established to deal with and then quickly and definitively decide the disputes between the municipal administrations and the former barons. Between 1809 and 1810, this commission issued sentences that settled the disputes between the towns of Bocchigliero, Calopezzati, Caloveto and Crosia and the Prince of Campana; between the town of Campana and the former Baron Sambiase; between Corigliano and Rossano and the Duke, former Baron of Corigliano; between Cropalati, Longobucco, Paludi and Rossano and the Prince Borghese; between Scala Coeli and the former Barons Sollazzo and Abenante; and finally between the town of Umbriatico and the Prince of Pallagorio.[12] The proceedings of the trials and the sentences themselves contain useful information about the state of the forests, about the use of the land and, mainly, about the exercise of civic rights of use in the Sila areas, where, for centuries, the people gathered firewood and pastured their flocks during the periods of the year when it did not harm the vegetation, gathered chestnuts and acorns *allo sbarro* and planted crops.

Still another question is related to the civic rights of use: the expropriated country estates of the clergy and nobility could not be entirely aggregated to the royal demesne, since – on the legal level – this specified protection of the rights acquired by the population through centuries of civic use. Thus it was decided that a portion (usually one-third, but in some cases even half) of the lands taken from the clergy and nobility should be assigned to the towns. Within the framework of these measures, land agents were sent to the sites to measure and appraise the estates and then ascertain the exercise of civic rights of use. The reports of these agents, the decisions and requests of the municipal administrations, the procedures of reintegrating the municipal property, the resolution of land ownership issues among different municipalities, which in the meantime had separated and obtained administrative autonomy, the settlements reached between the parties, and the orders issued by the *Intendente*, or Officer, of the Province of Calabria Citra in his capacity as *Commissario Reipartitore* (Commissioner for Land Division), comprise a documentary base of inestimable historical value. In particular, they also contain detailed information concerning the Sila forests.

From the *intendente*'s orders on the decisions of the suits between the municipal administrations and the former barons and the settlement of ownership claims between two or more municipalities, it is possible to obtain information useful for defining the status of the agrarian agreements, the exercise and quality of the civic rights of use, and the methods of cultivating the fields and grazing.

The most marked farming lands, such as those in the area of Bocchigliero, which were also quite limited as to size and extent, were planted in alternate years. During the planting years, the town collected the land tax and the citizens harvested wheat; during the fallow years the town sold the grass for 6 months and for the other 6 it was at the citizens' disposal. As to the civic uses, we see that they were limited to watering, spending the night and collecting firewood. However, in the same town of Bocchigliero other fields were planted every 5 years and during, the 4 years that they lay fallow, the grass was at the complete disposal of the flocks.[13]

In the documents concerning the town of Calopezzati, we find information about the forests: animals might sleep in the forested areas and the citizens might use the wood for farming tools and other civic purposes.[14] In the years of planting the *sectis segetis* citizens could use the pastures of wheat and grass until the first Sunday of October. In the following 2 years in which the crop alternation took place, they could graze from 1 May to the first Sunday of October. In the forests the citizens had the right to collect firewood freely and only in a few cases was it specified that the right to gather (or cut) was limited to the acquisition of wood for making farming tools.[15]

Olive, mulberry, oak and fruit trees were planted on the better lands, such as at Mirto di Crosia. Sometimes we can find data regarding the different qualities of the lands regarding which the *intendente* issued orders. One example concerns the fiefs of Dragonara and Malena in the area of Cropalati: Dragonara: 350 *tomolate*,[16] of which 250 cultivated with oak trees and 100 of divided and forest lands; Malena: 3000 *tomolate*, of which 200 for planting and the rest uncultivated, oak trees and forests.

The feudal lord Prince Borghese received the income from acorns on these lands and made the pastures available from 8 September to 24 December; subsequently the municipality sold the pastures and the *bruscia* of acorns, or *sbarro*, from 25 December to 7 September. In the planting years (the fifth year after 4 years of grazing), the town sold the *spicaggio*. The citizens might graze three ploughing oxen and two cows each for the entire year, take the profits from the acorns, collect wood for fires, for farming tools and for building.[17]

Thus we see a situation that was, on the whole, uncertain as to land ownership and utilization of the agricultural and forest resources, if we consider the effects of free grazing and the status of the forests that were open not only to collecting dry wood, but also to cutting wood for making farm tools, such as ploughs or carts and for building houses. Crop alternation was usually done on a 3-year basis, which meant that wheat for baking bread was only produced every 3 years, a custom that was 'inherited' from the 18th century. In the Campana land registry for the year 1742, it states that the Tacavallicata estate of 100 *tomolate* would be planted for the clergy 2 years out of every 6 and that for the other 4 would be used for grazing.[18]

The arable lands planted with oaks and fruit trees were used mainly for grazing.

The new French government inherited an economically difficult situation and developed a system of relationships based on the equilibrium between the interests of the state, which earned the income from the planting for 6 months, the municipal administrations, which rented the pastures for another 6 months, and the citizens, who exercised their rights for 12 months. This situation provided for the use of the lands, but it is obvious that no one had any real or direct interest in improving them. Protection of the forests was not viewed as an investment for the future, but only as a means of guaranteeing longer exploitation. The forests, which comprised the largest part of the areas of the towns located in the inland mountain zones, were used to procure wood and for grazing, but to a lesser extent, due to the distances, which were even more marked due to the lack of roads and transportation. The interests of the citizens were focused on the hills and the lands closer to the town centres.

The government pursued the same philosophy in relation to Sila. After having issued laws to protect the forests of the Kingdom, it took specific action in 1810 to protect the Sila forests, which were considered assets of national importance because of the wood they supplied for naval shipyards: the trunks of the centuries-old Corsican pines of Sila were indispensable to the Royal Navy.[19]

From a review of the laws and local documentation, we cannot see any policy of forest conservation in the global sense or for specific goals, such as protecting the area. Both the citizenry and the national government sought to limit the destruction of the forests only in order to increase the time they could be used and exploited.

The French succeeded only in establishing an economic policy capable of making changes in the social arrangements: the division of the former feudal and ecclesiastic lands, with assignments to the peasants and especially the most needy, which, if fully implemented, would have led to the development of a new class of small landowners interested in supporting the new regime and potentially committed to the improvement of the country estates and protection of the forests. When Napoleon finally fell from power and the Bourbons returned to the throne of Naples, much of the land expropriated by the preceding government was returned to the clergy; even the nobles regained power and influence, but they did not retrieve their feudal rights, which the Enlightenment had finally swept away from southern Italy, albeit with centuries of delay.

5 The economy of the Sila area during the Bourbon period (1816–1860)

The local administrations went ahead with their requests and petitions to obtain the assignment of a larger portion of the lands over which they claimed rights for civic uses. By leasing land to the peasants, the Decurionates hoped to meet their constantly increasing budget needs. In the meantime, the usurpations of the Regia Sila and hence the Sila Greca forests continued. The situation risked becoming explosive when the central government decided to appoint a Commissioner to verify the situation and draw up plans for restoring conditions of legality that would also protect the forests. The commissioner who did the best for restoring legality, for safeguarding the rights of the citizens and protecting the forests was Pasquale Barletta, whose work during the 1850s was both intensive and incisive.

During the Bourbon period, the population of Sila Greca (including Acri, for the reasons outlined earlier) grew from 41,675 in 1815 to 60,170 in 1861, with real and percentage increases of 18,495 and 44.38%, respectively (Izzo, 1965). The population increase went hand in hand with expansion of the area used for farming, and this was to the detriment of the forests. The maquis, where there was no vegetation, were turned into farmlands; then the hillsides and mountains closest to the towns were deforested. The need for wheat for bread-baking was the basis for the foundation of the *monti frumentari*, non-profit-making charitable organizations that were created to advance the seed needed for planting to the farmers. The government promulgated laws to protect these public charities, but it was mainly the Church that favoured and promoted them. From a review of the records, we can find important information concerning the creation of the three *monti frumentari* in the towns of Sila Greca on the basis of a large endowment made by the clergy, specifically the Archpriest of Pietrapaola, Don Luigi Nicoletta. In 1835 he made a donation to the town of Pietrapaola of 800 *tomola* of full measures of grain for the creation of a *monte frumentario*; Mandatoriccio made a donation of 500 *tomoli* of grain in 1838;[20] and the *monte frumentario* of Paludi received 400 *tomoli* of grain in 1841.[21]

In view of their role, the *monti frumentari* were recognized by the government. Even those of Pietrapaola,[22] Mandatoriccio[23] and Paludi[24] were officially recognized.

The development of farming was also favoured by the creation of two *casse di prestanze agrarie* (agricultural lending agencies), which were established in the Province of Calabria Citra in 1853, one in Cosenza and the other in Rossano.[25]

In this context, the forests of Regia Sila were exposed contemporaneously to a reduction in surface area in favour of farming, usurpation by powerful, more or less noble families and the exercise of rights of civic use, which, after fruitless attempts at liquidation, were implemented without any administrative controls.

The situation of the Sila led, in 1833, to the creation and appointment of the Civil Commissioner, with broad powers. The owners of all the lands in the Sila district were asked to submit their legal deeds of title; the exercise of rights of civic use was once again sanctioned by a rescript dated 1841. In 1847 the Civil Commission for Sila Affairs appointed Cavaliere Pasquale Barletta, who worked decisively and competently and with autonomy of judgement. He checked the entire, complex situation and drafted solutions that remained on paper only, because of the well-known national political events. In the 1850s, the economic damage caused by continuous and uncontrolled deforestation raised the alarm, launched by the Real Società Economica of the Province of Calabria Citra, which called for urgent measures to safeguard the integrity of the Sila forests (Greco, 1864).

6 Sila becomes a national issue of the Kingdom of Italy (1861–1914)

With the unification of Italy, the Sila issue came to the attention of the government. In a short time, Garibaldi's plans to satisfy the expectation of the citizens and local administrations of the Sila forests went up in smoke. Prospects for profound changes were contained in the decree issued by Garibaldi on 31 August 1860 from Rogliano: 'In the name of Italy, the poor inhabitants of Cosenza and the hamlets shall freely exercise the right to graze their flocks and plant the State lands of Sila. And this shall be effective until definitive provisions are issued', but the central governments (right and left) thwarted the promise of free grazing and planting on the Sila lands – in fact, they favoured the interests of the large landed estates.

Among other things, the government promoted a law which, by eliminating the prohibition of importing wheat into the former Kingdom of the Two Sicilies, exposed agriculture to foreign competition.[26] The southern farmers, including those in Calabria and specifically those in the Regia Sila area, had to increase output in order to counter the drop in wheat prices caused by the liberalization of the market. The area dedicated to farming increased, crop alternation periods tended to shorten, oaks were progressively cut down, planting and grazing, which up to then had been the basis of the local economy, focused more and more on planting, and the forested areas were being consistently transformed into farm grazing lands. The property of the clergy was expropriated and alienated. Starting in 1867, country estates were sold *en bloc* to the highest bidders in the magistrates' courts.[27] There was no political plan for the creation of small and medium landowners who would be potential supporters of the new regime. But the masses in the south were controlled by the army: political support was not needed. What was useful, however, was the financial liquidity that could be obtained in the south by alienating Church property and using the money for the economic development of the north.

The local economy regressed to bartering. The new landowners, after having used all their savings to purchase the former Church lands, were strangled by the increasing land taxes, which rose at dizzying rates from year to year. The agrarian agreement deteriorated until barely one-fifth of the product was destined to the farmer; the lands which for centuries had guaranteed the local population a minimum level of survival through the exercise of civic rights of use disappeared. Encroaching on private property became a crime and a dangerous enterprise, due to the presence of armed guards; the administration of the Sila forests changed to the detriment of the landless farmers. Ironically enough, the *monti frumentari* were replaced by a tax on flour, which made milling more expensive, with a negative impact on bread-baking. Taxes on livestock, collected by the towns, which were struggling with budget problems, actually cut back on breeding, which, for centuries, had been an important item in the local economy. Compulsory military service, which recognized no attenuating social factors and replaced the Bourbon method of recruitment, which had been based on a lottery system with a possibility of finding replacements, removed young men from primary activities. The Church could not intervene materially and the municipal administrations, without adequate resources, had to provide the structures required by the new public organization, such as the appointment of the district doctor and midwife, supporting local schoolteachers, maintaining the roads and building cemeteries.

This new situation pushed the population of Sila Greca in three main directions: banditry, worker associations or unionism and emigration.

Above all, the reasons and opportunities that prompted the people to turn to weapons and violence increased. In the area, and especially in Corigliano, there was an anti-government tradition, which had already manifested itself in the spring of 1806, and the Sila forests were a place of refuge for the bandits and outlaws.

The farmers and artisans created a common front in face of the insupportable socio-economic situation; helping each other in the most critical situations seemed the obvious thing to do, and it was also propagandized by the new socialist theories. The towns of Sila Greca were interested in this type of cooperation and associations; the Società Operaia di Mutuo Soccorso di Crosia, the Unione Cooperative Agraria e Lavoro di Paludi and, in nearby Acri, the Società Operaia dell'Independenza were established during these years (Falcone, 1884).

However, the main release valve for the crisis of the latter half of the 19th century was emigration. America and Africa were the destinations of the illiterate farmers and artisans. Opportunities for jobs were created following the construction of the Suez Canal. From the files of the notaries, we can obtain an idea of property transactions and we can understand the context in which the emigrants sold their land to put together enough money to cross the Atlantic. Often they sold a plot of land that their ancestors had desired for generations and finally obtained in the 19th century after years of bitter struggles.

The demographic data for the Sila Greca area are particularly significant. The total population rose from 60,179 in 1861 to barely 62,430, with real and percentage increases of 2260 and 3.76%, respectively. Between 1861 and 1901, 12,164 people emigrated from the Rossano district; the yearly figure was below 1000 until 1886 and then it progressively increased, to peak at 1644 in 1901. The mean, calculated on a population sample of 10,000, was 156.2 during the 5-year period from 1887 to 1890, rose to 177 between 1881 and 1895 and reached 195.3 in the last 5 years of the 19th century (Izzo, 1965).

The citizens tried to avoid emigration by petitioning to be allowed to plant the forests. The petition that the inhabitants of Cropalati submitted to the Minister of the Interior in 1891 was particularly significant:

> Our town possesses approximately 1000 hectares of forested land[28] . . . we are writing to Your Excellency because we are all suffering from hunger, when we do not have work, and for this reason everyone is emigrating, to suffer who knows where. The children they leave behind are left in the streets; in order not to suffer the slavery of those landowners from whom some of us rent land . . . we always work, but never collect, because the meagre crops are not sufficient for the owner of the land. And therefore, our brothers are emigrating because the sowable lands are not sufficient for everyone to be able to work and live. Therefore, we beg Your Excellency to allow us to plant the forests, to allow the poor, who are dying of hunger, to live. And thus, none of our brothers will emigrate. We beg Your

> Excellency to look at us, and if possible divide up this land, so we can plant it, we undertake to pay an annual tax, each of us, to our town.[29]

The government, however, was sensitive to other interests. It took action on the Sila issue in 1876 and settled the centuries-old disputes between the towns and the landowners with a solution that called for partial devolution of the Sila territories to the state, with recognition that three-quarters belonged to private owners, with the liberalization of the exercise of rights of civic use.[30] The state forests were de facto broken down; they were exposed to destruction on the part of the landowners, who could hardly believe that, at a single stroke, they were freed from the shackles of rights of civic use, from the control of the central government and from the conflicts with the local governments. With the fear that the miracle would be short-lived, they sold huge tracts of centuries-old forests at any price, and the forestry industries of central and northern Italy quickly began to exploit them.

The construction of the Sila roads, a consequence of the law of 25 May 1886, brought Sila Greca out of centuries of isolation, with positive effects. However, it exposed the forests to greater exploitation. The construction of the railway at the beginning of the 20th century did the rest.

Emigration rose markedly during the first decade of the 20th century. However, it had positive effects on the local economy because of the hard currency that was sent from abroad. These moneys were used to build houses and purchase land. The emigrants who returned home also brought with them new farming tools; thus traditional subsistence farming was supplemented by farming for sale and profit.

7 The difficult years during and after the First World War (1915–1921)

In some cases the local administrations finally divided up the state lands and regulated the use and exploitation of municipal lands, in so far as rent and protecting the forests was concerned (the forests were already suffering the effects of abandonment and mainly uncontrolled deforestation), and the grazing lands – for goats as well – in spring. Some towns approved regulations for renting and using the municipal country estates and the forested lands in particular. The town of Corigliano Calabro issued the *Regolamento degli usi civici sui demani comunali inquotizzati* (Regulations for the exercise of civic rights of use of two undivided municipal lands) as early as 1900.[31] Acri drew up the *Regolamento per l'esercizio dei diritti d'uso nei beni demaniali*[32] (Regulations for the exercise of civic rights of use of municipal lands), which was approved by the Cosenza Ripartimento of the Government Forestry Service; and later it approved the *Regolamento per la concessione in utenza delle terre demaniali*[33] (Regulations for concessions to use municipal lands). The local governments had to face a difficult situation: the population, especially during the war, pressed for intensification of planting. The needy citizens were interested in a piece of land, but even the emigrants always hoped to return. The mutual aid associations, which accentuated their working-class nature, pushed for intensive farming to the detriment of the forests and grazing, which was generally done by the large landowners even on municipal land that was rented for a few lire.

The general malaise of the population risked becoming rebellion in Paludi, where, to combat the usurpations and the terrible administration of the municipal lands, the citizens turned to the Minister of Agriculture in 1916, and the following year they appealed to His Royal Highness Crown Prince Umberto II, future King of Italy;[34] in 1920 the situation was precipitated and the municipal lands were occupied.[35]

On the issue of protecting the forests, the work of the Azienda del Demanio Forestale dello Stato was significant, starting in 1910, not only because of the control it exercised over all the forest assets of the area, but also for the direct actions it took on the areas which it had fully at its disposal. The Azienda Forestale di Cosenza directly managed 2788 ha of stands, 217 ha of coppices and 204 non-productive hectares; and it extended nurseries for the production of plants for reforestation over 9 ha.

Some of these lands were located in the Sila Greca area and specifically in countryside around Longobucco: the Destra di Golia estate, former property of the Cistercian monks, which had been ceded by the Arnone family and comprised about 219 ha; the Santa Barbara estate, of roughly 709 ha, ceded by the Campagna family; and the 64 ha of the Acquacalda estate ceded by the Barracco family.

Alerted by the Ministry of Agriculture, the Prefecture of Cosenza set about guaranteeing public order and preventing the occupation of the lands. The Azienda del Demanio Forestale dello Stato, with headquarters in Cosenza, was forced to intensify its surveillance and protection of the forest.

In many cases, the situation was quite poor so that, for the energetic reforestation that was also necessary for soil conservation, the towns concluded transfers of large tracts of forested land to the Azienda Forestale. Acri, with a procedure that was begun in 1919 and concluded in 1923, sold the Galluzzo Pine Grove to the Azienda Forestale, maintaining that it could not be conveniently controlled and kept by the town and therefore was subjected to systematic and continuous acts of vandalism on the part of a few, heedless citizens who lived in the vicinity . . . and in consideration of that fact that due to these vandalistic acts the superb pine grove risked destruction.[36]

In some cases the municipal governments sold the forests they could not conserve and protect to the Azienda Forestale dello Stato. But, in many other cases, it was private citizens who sold entirely exploited forests. In this context, the government repurchased, at presumably higher than market prices, those forests for which, in 1876, it had recognized the right of ownership of many usurpers and it repurchased them when they were already degraded and required massive, expensive reforestation.

8 From 20 years of the Fascist period to the Second World War (1922–1945)

During the 20-year Fascist period, forests and those of Sila Greca in particular were directly affected by general politics and the national economy. Some socio-economic components which pursued the reduction in the forested surface continued and were accentuated, but new factors also came into play that pushed in opposite directions and hence in favour of the forest.

Emigration decreased considerably from the end of the First World War to 1930: the immigration quotas imposed by the USA and Canada became more and more rigid. Italian foreign policy subsequently put a freeze on emigration in the 1930s. All this, along with the government's population policy, led to a further increase in the natural population rise.

This time the forest was threatened not only by crops, but by market demands. Railway ties, which were usually made of Turkey or Adriatic oak, were now being made of the cheaper, centuries-old Corsican pine of Sila. The market also required poles for electrical power and telegraph wires between post offices. In Catanzaro and Cosenza, two plants were opened for the extraction of tannic acid; sawmills that benefited from the advent of electric power produced lumber for crafts; charcoal production also rose in relation to the growing needs of families, as living conditions improved mainly in the cities. It is in this context that we can explain the role of the forest in the local economy and the interest of northern companies in exploiting the forests of Sila Greca.

Thus, in addition to guaranteeing the wood supply for the needs of the citizens in the nearby towns, the forest began to become a source of income, in that it created secondary industries and encouraged investments by private parties, who considered it advantageous to employ capital in improving the forests.

There were also other positive factors, alongside the market's interest in the forest. The Ispettorato Ripartimentale delle Foreste intensified its protection of public and private forests and at the same time carried out reforestation in Sila Greca to restore and salvage deteriorated and destroyed areas. Between 1924 and 1927 the national parliament passed laws in favour of reorganizing the exercise of rights of civic use, with subsequent positive effects on forest protection.[37] During the 1930s farming was aimed not so much at increasing the cultivated areas as at, mainly, increasing production by intensifying cultivation, reducing alternation intervals, and by using new agricultural techniques and mechanization.

9 The post-Second World War period and the start of restoring and rebuilding the forests

After the Second World War the government of the Republic of Italy also inherited the issue of the Sila forests, on which it took action as early as

1947,[38] because of land occupation, which often developed into demonstrations that upset the public order (Alicata, 1951).

Emigration resumed after the war, and practically resembled a mass exodus: nearly 50% of the populations of the Sila Greca towns moved to the USA, Canada and mainly Australia first and then to other parts of Europe and northern Italy. Once again property sales increased considerably: in 1953 a ticket to Australia cost 400,000 lire, an enormous amount that was the equivalent of a teacher's yearly salary.

Along with mass emigration, there was yet another phenomenon; labour shifted from primary activities first to industry and then to the tertiary or service sector. The amount of farmland used decreased in every municipality and thus the area that could be potentially restored to the forests increased.

After the tragic floods of 1953, the disastrous effects of which were partly due to two centuries of deforestation, the parliament passed a law in favour of Calabria.[39] In Sila Greca over 15,000 ha of land were reclaimed as forest through massive reforestation and recovery of degraded forests, which markedly improved the forest landscape, with positive effects on soil conservation as well.

10 Conclusions

The review and interpretation of the documents have made it possible to verify the extent to which the historical–political context that characterized the Sila Greca area in the past two centuries influenced the consistency of the forest assets.

The periods to which the progressive and consistency variations in the forest cover are related are the French domination, the Bourbon restoration, the Kingdom of Italy and the two world wars. The first is characterized by the transfer to the royal domain of all former feudal and church properties. Therefore, this phase protected the forests, since it annulled exercise of rights of civic use and led to the utilization of the forests, albeit to satisfy military requirements.

The second period, characterized by the return of forest lands to the church institutions and by the division of the former feudal lands between the royal domain and the municipalities, and a significant population increase led to a marked decrease in forested area, due to the growing needs of the local populations, who took lands from the forest for farming. In this period the Regia Sila continued to produce wood for the shipyards of the Royal Navy.

The third period, still marked by population growth, as emigration had not yet peaked, was characterized by yet further reduction in the forest area, due to the utilization of the forests by the former usurpers, whose ownership was recognized on the basis of the law of 1876. This was compounded by the construction of roads, which, on the one hand, brought Sila Greca out of centuries of isolation but, on the other, exposed the forests to further and indiscriminate exploitation.

The fourth period is characterized by regulation of forest use and protective measures resulting from the creation of the Azienda del Demanio Forestale dello Stato, which purchased lands in Sila to protect them and areas to reforest. In the period between the two world wars, certain socio-economic factors that led to yet further reductions in the forest cover became accentuated, however, new factors also came into play which gave a thrust in the opposite direction and hence in favour of forest conservation (law for regulating the exercise of rights of civic use).

The effects of this progressive reduction in forest cover were manifested in all their severity by the floods that struck a large part of Calabria during the 1950s and which led to the passage of special laws. In this context approximately 15,000 ha of forest were recovered in Sila Greca through reforestation and work to salvage the degraded forests; this markedly improved the forest landscape and created already noticeable effects on soil conservation.

Notes

1 Archivio di Stato di Napoli, Fondo Catasto Onciario, the Catasti di Mandatoriccio (Vol. 6008), Campana (Vol. 5991), Bocchigliero (Vol. 5986) and Longobucco (Vols 5, 6003–6007) were particularly useful.

2 Archivio di Stato di Cosenza, Cadastre Français; Ufficio Tecnico Erariale of Catanzaro, historical archives.

3 Archivi di stato di Napoli, Cosenza and Catanzaro; Archivio Commissariato Regionale Usi Civici della Calabria.

4 Law no. 14 dated 19 January 1807; Royal Decree no. 922 dated 4 May 1911.

5 Royal Decree no.1875 dated 25 January 1820.

6 Royal Decree no. 2378 dated 7 October 1834.

7 Law no. 267 dated 26 June 1902.

8 Royal Decree no. 869 dated 9 April 1928.

9 Law no. 162 dated 29 April 1934.

10 Royal Decree no. 965 dated 10 June 1937.

11 Law no. 130 dated 2 August 1806.

12 *Bollettino della Commissione Feudale,* Naples, years 1809 and 1810.

13 Order of the Appointed Provincial Councilor for the division of the lands in the suit between the town of Bocchigliero and the Administration of the Royal Demain, issued in Cosenza on 18 March 1819.

14 Order of the Intendente of Calabria Citra in the suit for the division of the ecclesiastic lands between the town of Calopezzati and the Mensa Vescovile of Rossano, issued in Cosenza on 10 August 1813.

15 Order of the Royal Commissioner for the division of the lands of the Province of Calabria Citra in the suit between the town of Calopezzati and the ex-feudal lord Prince Campagna, issued in Cosenza on 5 August 1811.

16 Tr. note: the *tomolo* was a unit of land measure used in central and southern Italy.

17 Order of the Royal Commissario for the division of the lands of the Province of Calabria Citra in the suit between the town of Corpalati and ex-feudal lord Prince Borghese issued in Cosenza on 15 July 1811.
18 *Catasta Onciario di Campana*, 1742, *Rileva*, of. 116, Part 58, Archivio di Stato Naples, f. Onciario, vol. 5991.
19 Royal Decree no. 689 dated 2 July 1810.
20 Chiarella Nicola, notary in Mandatoriccio, ibid., no. 1047, document dated 22 August 1838.
21 Chiarella Nicola, notary cited, document dated 11 October 1841.
22 Royal Decree no. 3269 dated 22 January 1836.
23 Royal Decree no. 5269 dated 12 April 1839.
24 Royal Decree no. 7325 dated 17 March 1842.
25 Royal Decree no. 227 dated 18 April 1853.
26 Royal Decree no. 186 dated 18 August 1861 that liberalized grain trade throughout the Kingdom.
27 Law no. 3848 dated 15 August 1867 for the liquidation of the church property.
28 The lands in question were the Bosco Maleno, Dragonario, Difisella, Pagliariccio and Manco.
29 Ministry of the Interior, Div. 5121–5120. prot. 4951 dated 26 November 1891, in the *Archivio Commissariato Regionale Usi Civici della Calabria*, Cropalati folder.
30 Law no. 3124 dated 25 May 1876.
31 Resolution of the Royal Commissioner of Corigliano, who for emergency reasons assumed the powers of the Council, dated 18 September 1900.
32 Resolution of the Town Council of Acri, dated 8 March 1912.
33 Resolution of the Town Council of Crosia, dated 30 July 1921.
34 Archivio Regionale Usi Civici, Paludi folder.
35 Communication to the Prefect of Cosenza by the President of the *Cooperative di Lavoro e Agraria fra Agricoltori e Combattenti di Paludi*, in which we read that, on 23 March 1923 in the main square of the village, a rally was held in the presence of the Maresciallo of Carabinieri of Cropalati concerning the 'lands that are under the forest restriction and the occupation of many hectare of land . . . by a group of citizens'. Prefecture of Cosenza, Uff. Demaniale, prot. 219–226 March 1920.
36 Town of Acri, resolution adopted on 20 September 1919 by the Royal Commissario; on the same subject, see the resolutions of the town council dated 24 March 1923 and 31 May 1924; the Azienda Forestale dello Stato purchased the Galluzzo Municipal Forest for the sum of 265,000 lire to launch a 'reforestation project'.
37 Royal Decree no. 751 dated 22 March 1924; Royal Decree no. 1484 dated 28 August 1924; Royal Decree no. 895 dated 16 May 1926; Law no. 1766 dated 16 June 1927.
38 Laws 1629 of 1947 and 230 of 1950.
39 Law no. 1177 dated 26 November 1955.

References and further reading

Alicata, M. (1951) *La Riforma Fondiaria in Calabria*. Rinascita, Rome, anno VIII, no. 2.

Barletta, P. (1852) *Stato de'Demani della Sila Badiale di San Giovanni in Fiore Occupati Senza Concessione da quel Comune come Copro Morale.* Tip. del Fabbricatore, Naples.

Barletta, P. (1852) *Stato delle Contrade della Sila Badiale Verificate negli Anni 1850–51.* Tip. del Fabbricatore, Naples.

Barletta, P. (1854) *Verificazione dei Demani della Sila.* Tip. Migliaccio, Cosenza.

Ciasca, R. (1957) *Le Trasformazioni Agrarie in Calabria dopo l'Unità.* Atti del I° Congresso Storico Calabrese, Tivoli, Chicca.

Falcone, N. (1884) *Acri e la Società Operaio dell'Indipendenza.* Tip. dell'Avanguardia, Cosenza.

Gadda, A. (1881) *La Sila delle Calabrie.* Tip. del Risorgimento, Cosenza.

Galasso, G. (1957) *La Riforma Agraria in Sila e nei Territori Jonici Contermini.* Nord e Sud, Naples, anno IV, no. 3.

Greco, V.M. (1864) *Rendiconto della Real Società Economica della Provincia di Calabria Citra.* Tip. Migliaccio, Cosenza.

Izzo, L. (1965) *La Popolazione Calabrese nel XIX Secolo.* Edizioni Scientifiche Italiane, Naples.

Marini, C. (1844) *Sulla Selva Brezia e sulla Interpretazione ed Applicazione de' Reali Decreti 5 ottobre 1838 e 31 Marzo 1843 a le quistioni all'Agro Silano.* Tip. Migliaccio, Cosenza.

Martire, F. (1866) *La Sila di Calabria.* Le Monnier, Florence.

Palumbo, M. (1910) *I Comuni Meridionali Prima e Dopo le Leggi Eversive della Feudalità. Feudi. Università. Comuni. Demani.* Tip. L'Unione, Montecorvino di Rovella.

Pancaro, C. (1889) *Atti, Progetti, Tavole Dimostrative e Pianta Corografica della Sila con Prefazione da Servire al Consorzio Obbligatorio Stabilito con Legge 25 Maggio 1886 per la Costruzione delle Strade Silane.* Tip. dell'Avanguardia, Cosenza.

Patrone, G. (1934) *La Percentuale di Massa Utilizzabile nella Lavorazione delle Traverse Ferroviarie Normali di Pino Laricio Calabrese.* Annali del R. Istituto Sup. Agrario e Forestale, Florence.

Ministero di Agricoltura, Industria e Commercio. Direzione Generale delle Foreste (1915) *Relazione sull'Azienda del Demanio Forestale dello Stato: I° luglio 1910–30 giugno 1914, Roma Capaccini.*

Rossi-Doria, M. (1934) *La Calabria Agricola ed il suo Avvenire.* Il Ponte, Florence, VI, no. 9–10.

Taruffi, D., De Nobili, L. and Lori, C. (1908) *La Questione Agraria e l'Emigrazione in Calabria.* Barbera, Florence.

Zurlo, G. (1852) *Relazione Ufficiale al Cavalier Giovanni Acton Ministro di Sua Maestà Siciliana, Intorno allo Stato in cui Erano le Regie Sile di Calabria nell'Anno 1792.* Tipografia del Ministero dell'Interno, Naples.

Zurlo, G. (1862) *Stato della Regia Sila Liquidato nel 1790 da Giuseppe Zurlo Giudice della Gran Corte della Vicaria.* Stamperia Nazionale, Naples.

Economic Culture and Forest Policy in Italy, 1861–1915

23

Tiziano Raffaelli

Dipartimento di Economia dell' Impresa, della Tecnologia, dell' Ambiente, viale Sant' Ignazio 17, 09123 Cagliari, Italy

The influence of economic culture on Italian forest policy between 1861 and 1915 is examined. To encourage production, free traders favoured private ownership while for the prevention of 'public damage' they approved of state interference. The 'historical school' maintained that state ownership could solve the conflict between the two aims. Both views were unable to enforce adequate forest policies.

Forest legislation during the period under review is marked by two main acts which seem to obey opposing economic principles.

In 1877 the Italian Parliament passed an Act that substituted the various norms of preunitarian Italy.[1] Significantly, it took 16 years to find a satisfying solution to unifying laws of forest management, which were widely different, ranging from the almost absolute liberty in the exercise of property rights of Tuscany to the restrictive measures of the Papal States, Modena and Parma.[2] The Act of 1877, proposed by the Minister of Agriculture, Majorana, a member of the Società Adamo Smith, was basically inspired by *laissez-faire* principles: it favoured private ownership of forest land and prohibited only those activities which clearly tended to provoke 'public damage', identified with landslides and flooding.

In contrast, in 1910 a new Act of Parliament, proposed by the Minister of Agriculture, Luigi Luzzatti, a member of the Associazione per il Progresso degli Studi Economici, whose political and methodological aims are usually associated with the historical school, instituted the Azienda Speciale del Demanio

(eds M. Agnoletti and S. Anderson)

Forestale di Stato, a state enterprise for the management of forest resources.[3] The avowed aim was to promote both direct public production of timber and preservation of forest land conceived as a 'public good'. The change was described by Luzzatti himself as a switch from 'repression' to 'prevention'.[4]

Historians are aware of the connection between this change in legislation and the shift in economic culture that took place in the same period: the free trade hegemony of the 1860s and 1870s was challenged and quickly replaced by a new wave of interventionist and protectionist ideas, which accompanied the spread of 'historical Germanism'. As far as forest policy is concerned, historians tend to credit this cultural change with positive results. Its only recognized fault is that it happened too late. This assessment of historical research is confirmed by the fact that most criticisms of the 1877 law expressed by later reformers have survived the test of time and are still widely shared.[5] On the other hand, only too evident failure in later years to carry out preservation and proper management of forests and woodlands has been ascribed to lack of financial resources.

Though well established, this assessment of the relative merits of the two Acts is highly questionable.

From a theoretical point of view, the main principles of economic policy that were relevant to forest management were known in 1877 as well as in 1910. The fact that forests were necessary for deeper reasons than merely that of providing material for burning, building and other economic purposes was recognized by almost all the participants in the debates. In particular, the hydrogeological functions of forests and woodland were well known and indeed provided the main reason for legislation in 1877.[6] The change was one of attitude, rather than theoretical knowledge.[7] Moreover, the Act of 1877 made ample provision for state intervention, giving detailed prescriptions and regulations for forest management and even allowing public expropriation for planting new forests in order to avoid flooding and other calamities.[8]

It is often stated that the law simply dismantled any form of public intervention and control in forest management, that it was an application of the ideology of free trade, typical of the liberal government of the Left as witnessed by its opposition to railway nationalization. However, it must be noticed that criticisms of the 1877 Act were raised against its restrictions more than against its *laissez-faire* inspiration. Its lack of attention to sheep-farming and agriculture, contrasting with the need to avoid unnecessary prohibitions, was a much more widespread motive of concern than its inability to protect lower lands from flooding and mountains and hills from landslides.[9] These problems emerged abruptly only when floods and landslides actually took place, as in 1882 and 1907, for example.[10]

Criticisms of excessive restrictions on property rights were immediately expressed by economists of the Tuscan free-trade school, like Pareto and Peruzzi. The latter, in particular, led the Parliamentary opposition to the law on uncompromising *laissez-faire* principles. He attacked his friend Majorana, thereby proving that there were at least two varieties of free-traders and that the law was not the offspring of dogmatic positions.

In the debate at the Accademia dei Georgofili, Peruzzi suggested that the adverse effects of deforestation could be avoided by facilitating direct transactions of property rights between the owners of forests and woodland and the owners of lowlands that were at risk of being flooded or damaged by deforestation.[11] This anticipation of modern solutions to 'transaction costs' problems can be considered the purest expression of Tuscan economic *laissez-faire*.

Significantly, Peruzzi's worries about restrictions on private enterprise were shared by economists of the 'historical school', like Lampertico, whose main concern was state encouragement to economic activity.[12]

The law of 1877 was far from embracing the uncompromising position of Peruzzi. It denied compensation for the restrictions it imposed on private property and, while releasing about one-quarter of previously restricted lands, it made restrictions on the remaining three-quarters tougher, which was indeed the main cause of complaint against the law itself.

Where the law proved to be completely inadequate was in the way it was applied. In 1887, one of the main agrarian economists of the period, Ghino Valenti, who, as part of the Inchiesta Iacini, carefully studied the conditions of the hills and mountains of Marche, wrote that restrictions on property rights had been applied by inaccurate surveys. He also noticed that the foresters were few and badly prepared, while the bureaucracy kept its control over the whole process by issuing senseless prescriptions. According to Valenti, the provisions of the 1877 law for expropriation were the same as the French and Swiss provisions, the main difference was that the Italian Parliament had allocated no funds for expropriation and reforestation, whereas 'creating new forests needs money more than laws'.[13]

It seems that what was needed at the time was a better and more detailed consideration of long- and short-term interests in forest land utilization and a way of balancing them through state intervention. This would have required improvement in state administration and more funding. Indeed, these two objectives were central to Luzzatti's reform in 1910.

The reform was welcomed with a broad consensus and Valenti himself added his own voice to the chorus.[14] However, the fundamental problems of forest management were set aside more than solved. Luzzatti's project started out from the assumption that there was no sharp contrast between economic and extraeconomic interests in land management. The reform avoided the dilemma between profitable land utilization and long-term requirements of soil and water management. It implicitly assumed that state-owned forests could pursue both objectives simultaneously, achieving good economic results while at the same time responding to the other public needs.

Luzzatti was in touch with forestry experts, attended many of their conferences, held in imitation of the German and American meetings, and maintained that forest management could soon become economically viable almost everywhere, provided new scientific principles were adopted. From this point of view, state ownership was conceived as capable of achieving the results that private enterprise had failed to accomplish.

Economists were then indirectly supporting this view. Those who were influenced by the German historical school, like Toniolo, thought it necessary to extend state intervention in any direction, while supporters of free trade, like De Viti de Marco, admitted that forests provided an exception to the general rule that private property was preferable to public property.[15] Forest management immobilized huge amounts of capital in the form of timber, was characterized by great internal economies, which the state could exploit better than anyone else, and by routine rules of conduct that made continuous attention by the owner unnecessary.[16] However, De Viti's principles of public finance maintained a clear demarcation between the economic functions of forests and their nature as 'public utilitites'. This latter function required costly management, either directly by the state or indirectly through the imposition of restrictions on private owners.

Without explicitly denying this distinction, the rhetoric behind Luzzatti's Act contributed to the confusion. The concept of 'public utility' was almost forgotten and the article of the Act which was most widely criticized was precisely Article 27, which courageously extended previous restrictions on property rights to chestnut woods.[17] Paradoxically, the 1877 Act was less ambiguous on this issue and assumed that prevention of landslides and floods implied expenses that were different from economic investments capable of repaying themselves directly, sooner or later.

The ambiguity of the principles underlying the new legislation became clear in a few years, when it appeared that the objectives of timber production and land preservation did not coincide. In southern Italy in particular, forestry experts realized that such a coincidence was wishful thinking. If the Italian government wanted to prevent future disasters, it had to spend with no prospect of direct money returns. From an economic point of view, in most cases, woods and forests were to be considered as a 'necessary evil', according to a concept that economic science had taught throughout the period but which was hard to digest for political leaders and for public opinion as well.[18]

The positivist culture which prevailed at the turn of the century and inspired the new legislation made forests a proper object of 'scientific management' but failed to accept that different human aims made it necessary to choose between alternative uses of land and economic resources.

Notes

* This chapter is a shortened version of Raffaelli, T. (1997) Dalla repressione del 'danno pubblico' alla produzione del 'bene pubblico:' cultura economica e politiche forestali (1861–1915), *Pensiero Economico Italiano* 5(2), 41–79.

1 L. 20.6.1877, no. 3917, 'Legge forestale'.

2 On the laws of the Italian preunitarian states, see Grisolia, F. (1897) Foreste e boschi, *Digesto Italiano* 11, 580–739 (UTET, Turin).
3 L. 2.6.1910, no. 277, 'Provvedimenti per il Demanio forestale di Stato e per la tutela e l'incoraggiamento della silvicultura.' On the rivalry between the Società Adamo Smith and the Associazione per il Progresso degli Studi Economici, see Faucci, R. (1981) *La Scienza Economica in Italia (1850–1943)*, Guida, Naples, ch. 2; Faucci, R. (1995) *L'Economista Scomodo. Vita e Opere di Francesco Ferrara*, Sellerio, Palermo, ch. 6; Salvestrini, R. (1965) *I Moderati Toscani e la Classe Dirigente Italiana*, Olschki, Florence, ch. 5.
4 *Atti Parlamentari*, XXIII legislatura, Camera dei Deputati, ddl. 346, 11.9.1910.
5 The standard criticism of the Act of 1877 was stated by Luzzatti, Trifone and other reformers of forest legislation (Luzzatti, L. (1910) Indirizzo di saluto, *Atti del Congresso Forestale Italiano. Bologna 1909*, Pro Montibus et Silvis, Bologna; Trifone, R. (1957) *Storia del Diritto Forestale in Italia*, Coppini, Florence). This criticism has been accepted by modern historians (Mura, A. (1973) *Ordinamento Forestale e Problemi Montani*, Giuffrè, Milan; Zanzi Sulli, A. (1996) Origine ed evoluzione di una cultura tecnica forestale dello Stato unitario, in: *Atti del Convegno Nazionale della SISE*, Monduzzi, Bologna, pp. 637–652). More balanced, though similar, is the interpretation given by Del Vecchio, B. (1980) Sui moderati e la questione forestale (uno scritto poco noto di Bettino Ricasoli), in: Società Toscana di Storia del Risorgimento (ed.) *Agricoltura e Società nella Maremma Grossetana dell'800*, Olschki, Florence, pp. 285–312.
6 Italy was well advanced in hydrogeological research. The works by Mengotti and others were very influential and Marsh's famous book, *Man and Nature* (1864), written while the author lived in Italy, had a second edition published first in Italian (Barbèra, Florence, 1870) and was often quoted in the Parliamentary debates (cf. Vecchio, B. (1994) Un documento in materia forestale nell'Italia del secondo Ottocento: i dibattiti parlamentari, 1869–1877, *Storia urbana* 18, 177–204. For the relationship between Marsh and the Italian culture, see also Hall, M. (1998) Restoring the countryside: George Perkins Marsh and the Italian land ethic (1861–82), *Environment and History* 4, 1–14).
7 Even purely economic issues, like the incentive given by huge immobilized capital to cut trees before their maturity, were scientifically investigated before 1877. Forest economics was deeply indebted to engineering science and was therefore able to find complex mathematical solutions to economic problems before the so-called marginalist revolution. In Italy in particular, the lead was taken by Turin – under the influence of French engineering schools – with G. Borio (*Elementi di Economia e Stima dei Fondi Agrari e Forestali*, 2nd edn, UTET, Turin, 1868, §§ 202–211) and later on with C. Tommasina.
8 It is interesting to note that in Italy, and even in Tuscany, expropriation was widely accepted by *laissez-faire* economists, at least in principle. While in Germany *laissez-faire* forest economists, led by Pressler, opposed expropriation, in Italy they preferred it, when necessary, to state interference through regulations imposed on private owners.
9 Indeed, the parliamentary proposals of the period, presented under the title 'Modificazioni alla legge forestale del 20.6.1877, n. 3917,' aimed at reducing rather than increasing restrictions on forest lands: *Atti Parlamentari*, Camera dei Deputati, ddl. 18.1.1887, no. 141; ddl. 7.12.1891, no. 291; ddl. 26.11.1892, no. 149; ddl. 13.12.1893, no. 75; ddl. 12.12.1896, no. 364.

10 After the floodings that took place in Veneto in 1882, the Italian Parliament was unable to reform the law of 1877, but managed to pass an Act for reforestation, though without allocating a reasonable amount of money for the purpose (L. 1.3.1888, no. 5238, 'Disposizioni atte a promuovere i rimboschimenti').

11 Conferenze intorno alla legge forestale, *Atti della Reale Accademia dei Georgofili*, series IV, vol. 6, 1–306; see in particular pp. 197–198.

12 Lampertico, F. (1876) *La proprietà. L'Economia dei Popoli e degli Stati*, Vol. III, Fratelli Treves, Milan.

13 Valenti, G. (1914) Il rimboschimento e la proprietà collettiva nell'Appennino marchigiano, Mancini, Macerata 1887; reprinted in *Studi di Politica Agraria*, Athenaeum, Rome.

14 Luzzatti's law 'per la prima volta porta con sé i mezzi destinati ad attuarla', Valenti, G. (1911) L'Italia agricola dal 1861 al 1911. In: *Cinquant'Anni di Storia Italiana (1860–1910)*, a cura dell'Accademia dei Lincei, Hoepli, Milan; reprinted, in *Studi di Politica Agraria*, op. cit., p. 489.

15 Toniolo, G. (1909) *Trattato di Economia Sociale*, Vol. I, Libreria Editrice Fiorentina, Florence, p. 167; De Viti de Marco, A. (1894) Entrate patrimoniali e demanio, *Giornale degli Economisti: Principi di Economia Finanziaria*, Einaudi, Turin, 1934.

16 De Viti's principles were restated by Einaudi, L. (1932) *Principi di Scienza delle Finanze*, La Riforma Sociale, Turin, pp. 15–18.

17 The law of 1877 had put *ex lege* restrictions on lands situated 'above the limit of chestnut cultivation' (Art. 1), a limit suggested by forest scientists but widely questioned.

18 Valenti defined protective forests a 'necessary evil' in '*L'Agricoltura italiana* . . .', op. cit., p. 488. The definition was taken over by L. Piccioli, then Forest Officer in Catanzaro ('La legge Luzzatti sul Demanio forestale di Stato e la mutabilità parlamentare,' Comunicazione del 3.6.1914 al Congresso forestale italiano di Napoli, Moretti e Percotto, Udine, 1914). Piccioli soon realized that the two objectives of land preservation and economic management of forests could coincide only in particular conditions, like those prevailing in north-eastern Italy.

Forest Policy and Public Forest Production in Spain, 1855–1936

24

*Domingo Gallego Martínez,[1] José Ignacio Jiménez Blanco,[2] José Antonio Sebastián Amarilla,[2] Juan Francisco Zambrana Pineda[3] and Santiago Zapata Blanco[4]

[1]Zaragoza University; [2]Facultad de Ciencias Económicas y Empresariales, Universidad Complutense, Campus de Somosaguas, 28223 Madrid, Spain; [3]Malaga University; and [4]Extremadura University, Spain

This chapter deals with Spanish forest history between 1855 and 1936. Specifically, it concerns forest policy and forest production, neither of which can be considered separately. Forest policy is divided into four stages, and the main features of each are analysed. These features could be arranged in two principal subjects: privatization of public land properties and management of the public lands that remained (about 6.6 Mha). Forest production is analysed by production trend, forest product uses and changes in use and the productive features of the different administrative types of forests.

Spain is a country where forest area is of great importance. Of a total surface of 50.5 Mha, 6 Mha are unproductive, 21.8 Mha are cultivated and 22.7 Mha are occupied by forests. Approximately, 52% of the productive surface is covered by forests.

This does not mean that all this territory is occupied by trees. In fact, only 49% is defined as woods. The rest includes pasture lands (*dehesas*), shrub lands and bush lands. One of the main features of today's Spanish forest lands is the poor condition of the vegetal layer. All this territory, however, is called forest.

* These authors constitute the Grupo de Estudios de Historia Rural.

At the beginning of the 19th century, the forest surface was even greater. Since then, there has been a net growth in both cultivated and unproductive areas. In that period, part of these lands were owned and used by individuals, but the majority had a communal use, although it was not always clear to whom they belonged.

The aim of this chapter is twofold. We intend to analyse Spanish forest policy and public forest production during the second half of the 19th century and the first 30 years of the 20th century. The two analyses are not to be separated, since production results are used as a contrast to forest policy goals. Private forests cannot be analysed because information about them is scarce. The period chosen is explained by quantitative sources being available. The end of the period is determined by the Civil War.

1 Forest policy

Spanish forest policy between 1855 and 1936 had two main objectives. The first one was to define property rights over forest lands. Under the influence of the dominant liberal ideology, it sought to privatize most of them. However, for various reasons, this goal was not fully achieved. In spite of the intensive expropriation process that took place in the 19th century, in the end some 6.7 Mha were still publicly owned, the great majority of them in the hands of villages. In 1946, public forests represented 37.5% of the total Spanish forests. The second goal of the forest policy was to define the management system to be followed by the remaining public forests. The leading role of both these goals changed over time.

For the period analysed, the forest policy can be divided into four stages. The first one extends from 1855 to 1875. During this stage, state action consisted of regulating the privatization of forest lands and creating a public forest patrimony. In 1863 the first Spanish Forest Act was enacted and this set the framework for state activity in the forest sector up to 1936, although the management system created by this law was not applied during this period.

The second stage comprises the period between 1875 and 1900. It was then when public forest use schemes, drawn up by engineers belonging to the State Forest Service, were enforced. Although most forests belonged to villages, from 1863 onwards their management began to be supervised by the state. It is also during this stage that the surveillance of public forests was given to the Guardia Civil.

One of the outstanding aspects of the third stage (1900–1924) was that public forests were divided into two groups. The first group consisted of public utility forests, which remained under the rule of the Ministry of Fomento; in the second group we find those forests not fulfilling the conditions established for public utility forests and their management was entrusted to the Ministry of Finance. A second feature of this stage was the growing application of reforestation and planning to public utility forests, which makes evident the increasing importance of state intervention. However, planning could be carried out by individuals until 1918. Finally, in this stage, communal uses which were open to every villager at no expense lost their importance (Table 24.1).

As well as promoting the expansion of ordinary uses, management plans gave rise to an improvement of the average productivity per hectare of public utility forests, as may be inferred from Table 24.2, where production is more than proportional to the corresponding surface.

The fourth stage occurs from 1924 to 1936. Two facts stand out. First, the Ministry of Finance closed its forest section and these forests began to

Table 24.1. Public forest and public utility forest production in percentages and according to type of use (Spain, 1861–1932)

	Public forests			Public utility forests		
	Ordinary	Communal	Extraordinary*	Ordinary	Communal	Extraordinary*
1861/62	31.3	55.8	12.9	–	–	–
1870/71	17.3	56.6	25.7	–	–	–
1879/80	32.6	46.9	20.5	–	–	–
1903	–	–	–	46.0	44.2	9.8
1913	–	–	–	57.6	39.1	3.3
1920	–	–	–	68.3	26.5	5.2
1924	58.9	35.6	5.5	–	–	–
1932	69.5	24.6	5.9	–	–	–

*Most extraordinary uses were fraudulent.
Sources: Calculations have been carried out on the basis of GEHR (1991), tables 17 and 19; Dirección General de Agricultura, Minas y Montes (1915), pp. LII–LIII; Dirección General de Agricultura, Minas y Montes (1922), pp. XXXIII–XXXIV; Dirección General de Agricultura y Montes (1926), p. 30; and Dirección General de Montes, Pesca y Caza (1934), p. 29. The names of the institutions have not been translated to make it easier to consult the sources.

Table 24.2. Surface and production of managed and reforested public utility forests (expressed as a percentage) (Spain, 1903–1932).

	Managed forests		Reforested forests	
	Surface	Production	Surface	Production
1903	2.8	12.3	0.9	0.2
1913	8.5	32.3	2.4	1.1
1920	8.8	28.3	2.8	1.4
1932	11.3	38.0	4.0	2.1

Sources: Calculations have been carried out on the basis of GEHR (1991), tables 16 and 19.

be supervised by the Ministry of Fomento or by the villages owning them. Secondly, fast-growing species acquired a leading role in reforestation.

2 Public forest production

The study of public utility forests between 1901 and 1933 gives us a better understanding of the goals of forest policy and their main effects. The evolution of production is represented, as a whole, in Fig. 24.1, whereas the different uses can be found in Fig. 24.2.[2] In the latter, we have multiplied each product's effective production by the average values for each one in the 1912/13 data, which means that the curves represent production in terms of volume. In the former, the results obtained have been added up. As may be noted, total production went up continuously and steadily during the first 30 years of the 20th century, though in the years of the First World War a slight fall can be noted. With an annual rate of 2.5%, this increase is greater than the rest of the Spanish agriculture sector, which only reached 1.9%.[1] This steady rise shows that forest products made their way to the market easily and that the different individuals organizing public forest uses correctly sensed the market's impulses, all the more correctly when our estimation of its magnitude is very probably below its real value.

Figure 24.2 and Table 24.3 make it clear that timber and resin were the products responsible for this growth, not only because of their degree of intensity, but also because of their importance in total production, which agrees with other sources. Timber was a raw material that was scarce in Spain, as is shown by the fact that the majority of forest imports were tree trunks and boards, whose sustained increase during the first 30 years of the 20th century

Fig. 24.1. Public utility forest production valued according to 1913 prices (1901–1933). In 1000 pesetas (from GEHR, 1989: pp. 105, 109 and 110, tables 3.8, 3.13 and 3.14).

allow us to infer, for such a period, a constantly growing domestic market, pushed forward by the general development of the Spanish economy. Undoubtedly, public forest users took advantage of it.

Whatever the reason, the certainty about public forest production in the first 30 years of the 20th century is its forest character in the strict sense, and

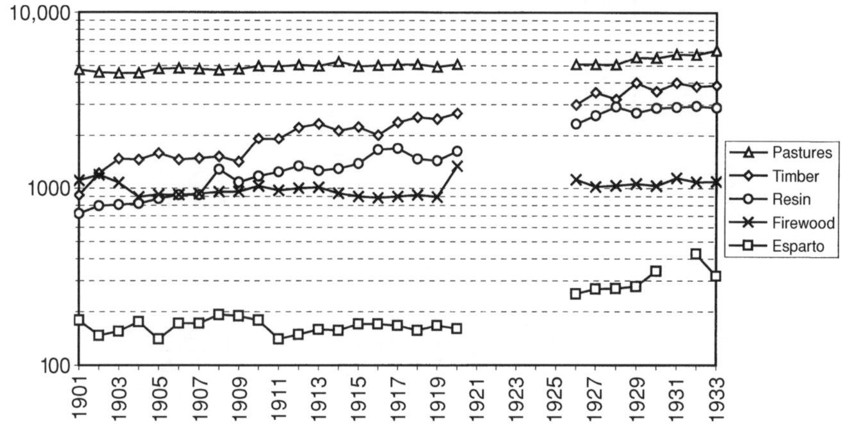

Fig. 24.2. Public utility forest production by use according to 1913 prices (1901–1933). In 1000 pesetas (from GEHR, 1989: pp. 105, 109 and 110, tables 3.8, 3.13 and 3.14).[2]

Table 24.3. Public utility forest production expressed in 1913 prices (Spain, 1901–1933). Five years' average in 1000 pesetas.

	Timber	Firewood	Pastures	Esparto	Resin	Cork and bark	Ploughing	Total
Average annual production in thousand PTA								
1901–05	1,324	1,037	4,622	160	804	211	0	8,157
1916–20	2,414	985	5,025	165	1,568	308	49	10,514
1929–33	3,837	1,082	5,767	273	2,853	1,459	296	15,567
Production average for each use as percentage of total production								
1901–05	16.2	12.7	56.7	2.0	9.9	2.6	0.0	100.0
1916–20	23.0	9.4	47.8	1.6	14.9	2.9	0.5	100.0
1929–33	24.7	7.0	37.0	1.8	18.3	9.4	1.9	100.0
Volume indices in hundreds cf. 1901–1905								
1901–05	100	100	100	100	100	100	–	100
1916–20	182	95	109	103	195	146	100	129
1929–33	290	104	125	171	355	691	407	191

PTA, pesetas.
Sources: GEHR (1989), pp. 105, 109 and 110, tables 3.8, 3.13 and 3.14.

Table 24.4. Production according to the different types of public forests, as a percentage of total production valued at current prices (Spain, c. 1920).

Types of forests	Timber	Firewood	Pastures	Acorn pastures	Esparto	Resin	Cork	Ploughing	Total*
Ministry of Finance (1)	3.1	5.7	58.1	1.4	3.0	1.5	0.6	23.8	97.2
District (2)	25.9	9.4	44.7	1.0	2.6	12.5	0.4	0.9	97.4
Managed (2)	40.3	3.0	10.2	3.2	0.0	37.6	3.6	0.5	98.4
Reforested (2)	19.4	14.1	46.4	0.0	10.9	0.0	0.0	0.8	91.6
Public utility (3)	30.2	7.6	34.3	1.6	2.0	19.9	1.4	0.7	97.7
Public forests total (4)	22.5	7.1	42.1	1.6	2.3	14.7	1.1	7.7	99.1

*The percentages do not total 100, because other uses, such as hunting, quarries and extraction of medical and aromatic plants, are missing.

(1) Forest year 1920/21.

(2) Forest year 1919/20.

(3) These percentages result from the addition of district, management plans and reforesting production.

(4) The percentages result from the addition of Ministry of Finance (1920/21) and public utility forests (1919/20) production.

Sources: The estimations have been carried out using books 23,469 and 23,470 from the Library of the Ministry of Finance and from Dirección General de Agricultura, Minas y Montes (1922), pp. 56–57, table 1.

this in spite of the very likely undervaluation of pastures. The typical forest products and by-products – timber, high forest firewood, resin, cork and bark – varied from 32% of public utility forest production in 1901–1905 to 55.4% in 1929–1933. On the other hand, forest uses such as underbrush firewood, pastures, esparto and ploughing changed, in the same years, from 68% to 45% of the cited product.

In Table 24.4, the relative weight (as a percentage) of each use in relation to overall production for each type of public forest in 1920 is presented. It must be noted that the forests managed by the Ministry of Finance were farther from forest activity as such and closer to agriculture and cattle raising, since pastures and ploughing totalled more than 80% of their production. District-managed lands were the most representative of all Spanish forests. Their high percentages of pastures and the fact that most of the firewood collected in them was brushwood made them similar to Ministry of Finance forests. However, the relative importance of their production of timber (over 25%), resin and high forest timber put them closer to forests resulting from Management Plan inspections, the most forest-like of all.

In managed forests, timber amounted to 40% of production value and at the same time nearly every other relevant use, such as resin, cork and high forest timber, was typical of forests. Although cattle rearing was still one of their main activities, it was, very probably, a residual and declining use.

Finally, reforested forests were mainly forest areas in the process of formation, only producing subsidiary products. Moreover, it is also clear that, in the period studied, reforestation aimed more at increasing the protecting function of woodlands, especially soil fixation, than at creating new productive surfaces.

Therefore, we can see that Spanish public forests adapted to the demand from forest markets, even if, more often than not, villages (the owners of the land) were against this movement. In this process, the role of the state should not only be analysed with regard to its direct influence in forest policy, but also in relation to its indirect activities, by selecting the beneficiaries among the different people or companies interested in public forest uses.

Notes

1 Public forest production growth rates are calculated from GEHR (1989), p. 110, table 3.14. In the case of agriculture production, the data are from the estimation of the agriculture production index which we carried out a few years ago (GEHR, 1987: pp. 420–421). In both cases, the rates were established from the average production, expressed in volume, of the periods 1903–1907 and 1929–1933.

2 The data from 1922 are missing because we have not been able to find the corresponding statistics, as well as those from 1922 to 1925, because in those years Ministry of Finance forests were also administered by the Ministry of Fomento and the source does not differentiate between this type of forest and public utility forests. The

figure for the production of esparto in 1931 is 0 (see Fig. 24.2). We have not shown in this figure the evolution of cork because its sharp variations could prevent us from seeing the development of the other uses.

References

Dirección General de Agricultura, Minas y Montes (1915) *Estadística general de la producción de los montes de utilidad pública, correspondiente al año forestal de 1912–1913*, Madrid.

Dirección General de Agricultura, Minas y Montes (1922) *Memoria sobre la estadística general de la producción de los montes de utilidad pública, correspondiente al año forestal de 1919–1920*, Madrid.

Dirección General de Agricultura y Montes (1926) *Estadística general de la producción de los montes públicos y apéndices a la misma, correspondiente al año forestal de 1923–1924*, Madrid.

Dirección General de Montes, Pesca y Caza (1934) *Estadística general de la producción de los montes públicos y apéndices a la misma, correspondiente al año forestal de 1931–1932*, Madrid.

GEHR (1987) Un índice de la producción agraria española, 1891–1935, *Hacienda Pública Española* 108–109, 411–422.

GEHR (1989) El sector agrario hasta 1935. In: Carreras, A. (ed.) *Estadísticas Históricas de España. Siglos* 19–20, Madrid, 91–129.

GEHR (1991) *Estadísticas Históricas de la Producción Agraria Española, 1859–1935*, Madrid.

Economic Policy and Forest Resources in Spain: the Case of the Province of Pontevedra, 1926–1956

25

Eduardo Rico Boquete

Universidad de Santiago de Compostela, Francisco Mosqueda 10, 4C, 28043 Madrid, Spain

During the first decades of the 20th century in Spain, the economic policy was a nationalist and protectionist policy, to achieve economic self-sufficiency. In this era (1900–1936), official reforestation began and, in 1926, by decree, a provincial consortium for reforestation was created and the species *Pinus pinaster*, *Pinus insignis* and *Eucalyptus globulus* were grown in the province of Pontevedra.

Franco's economic and forestry policy presents a lot of continuities with its predecessors related to the purposes and to the institutional and administrative apparatus used. It kept the purpose of production of softwood for the Spanish cellulose industry (Empresa Nacional de Celulosa (ENCE)) and other sectors. The strategic objectives were to guarantee the economic and political independence of Spain.

Between 1926 and 1956, more than 40,000 ha (100,000 acres) were reforested. In this process economic and industrial reasons were very important. The plantations offer a different evolution according to the different aspects which occur (climate, exposure, land). Problems arising are appearances of important plagues and illnesses, mycoses unknown here, and the spreading of fires.

1 Introduction

The industrial revival of wood pulp production by chemical methods, as cellulose, occurred subsequent to the Spanish Civil War. Until then, small

(eds M. Agnoletti and S. Anderson)

factories for man-made fibres, built in the 1920s, used pulp imported from northern countries.[1] A late period of prosperity had seen an important new dimension and annual increases were experienced in other countries, where cellulose production had already begun by the end of the 19th century. For example, if in 1924 world production was 6 million tons, a year later it was 7.5 million tons. This fact and the new applications which were daily being discovered for cellulose (man-made textiles, celluloid, varnish, paint, explosives, enamel, etc.) were good examples of the economic possibilities that cellulose production offered.

As is known, the economic policy developed by the different governments of the Restoration had a markedly nationalistic feel. This was shown in the attempts to apply a series of measures aimed at the protection and promotion of Spanish industries. The search for self-sufficiency and nationalization of raw materials would constitute another of the essential objectives of the following governments, more noticeable during Primo de Rivera's dictatorship but kept in place during the short republican period.

In 1901 the national market consumed 28,000 tons of paper; 15 years later the annual consumption exceeded 57,000 tons and was increasing. In those days, mechanical pulp was only produced by Papelera Española factories, producing about 15,000 tons a year, which was clearly less than the country's needs. Forest technicians were aware of this situation and put forward possible solutions to the problem while taking into account what had happened in more advanced European countries. They announced, adapted and fitted their suggestions into the prevailing economic situation, which was at the time of an increasing nationalistic nature. Forest engineer Octavo Elorrieta, for example, in 1916 affirmed that the problem of obtaining raw materials for mechanical pulp industries would always have a Spanish solution when the administration showed their will to solve it. But the situation in the cellulose industry was worse, because it simply did not exist; 'So, if we want to be self-sufficient and solve the paper problem by ourselves, we will have to begin obtaining chemical pulp.'[2]

Similar arguments would be expressed in the National Economics Congress in 1917 and in the National Engineering Congress, events marked by their nationalistic and supervisory features, where companies such as La Unión Resinera, Papelera Española, Central Papelera and others took a leading part. In the first meeting, an amendment proposed by O. Elorrieta to the report by Critóbal Massó called for 'economic nationalism' and asked the government to initiate the creation of a new Forestry Commission, which would be the starting-point to self-sufficiency in forest products. In the second event, reports by Antonio Ganuza, Gaspart Brunet and Nicolás Maria de Urgoiti picked up on the central objective of the meeting: the nationalization of raw materials.

The government's policy of promotion for the industry was based on the projects of investigation and experimentation which started to be formulated at the beginning of the century. The creation in 1907 of the Central Institute of Technical and Forestry Experiments, which followed the lead of other countries, had the objective of 'cooperating in the advancement of the forest industry'. This was the first step on an established path.

The National Institute for Forestry and Farming Investigation and Experiments was set up by Royal Decree (RD) on 22 October 1926. Its regulations (RD 24 March 1927) specified that its mission was, among other things, to cooperate in the advancement of the forest industry through the application of the results of its investigations and to verify the tests required by public and private corporations. In the organization and the objectives of the Cellulose Pulp Laboratory, the examples of large-scale cellulose investigation institutes were followed, such as those in Abo (Finland), Eberswalde and Darmstadt (Germany), Montreal (Canada), Milan (Italy), Bordeaux (France), Helsinfors (Sweden) and Madison (USA).[3]

In 1928, a law was decreed on 29 December that led to the end of the National Institute and set up the Forest Institute of Investigation and Experiments (IFIE); its rules were approved by Royal Order on 18 January 1930. The objectives and the cellulose section were kept. In this section the first tests were made and the first samples of viscose, man-made fibre and cellulose were obtained. They continued in their aim to reduce the costs of investigation, experimentation and application in order to facilitate private investment, which would guarantee the economic and technical success of the company, in short, to avoid risks and to ensure profits.

The main lines of investigation, in the case of cellulose, were directed at new species (*Pinus insignis, Eucalyptus, Pinsapo, Populus*). The economic profitability of specific production was studied (the suitability of the pinaster pine to make cellulose, required by the Council of Pontevedra), the most suitable methods were looked for to achieve a higher yield and better quality final product (analysis of the different systems of obtaining cellulose), tests were carried out with new machines (tests with the machines of Bertrand Ltd), engineers were sent to the most important centres of investigation (visits of Cejudo, Echeverría and others to the laboratories of Bordeaux, Grenolde, Eberswalde and Milan) and investigations for private companies were carried out (Santander Eucalyptus).

In relation to the country's industrial reality, the government's action was fixed on two main fronts: the announcement of measures to protect the industry and to initiate a policy of promotion. The first examples were the decree of the prohibition of wood export (RD 12–7–1921) and the decree of protection for the paper industry (RD 11–5–1928). At the same time, during the republican period, the commissions studying the pulp and paper problem would be created, and the companies and public companies interested would take part in them. So, in 1932 and 1933, the following organizations appeared: a Mixed Commission for the in-depth study of the manufacture and consumption of paper (Decree 3–12–1932) and a Mixed Paper Pulp Commission (Ministerial Order 27–11–1933).

The role of the Mixed Pulp Commission consisted of analysing the situation for making the necessary raw materials to manufacture pulp and the possibility of increasing its annual production and studying the possibility of building cellulose factories. What is more, they wanted to draw up an installation plan and a progressive development of this industry, which was to be carried out in a maximum period of 10 years, with the objective of giving a full supply to the paper industry.

The commission was made up of personnel from paper companies (Central de Fabricantes de Papel); representatives from the Forest Institute of Investigations and Experiments, the Department of Industry, workers, media companies and energy producers; the owners of pine woods and eucalyptus woods (Galicia, Santander and the south of Spain) and Provincial Corporations with high interests in the sector (Pontevedra, Guipúzcoa, Vizcaya and Navarra).[4] However, all their work was stopped when the Civil War (1936–1939) began.

2 Afforestation and industrial projects before 1936

The first projects which proposed the obtaining of chemical pulp from wood in Galician forests started in 1927, and in the 1930s negotiations to build a cellulose factory in Pontevedra were started.

The first information we have of project talks involved a group of entrepreneurs who had started negotiations and studies into the building of a cellulose factory. Their negotiations took them abroad to see similar factories, which seems to indicate that their efforts were serious. However, it seems that they had the will but not the money, and this meant that the project was suspended.

On the other hand, in 1910 the Forest District of Pontevedra had already started the afforestation of several woods with species of quick-growth trees, above all conifers, which are suitable for the manufacture of cellulose. Also, due to the success of this initial afforestation, Pontevedra Council started the afforestation of its woods with the same species in 1926, with the aim of creating an important mass of wood capable of supplying a big cellulose industry; it even considered the possibility of taking part in the future industrialization of wood products. The afforestation, initiated by Daniel de la Sola, President of the Council, was planned and carried out by engineer Rafael Areses. In this project, which foresaw the plantation of 25,000 ha in the province, there was space for a lot of species. However, the Council estimated the afforestation should have an economic aim, so it specified that the species used should be of quick growth and suitable for the production of paper pulp. It was not about making extensive woods with a great vegetative variety; it was about growing a mass of high-productivity trees, which would be easy to cultivate and produce high economic profits.

Afforestation was carried out after establishing a new collaboration agreement, called a Consortium, between the provincial council and the town halls involved. It is possible that the engineers and the council knew of the existence of similar agreements in some places in Italy (Consorcio Provinciale di Rimboschimento), which were possibly introduced here by the famous Florentine worker Aldo Pavari, who travelled to Spain in 1923.

The future industrial use of these plantations determined the site, with the first afforestation being carried out in forests near Pontevedra, due to its good communication and transport network and its important commercial harbour nearby.

The Managing Commission of the provincial corporation studied, in its meeting on 24 September 1927, an installation project for a cellulose factory required by the corporation and carried out by engineer Ignacio Echeverría, from the cellulose section of the IFIE.[5] The conclusion of the investigation was emphatic. The viability of the factory was undeniable and with it the possibility of the nationalization of cellulose becoming a reality. Echeverría planned to build a factory with a capacity to produce about 10,000 tons of chemical pulp, using as raw material the pinaster and *Pinus insignis* and the existence of an abundance of water with good land and sea links. The machinery would be built abroad, in specialized workshops in Sweden, Germany or England, and the assembly would also be carried out by foreign technicians. There would also be the possibility of making specialized machinery for wood (cutters) and auxiliary tools in Spanish workshops. In any case, the engineer established very clearly that this concept would take most of the foreseen budget, with the warning that tariffs and exchange variations could alter the final cost of the machines. In short, according to the forest worker, the industrial capital necessary to establish a factory capable of producing 33 tons of cellulose a day, caustic soda and an installation for the recovery of bleaches, could be divided into the headings shown in Table 25.1.

At the same time, Echeverría considered the possibility of completing the productive cycle with the transformation of the pulp into paper in the same factory, so a machine would be needed to make paper. This would mean an increase in the budget of 1,300,000 pesetas. This increase could be compensated by the higher added value of the final product.

Table 25.1. Echeverría's calculations of cost (in pesetas).

Machinery	3,480,000
Buildings, land and water piping	800,000
Boilers, engines and accessories	500,000
Customs, transport and assembly	900,000
Taxes	300,000
Unforeseen expenses	300,000
Total fixed capital	6,280,000
Total circulating capital	1,250,000

In the preliminary plan the fundamental parts of the future cellulose industry were present: the usage of conifer wood, the location of the factory near Pontevedra, the obtaining of the machinery abroad, the whitening of the pulp, the participation of the administration and, importantly, the concern about finishing the productive cycle.

As we have said, the plan could not be carried out, among other reasons, because of the lack of private and public capital and the political changes that occurred in the country with the disappearance of the dictatorship and the arrival of the republic (1931). However, the new Republican Provincial Administration did not abandon the forests which had been created or the aim of industrialization of forest products. So, in 1934, it asked the Forest General Management if IFIE could study the possible industrial applications of the plantations arising from the afforestation, which, including only that carried out by the council, consisted of 6000 ha. The corporation was worried in case the word got out to the market, as there could be a general fall in prices or such a saturation that it would force a complete stop to afforestation in private forests, a situation similar to what had happened with the forests in Santander a few years before.

At the same time, trying to continue the collaboration with IFIE through Ignacio Echeverría, the council put a proposal to the Forest General Management, suggesting that the said engineer could make a study trip abroad in order to find out more about the latest advances in the investigation field and in the industrial application of the production of cellulose. The civil conflict stopped the experiment now under way and it restricted the aspirations of the Provincial Administration. Nevertheless, at the end of the civil conflict, interest would be shown in achieving its aims again, either directly or through private companies.[6]

3 The industrial policy of the new state

The policy implemented in the Franco period in 1939 had, as a principal reference, that carried out by Primo de Rivera's dictatorship, to which the influence of the fascist powers' – Germany, Italy and Japan – autocratic policies were added. This meant an emphasis on ideas of nationalism and power and, as a consequence, all the industrial sectors were forced to cover their needs for raw materials, and they had to contribute in supplying the market with their final products, above all in the biggest deficit areas and products declared as in the national interest. The first legislative measures, statements from the highest responsible ministers, technicians and economic sectors, insisted on the need to create an economic and industrial structure with an autarchic content and to be perfectly regulated. It was a voluntary autarchy for war and peacetime situations and for normality in international relationships. In the legislative field we have examples in the creation of the Regulatory Commission of

Chemical Industries, which was established 'to achieve an autarchy that completely satisfies our needs, making us independent from foreign markets' (OM 30–10–39). The same could be seen in the Protection and Promotion Law for the national industry, proclaimed on 24 October 1939, and in the decree on 15 March 1940, which declared the manufacturing of man-made textile fibres as of national interest in order to reinforce this idea.

But it was not just about providing for national consumption; it was also about the possibility of turning the country into a big exporter of certain products and this is what the theory was from the start. In relation to cellulose and its by-products, it would soon be confirmed that, in the near future, we would be in a condition to make important quantities available on the international market. In this way, due to the high profits that its production produces, it would be possible to contribute effectively to adjusting the balance of trade. According to Alarcón de la Lastra (1940),[14] the challenge consisted of being able to generate a productive, cheap and quality surplus, which could be competitive on the world market. What Alarcón and others emphasized was that the proposed autarchy was not complete in nature. His stand-in deputy, Carceller (1943),[15] confirmed that autarchy should affect what he called 'industrial economic liberation': chemicals, textiles, machinery, paper and wood. It is not possible to forget the relationship between self-dependency concepts, the totalitarian character of the government, as proudly proclaimed by politicians, its military existence and the warlike situation in Europe; in other words, before taking part in a military conflict, autarchic control was imposed.

In relation to cellulose and the sector of man-made fibres, politicians and industrial technicians and foresters, such as Paris Eguilaz, Antonio de Miguel, Granell, Creus Vidal, Gay de Montellá, Echeverría, Robert Robert, Abollado, Navarro Sagristá and others, would keep the ideal autarchy during the first decade of the regime. In the same way, publications of an economic, industrial and forest character were enthusiastic over the new economic direction.

As in previous years, the government would try to stimulate the implantation of this industry through various methods, which would be established according to its principles and to its political and economic objectives.

On the matter of investigation, experiments with native and exotic forest species had been started by IFIE in 1927 with the aim of finding high and quality production levels that would be capable of generating high-quality cellulose. For this, the cellulose section in the IFIE was strengthened, now being directed by Ignacio Echeverría, and they would make contact with the Superior Council of Scientific Investigations (CSIC), to pool efforts towards the same goal in order to avoid a duplication of the investigation. At the same time, the creation of the so-called Regional Centre of Teaching and Investigation and Forestal Experiments was decided on. It was set up in Lourizán (Pontevedra), just opposite the site that, years later, would be occupied by Celulosas of Pontevedra. In the order of creation, on 25 February 1943 it was stated that the centre would be to complete the teachings of the forest school and to spread IFIE experimentation. The wet areas of the north and north-west of Spain 'were highly suitable for the study of quick-growth forest species, the increase of which is of the utmost national need'. So it was clear that they wanted to try out everything related to quick-growth species, from the creation of plantations to industrialization.

For the leaders of the new government and the Forest Administration, the increase in afforestation activity must be by the government and only the government. With its economic, technical and coercive capacity, it had the power to ensure creation in a short period of time and the necessary tree masses to produce cellulose in the best conditions. They had the example of what had been done during Primo de Rivera's dictatorship, with the contributions of the republican period and with what the Italian government was doing, especially with its afforestation work with black poplars in the Po basin. Now the main implemented measures would be exposed:

1. Re-creation of the Forestry Commission (PFE) and the recuperation of consortia as an ideal way to carry out afforestation, which made possible the concentration and centralization of raw materials under the government's control (D 26–8–1939; Law 10–3–1941). They did not pursue a wealth of land but rather a wealth of timber.
2. Prohibition on the export of conifer seeds, together with *Cupressus* and *Eucalyptus* seeds (OM 3–12–1941). With these the foreseen afforestation would be done; obviously the export of autochthonous leafy tree seeds was not banned.
3. Decree of 3–9–1941. The declaration of areas of forest interest in those areas suitable for the production of cellulose species, for example, Guipuzcoa and Huelva.
4. Law of 18–10–1941, about afforestation in rivers and on river-banks, which, apart from strengthening the fluvial river-beds, also produced wood in a short period of time.
5. Order of 25–1–1952, creating the National Commission for the Black Poplar, with the objective of promoting and rationalizing the plantation of black poplars everywhere; the work was carried out by the PFE (the Forestry Commission).
6. Order of 25–1–1952, to create a commission to draw up a coordinated plan of afforestation in forested areas or in areas suitable for afforestation with quick-growth species and a programme for cellulose industries and paper pulp production. This order was given by the government's president.
7. Order of 23–5–1956, creating a National Commission for Eucalyptus, with the same aim as for the black poplars.

On the other hand, the importance given to textile cellulose made the creation of the National Industrial Cellulose Plan possible and its manufacture was declared to be of national interest (D 15–3–1940), stipulating a series of benefits for those companies that began manufacturing it in a complete cycle and using national raw materials.

4 The regime's forest policy: afforest, afforest and afforest

The title of this section is the forest version of the known slogan of the dictator: 'produce, produce, produce'. Produce until 200,000 ha a year of afforested surface was reached – that was the wish he expressed to the Minister for Agriculture. The State Forestry Commission would be instrumental, ready to act, and the so-called public utility forests and others not catalogued were designed to bear the brunt of afforestation activity.

As stated before, the technicians and people responsible for the forest had assumed the autarchy orders and were ready to start. To achieve cellulose autarchy, it was necessary to carry out massive afforestation of public forests in the Atlantic and Cantabrian areas, that is to say, those areas in which quick-growth species could have a greater development and which had been under investigation for some years. All of the afforestation in these regions would be carried out quickly with the participation of the councils and town halls, using conifers and *Eucalyptus*.[7]

The Council of Pontevedra and the PFE, on whose Board of Directors was Daniel de la Sota, agreed on the envisaged problem and, in 1940, they signed an afforestation consortium with the objective of creating enough raw material to supply the cellulose industry. The engineer of the Council Forest Service did not forget that the use of pine woods set up by the public and private afforestation

> could give a patriotic service of great importance, in being the first Spanish test on a large scale in the liberation of our foreign dependency . . . it could be the starting point in establishing a precedent for the installation of a productive establishment of chemical pulp.[8]

A month later the council reminded the PFE that: 'The coastal area with its maritime climate is suitable for the quick-growth species and will offer an abundance of raw material for industrialization to produce pulp for paper.'[9]

The signing of the consortium between the Provincial Corporation and the PFE was at the end of 1940 and it had a series of clauses which meant the centralization and concentration of future plantations under control of the State Forestry Commission, because, according to the basis of the consortium, the created timber capital would belong to the PFE, which gave part of its value to the other consortium members in accordance with their contributions. That is to say, the Forestry Commission had all the power to do and undo what they wished in relation to the use of the pine and eucalyptus woods. In this way they could be sure that woodland work and fellings would be done in the way they wanted them done. In other words, the wood would be given to whoever they decided to give it to, without having an auction; from all of this Celulosas would benefit.

Afforestation started in 1940/41 in areas which had already been afforested in the 1920s, that is to say, extending the area of those plantations. During the first 10 years the afforestation process suffered some ups and downs, produced by budgeting difficulties and by problems arising between the council and the PFE. The consortium and the afforested hectares of the first decade are shown in Table 25.2.

The existing problems between the two institutions, which in some years meant the complete paralysis of joint afforestation work, were solved partly by the capacity of the Forest District to join hands with town halls, leaving the council out of it. Despite this the figures reveal important variations in the annual afforestation. At this stage, the meeting of the Galician Agricultural–Forestal Congress stands out, held from 25 to 30 October 1944 in Compostela. In Pontevedra's report it said: 'In the Province of Pontevedra, afforestation has been aimed for a long time at obtaining forest masses with quick-growth species, with a view to exploitation to obtain cellulose.'[10]

At the beginning of the 1950s, the Ministry of Agriculture showed interest in asking the National Institute for Industry (INI) for a copy of the Italian IRI. It planned to install a cellulose factory in Galicia with a capacity to consume more than 150,000 m^3 of wood a year. For this, the INI created, in 1951, the Cellulose Management Commission (CGC). This organization, directed by forest engineers, would be chosen to plan and install a big cellulose factory in Pontevedra. Nevertheless, the government, through the National Forestry Commission, had already started a vast programme of afforestation in forests near the coast. Even so, after the creation of the CGC, the afforestation phenomenon experimented with big increases, because it had asked Forestry for strong action to afforest and to make an inventory of the existing plantations.

Table 25.2. Pontevedra: consortia and afforestation, 1941–1950.

Year	Consortium (ha)	Afforestation (ha)	Total afforestation (ha)
1941	13,695	1,079	1,079
1942	26,353	2,824	3,903
1943	3,987	2,127	6,030
1944	11,719	2,514	8,544
1945	2,331	3,305	11,849
1946		4,777	16,626
1947	1,849	1,030	17,656
1948	1,659	843	18,499
1949		1,067	19,566
1950	3,756	595	20,161

Source: DGCONA Archive, Consortium Section.

What was done in the years following the creation of the CGC was the formulation of the cellulose project, from 1951 to 1956, as shown in Table 25.3.

It is easy to appreciate the magnitude of the forestry partnership, concentrated in a few key years. From 1953 it can be said that all the effort was concentrated on the afforestation of the lands; that is to say, what the PFE was really interested in was obtaining the timber capital. At the same time they tried to involve private initiative in afforestation work, with the promulgation of a law on 7 April 1952 about assistance for free initiative of afforestation in public and private lands. Even so, the results were not very good or at least not as good as was expected.

From 1940 to 1971 afforestation was carried out using two conifers and a leafy tree: *Pinus pinaster*, *Pinus insignis* and *Eucalyptus*. Other species, such as various pines and acacias, were not used. Of course, those other trees were of slow growth and therefore of small profit in a short period of time, and so would not be planted. Until 1955 the Forestry Commission had reforested 26,255 ha, of which 22,908 ha had been afforested with pinasters and the rest, 3347 ha, with *P. insignis*.

For example, let us take a look at the list of plants produced in the fixed and temporary nurseries installed in the province (Table 25.4). In Table 25.4, we can see how *Eucalyptus* accounted for 90% of the leafy trees obtained in that year, a percentage which was maintained and exceeded in the following years, as mentioned before. This increase would influence, years later, the progressive increase in the use of eucalyptus in the Lourizán factory.[11] At this moment, as some engineers reported, nearly all the eucalyptus obtained in Galicia was sent to factories built elsewhere. This was not the only inconsistency they criticized: 'It is an economic incongruity that the wood factories in Galicia are, in a way, basic and not transformer industries.'[12] That is to say, the productive cycle should be completed in the same region because afforestation should be done with species that are 'in keeping with the tendencies of present industrialisation', the most appropriate thing being to

Table 25.3. Pontevedra: consortia and afforestation, 1951–1956.

Year	Consortium (ha)	Total consortium (ha)	Afforestation (ha)	Total afforestation (ha)
1951	2,297	2,297	1,153	1,153
1952	11,043	13,340	6,159	7,312
1953	41,338	54,678	7,299	14,611
1954	588	55,266	8,255	22,866
1955	735	56,001	4,870	27,736
1956	450	56,451	3,392	31,128

Source: DGCONA Archive, Consortium Section.

Table 25.4. Pontevedra: plants produced in 1955.

Species	Amount
Pinus pinaster	16,704,000
Pinus insignis	5,400,000
Pinus sylvestris	5,878,000
Pinus laricio	135,000
Cedrus deodara	240,000
Cupressus macrocarpa	300,000
Eucalyptus globulus	979,000
Quercus rubra	80,000
Total	29,716,000

Source: DGCONA Archive, Consortium Section.

promote the industries capable of using the wood in a complete productive cycle.[13] The afforestation done by the Forestry Commission affected all the town councils but with varying intensity. Of the hectares planted by the PFE up until 1955, 27.25% corresponded to the provincial capital and to ten bordering municipalities; this proportion is 40% for the plantations of *P. insignis*. The plantation by the PFE was 97% in the forests of public utility; the other 3% was in non-catalogued lands, which were free to be used by town councils. The private owner's participation was almost non-existent, because the conditions offered by the PFE were not very appealing and they meant the creation of a joint ownership in consortium forests.

In short, when the INI approved the project of a cellulose factory in Pontevedra (1956), the obtaining of the raw material was guaranteed and conifers and eucalyptus increased in their expansion in the following years, which has produced a deep and irreversible change in the floral composition of the province.

5 Conclusions

From the first years of the 20th century, Galicia appears to be one of the areas which has had the best conditions to become a big area for expansion of the cellulose industry and its by-products. In Galicia, Pontevedra seems to have been the ideal place to build it: good communication links, excellent forest mass growth and abundant water resources are its best resources.

The work of the council of Pontevedra, afforesting with the most suitable species and in great mass, promotion and investigation, together with the presentation of the first project of a cellulose factory, are good examples of the good will of the public and private authorities in the area, and it shows what would happen three decades later. Once the objective conditions were created, the installation of the factory was just a matter of time, although in this case the economic and political events in the 1930s and the regime after the civil war acted as delaying elements.

When the civil conflict ended, the self-styled New State had the aim of achieving autarchy, which would guarante economic and political independence, showing itself a loyal follower of the economic theories of its mentors, German Nazism and Italian fascism. The achievement of this objective was especially desirable in those productive sectors qualified as strategic, that is to say, industries for national liberation, as they were then called. That is how cellulose was qualified, because of its multiple and very important industrial and war applications as it was the basic material for the production of man-made textiles, paper, varnish and nitrocellulose. At first, they tried to stimulate the obtaining of textile cellulose with the objective of eliminating, little by little and then completely, the importance of cotton, wool and any other natural fibres which could be substituted by an artificial fibre. This is how, in 1940, Industry General Management formulated the National Cellulose Textile Plan; they increased support and help to private industries, which was not enough to achieve the plan's success.

On the contrary, what they achieved was a serious failure because, by the mid-1950s, the indices for consumption and production of all kinds of cellulose were lower than 20 years before and, even worse, it could be seen that if things carried on like this, the situation would not be easy to overcome in the near future. The confirmation of this failure was the best argument the INI had to justify the necessity of its Multiple Cellulose Project.

In this situation, Franco's authorities concluded that the creation of an important cellulose industry in Spain would only be possible with the direct action of the government. And that was the mission given to the INI, to carry out the constitution of the Cellulose Management Commission and Technical Council. The Commission, at the beginning of the future New Company Management, was directed by a forest engineer, who had been the General Forest Director, and it was formed of industrial and forest engineers linked to an organization structure of the Forestry Commission and Industrial General Management, who were supporters of substituting imports. The commission conceived Celulosas de Pontevedra as a factory of great productive capacity: public, modern and competitive abroad, antimonopolistic, non-integrated and the supplier of pulp to a group of paper industries that the commission had qualified as non-productive, old and parasitic.

From 1940 and as part of the measures taken to reach autarchy, the regime decided to carry out a big campaign of afforestation in Galician forests with quick-growth species to supply the necessary raw material for industries that would probably be built in the near future. In the case of Pontevedra, it meant carrying on with the afforestation plan started by the Council in 1927, with the same aims but with greater intensity, which was increased at the beginning of the 1950s, when the possibility of the government building a cellulose factory was starting to be seen. The tendency to create big monospecific forest plantations of intense productivity increased, with the predominance of the conifers, which were treated as real forest crops of low-quality wood.

Finally, the projects of afforestation and the projects of setting up cellulose industries were always interrelated, feeding off each other.

Notes

1 The most important were the following: La Seda de Barcelona, La Sociedad Anónima de Fibras Artificiales de Blanes and La Sociedad Española de la Seda Artificial de Burgos. They imported 3500 tons of cellulose a year and they produced 1500, 1300 and 500 tons a year respectively. In 1906 La Sociedad Española de la Seda Viscosa was created in Barcelona (Tamames, R. (1985) *Estructura Económica de España*. Alianza Editorial, Madrid.).

2 Elorrieta Artaza, O. (1916) El papel y los montes españoles, *España Forestal* 17, 185–187.

3 Apart from following its organizing model, in the tests the rules and criteria of Reichauschuss für Liefer Bedingung (RAL) and the Bureau Internationale pour la Standardisation des Fibres Artificielles (BISFA) would be used.

4 Something to take into account is the relationship between public organizations and the sector companies. Thus, Fernando Baró, Simeón de Pedro and José Lillo, members of IFIE, also belonged to the technical body of Papelera Española. O. Elorrieta, in 1928, the first Forest General Director, was always linked to wood companies, would be the founder and president of the Spanish Chamber of Wood, a pressure and defence group for the interests of the employers' organization of the sector members of the National Economy Council.

5 Pontevedra's council archives, session records 24-9-1927. This forest worker would be the most important specialist in cellulose and his biography is a good example of this. The first study was done abroad after being given a grant by the Allowance Board and this gave rise to his first book. He later carried out investigation work for IFIE and became a teacher of the subject in a specialist school for wood engineers.

6 Echeverría was one of the most important international technicians, because, in 1929, he had already been in Edinburgh (Scotland) checking out the functioning of the machinery ordered by the cellulose section of IFIE from Bertrands Ltd. He also visited the Tullis Russel Co. cellulose factory and the Imperial Institute Laboratory of London (AGA, section of Agriculture, Ignacio Echeverría – personnel dossier – C^{a} 5924).

7 This tree had been defined by Aldo Pavari (1938) as the autarchic tree *par excellence* (Pavari, A. (1938) Selvicoltura naturalística e selvicoltura autarchica, *Estratto dagli Atti della Reale Accademia dei Georgofili*, XVII, Florence).

8 DGCONA Archive, Consortium Section, Pontevedra 1, Council of Pontevedra's Triennial Work Plan, 20-10-1939.

9 DGCONA Archive, Consortium Section, Pontevedra 1, Letter from the Council of Pontevedra's President to the PFE General Director, 22-11-1939.

10 An Immediate Intensification Plan for the Development of Afforestation up to a minimum of 5000 hectares a year in the Pontevedra Province. Development of the wood industry and promotion of fluvial fishing, originally typed, p. 2.
11 Prada Blanco, A. (1991) *Montes e Industria. O Circuito da Madeira en Galicia,* Fundación Caixa Galicia, Vigo.
12 Consejo Económico Sindical Provincial (1968) *Ponencias y conclusiones del III Pleno de Pontevedra,* Vigo, p. 313.
13 Consejo Económico Sindical Provincial (1968) *Ponencias y conclusiones del III Pleno de Pontevedra,* Vigo, p. 292.
14 Alarcón de la Lastra (1940) *El Triunfo Nacional y su Repercusión en las Orientaciones de la Industria y el Comercio.* Conferencia pronunciada en el Instituto de Estudios Políticos, Madrid.
15 Carceller Segura, D. (1943) *La Situación Económica en España,* Dirección General de Comercio y Política Arancelaria, Madrid.

26 The Impact of Five Forest Commissions on the History and Practice of Forestry in British Columbia, Canada

Paul Mitchell-Banks

Central Coast Consulting, 4439 West 7th Avenue, Vancouver, BC V6R 1X1, Canada

Industrial forestry is a relatively new activity in British Columbia, having started approximately 125 years ago. The Fulton Commission (1909–1910) was established as a result of growing public concern about the alienation of forest land from the Crown (government) and public fears over deforestation and inadequate forest management. This commission resulted in the establishment of a Provincial Forest Service, the establishment of forest inventories, as well as the design of new tenures (licences) by which the government assigned harvesting rights to the private sector.

The First Sloan Commission (1943–1945) resulted from growing industry frustration about secure fibre (timber) supplies for their mills, along with concerns about the unbalanced pattern of timber harvesting and inadequate silviculture and forest management. This commission was noted for significant tenure reform and incorporating the concept of sustained yield into forest management.

The Second Sloan Commission (1955–1956) was established to monitor the changes from the first Sloan Commission and determine where policy should be modified. There was once again criticism of inadequate forest management. There was concern over forest industry concentration.

The Pearse Commission (1974–1976) was established as a result of continuing concern about the inadequacy of forest tenures and the ability of the Ministry of Forests to effectively manage the forest resource. Arguably the most important of all of the forest commissions in British Columbia, Pearse was critical of the degree of industrial concentration and the inadequate forest management being undertaken. Pearse argued for the Ministry of Forests and Ministry of the Environment to be joined, and for practices such as integrated resource management to be undertaken.

The Forest Resources (Peel) Commission (1989–1992) resulted from growing public concern over the state of the forests, the concentration of harvesting rights and processing facilities in the hands of the major forest companies and a move by the government to convert the majority of the tenures to a land-based and long-term licence. This commission made a number of recommendations addressing the lack of competition within the forest industry, inadequate forest management and the need to undertake large-scale strategic land-use planning.

While each one of the forest commissions has made significant contributions to the understanding of the forestry challenges faced in British Columbia, there has been a reluctance on the part of the government and a strong resistance on the part of industry to implement a number of critical recommendations. This chapter will review the history of the development of forestry within the province and critique the commissions and the resulting government policies.

1 European settlement: the period 1865–1909

At the time of early European settlement in British Columbia, the forest seemed to be eternal and much effort was spent in harvesting the trees to stimulate the economy as well as to convert some forested land into farmland (Godwin, 1994). A concerted effort was made by the colonies, and later the provincial government, to encourage settlement and economic development. The principal policy tools employed by the governments were Crown land grants in the earliest years of settlement, and later the granting of resource rights without alienating the land through a Land Ordinance issued by the Governor of Vancouver Island (Pearse, 1976).

The 1865 Ordinance empowered colonial administrators to issue tenures known as timber leases to individuals or companies engaged in lumbering. The scale and scope of these timber leases were left to official discretion, and initially the Crown retained no financial interest in the timber (Pearse, 1976). Between 1865 and 1905, three other forms of tenure were devised: timber licences; pulp licences and pulp leases – designed to attract investment into the pulp industry (Pearse, 1976).

In 1883–1884, a grant of 1.9 million acres of forested land on Vancouver Island was given to support the construction of the Esquimalt and Nanaimo (E&N) Railway. The 'railway belt' grant of 14.5 million acres to the Dominion (National) Government was made at the same time as the E&N grant, though this land was to support the completion of the transcontinental railway through British Columbia. As part of the federal land settlement policy, the Dominion government issued 'timber berths' – forest tenures giving rights to cut the timber without alienating the land from the Crown.

Timber and pulp leases, timber and pulp licences and timber berths comprise what are commonly referred to as the 'old temporary tenures'. Each form of old temporary tenure was designed to meet a particular need, but they have main elements in common. The tenures conferred rights to harvest the existing crop of original old growth and, as this crop was removed, the tenures reverted back to the Crown. To maintain their tenure rights, the tenure holders had to pay annual rents (for timber and pulp leases) or renewal fees (for timber and pulp licences and berths) and royalties on all timber harvested. Since 1907, timber cut from the old temporary tenure areas has been subjected to provincial export restrictions (Pearse, 1976). Finally, the old temporary tenures could not be transferred without ministerial consent (Pearse, 1976).

Timber speculation began to increase around 1905. In the USA, timber was being rapidly exploited and loggers began to travel further north in search of virgin stands. Amendments were made to timber licences (also referred to as special timber licences) to make them more attractive to speculators (Pearse, 1989). The use of these new timber licences, which were renewable annually and completely and freely transferable, led to a large increase in timber-staking activity. Over a 4-year period, staked claims rose from 1500 to over 15,000, a tenfold increase. While the government welcomed the revenue, concerns grew about the timber commitments being made (Shelford, 1993). By 1907, approximately 10 million acres of Crown forest land had become committed under the four forms of forest tenure, with 90% of this being under timber licences. The government believed that the volume of timber under tenure would satisfy industry requirements for years, and suspended any further tenure allocations (Pearse, 1976).

2 Fulton Commission, 1909–1910

The forest was not endless, and the markets were certainly not. Growing public concern over the state of the forests and the problems prevalent throughout the forest industry were driven by strong memories of an earlier period when hysteria over another public resource had created significant social, economic and political changes. The timber staking was similar to the earlier gold rush, which:

> had precipitated profound political changes in the region half a century earlier. Alarmed by these developments, the government appointed its first Royal Commission of Inquiry into forest policy, which produced the Fulton Report of 1910. With the meagre information available to them, the Commissioners estimated that the province had already alienated two-thirds of its merchantable

> timber, and after concluding that this would meet the needs of the industry for several decades they prudently recommended that the remainder be held in reserve. To meet minor and special needs, they proposed competitive, short-term timber sales, variants of which are among the most important tenure forms in use today. Many other significant changes resulted from the recommendations of this influential Commission, including the passing of the first Forest Act in 1912, which provided for a provincial Forest Service and embodied the first significant efforts toward forest protection and management.
>
> (Pearse, 1976)

The Fulton Commission and its report represented a remarkable change in attitude towards the forests. Until the establishment of the Forest Service there was very little comprehension of the extent or status of the resource. Fulton was the first to recognize the need for a good forest inventory as a basis for responsible management. He also suggested that forest revenue should be treated as capital and reinvested in the forest land base (Shelford, 1993). Successive governments have rejected this, with forest revenues funding the provincial general revenue account.

3 British Columbia's first Forestry Act, 1912

The Fulton Report recommendations were well received and many were incorporated into the first Forest Act, passed in 1912. The Act created the Forest Service and also introduced new licensing arrangements, with the most important being the creation of the timber sale licence (Pearse, 1976). Areas of timber that were to be sold were surveyed, cruised (inventoried) and the timber was classified as to quality, species, etc. Timber sales were advertised in the *British Columbia Gazette* for a minimum of 3 months and competitive bids were made by way of sealed tenders accompanied by a deposit of at least 10% of the bid price (Pearse, 1976). The timber sale licence introduced another legislative change to forestry in that each licence was subject to special conditions, which addressed proper harvesting, protection and forestry practices. The terms of the timber sale licences varied, but the most common term was for 2 or 3 years (Pearse, 1989).

4 The period 1912–1942

Between 1912 and amendments to the Forest Act in 1947, timber was cut from much of the land that was alienated prior to the Fulton Commission and the majority of this land reverted back to the Crown through timber removal, default on taxes and rentals or for other reasons (Pearse, 1976). During these 35 years, which saw two world wars and the Great Depression, there was little change in tenure policy. By the middle of the 1940s, the only significant method for obtaining new timber rights was through the short-term timber sale licences, a system that was in widespread use. By 1945, the timber sale licences accounted for over half of the interior lumber harvested and approximately 25% of the entire provincial harvest (Pearse, 1989). While the government had prohibited the transfer of forest land into private ownership, there was no legal way to regulate or restrict the annual timber harvest, which had increased significantly. There was also no legislation addressing the restoration and maintenance of the productive capability of logged-off lands to ensure a sustainable forest industry (Drushka, 1995).

5 The first Sloan Royal Commission on Forestry, 1943–1945

Industry found the system frustrating and considered it inadequate to provide the secure feedstock supply that they required to invest in new processing facilities. This, along with concerns about the unbalanced pattern of timber harvesting and inadequate silviculture and forest management (Pearse, 1976), led to the first Sloan Royal Commission on Forestry in 1943–1945.

The first Sloan Commission is noted for significant tenure reform and incorporating the concept of sustained yield into forest management. To achieve this, Sloan proposed two new types of sustained yield management units. The 'private working circle' enabled owners of Crown-granted land and old temporary tenures to combine their holdings, to which would be added additional Crown timber land. This would, theoretically, create a coherent management unit that would be subject to sustained yield management by industry, with the Forest Service ensuring that public concerns and safeguards were addressed.

The second sustained yield management unit was a 'public working circle', which incorporated all provincial Crown forest lands not taken up in the private working circles, with the land being managed by the Forest Service and serving the needs of the smaller and unintegrated forest enterprises (Pearse, 1976). The government incorporated most of Sloan's recommendations in amendments to the Forest Act passed in 1947 (Pearse, 1976).

Beginning in 1942 and throughout the first Sloan Commission, C.D. Orchard, the province's Chief Forester, had an agenda to bring about sweeping changes to the provincial Forest Act (Drushka, 1995). Orchard proposed a tenure called a forest management licence (FML), which would see the land still owned by the Crown but the timber harvesting rights being sold to a forestry company, which would also be subject to reforestation and management obligations. The intent was to open up large tracts of forest land to the companies without either alienating the land from the Crown or having the inadequate reforestation and forestry management practices of the past continue.

In his Royal Commission report, Sloan made reference to the debate of private versus public forest land ownership:

> The License area is a Tree Farm; who owns it, who manages it, now or in the future, is of secondary importance, provided it is managed with ability, interest, and imagination.
>
> (Mahood and Drushka, 1990)

It is unlikely that Sloan anticipated the concentration of corporate control over the FMLs. In his report he makes reference many times to the importance of the forest being managed to address community concerns and the dangers of overharvesting:

> Areas of reverted land situated in or near settled communities could also be managed on a sustained yield basis as public working circles by municipal authorities, subject to regulations designed to prevent improvident future management and transactions in relation thereto. . . .
>
> Ghost towns in the interior bear distressing and silent witness to the past policy of too many mills cutting out areas that could have supported in perpetuity, on a system of planned management, the potential capacity of probably half of them.
>
> (Sloan, 1945)

Sloan was sensitive to the growing public concerns of forest industry concentration and control of the resource, with MacMillan Company's controversial purchase of the Victoria Lumber and Manufacturing Company in 1944 undoubtedly giving him pause to think. He directly addressed this in his final public session:

> The position [MacMillan takes] is that this is our timber, and we propose to do what we like with it. There is no law in the country to prevent me from buying a mill, closing it down, and disposing of my timber as I see fit. That is the privilege the buyer has. Now should the present state of the law be continued from the point of view of an economic policy is a question including many factors. Should a small community suffer because of modern trends concentrating industries in large cities – that is something to be considered.
>
> (Drushka, 1995)

Sloan raised concerns over very large-scale or regional forest control in his report. These concerns anticipated the potential for conflict between regional and local forest planning concerns and needs, and the potential for local needs not to be met within a system of forest management and control that was ultimately driven by regional needs:

> I cannot subscribe, with respect, to the suggestion that the entire coast be treated as one working-circle and that the overall cut be kept within the yearly or periodic increment of that production unit then the whole Coast will be on a sustained yield basis. If, on the other hand, individual working circles are over cut the Coast will not be on a sustained yield basis notwithstanding the fact that the overall Coast production remains within the total Coast increment.
>
> (Sloan, 1945)

Mahood and Drushka (1990) argue that Sloan believed that the practice of long-term dedicated forestry found in the private forestry of Europe and the USA would also develop on public land held under tenure.

6 The 1947 Forest Act

On 3 April 1947 the amendment to the Forest Act was passed and forest management licences became the only means by which the forest companies could enter into management arrangements on Crown land and gain access to the private working circles.

In 1948, an amendment was made to the Forest Act to introduce the farm wood-lot licence. This forest tenure was to provide sufficient Crown forest land for farmers to yield a maximum of 10,000 cubic feet per year, including that timber harvested from the farmer's private land. The licence was intended to provide timber for winter harvesting employment for the farmers and to encourage management of their timber lands. The licence was not transferable and was subject to stumpage. It was a good idea but one that did not prove to work effectively in practice (Ainscough, 1976).

7 The period 1948–1955

In time, FMLs were to be known as tree farm licences (TFLs). Initially the FML/TFLs were given perpetual terms, to reflect the belief that a long-term tenure was essential to provide an incentive for forest management. Those issued after 1956 (second Sloan Commission) were given 21-year terms (Ministry of Forests, 1991). The forest companies immediately began to submit applications and to begin the process of lobbying and jockeying for FML awards – MacMillan, a critic of the FML system, had no choice but to seek awards of the tenures he disagreed with (Drushka, 1995). The award of forest management licences appeared to have been a fairly arbitrary and controversial procedure:

> Applications submitted to the Minister were advertised, and sometimes public hearings were held, but the legislation gave unqualified discretion to the Minister to decide how many licenses were to be issued, what size they would be, and which of competing applicants would be successful. Although the Crown lands committed under these licenses would otherwise have been available for development under Timber Sale Licenses, there was no provision for competition for these allocations. Because there were many conflicting applications and no clear criteria governed Ministerial discretion, awards sometimes generated a good deal of criticism from unsuccessful applicants as well as from the independent logging

> industry which was concerned about its narrowing opportunities to acquire harvesting rights.
>
> (Pearse, 1976)

In his 1945 report, Sloan recommended a follow-up Commission in 10 years, to which he was subsequently appointed in 1955. In the 10 years between the two Sloan Commissions, the government had awarded 23 forest management licences, given preliminary approval to 18 and received applications for a further 28 (Pearse, 1976). By 1955, Orchard's and Sloan's visions of a few hundred modestly sized FMLs had not transpired, and there existed a handful of some very large licences, including in some cases land reserved in public working circles, which was meant to be managed by the Forest Service to serve the needs of small operators and unintegrated forestry companies (Drushka, 1995).

8 The second Sloan Royal Commission on Forestry, 1955–1956

The second Sloan Commission ran from 1955 to 1956 and produced a report in 1956 that addressed primarily administrative concerns. While it was longer than the report of a decade earlier, its recommendations were not nearly as far-reaching (Pearse, 1976). In his second Royal Commission, Sloan continued to support the FML/TFL tenure type. He was harshly critical of the more than 2 million acres of forest land that had been cut down and left in a not satisfactorily restocked (NSR) state, and noted that most of this derived from timber sales administered by the Forest Service, which carried no reforestation requirement. As Fulton had in the first Royal Commission, Sloan made strong recommendations that all cut-over forest land be restocked and that young growth be protected from fire (Shelford, 1993). Sloan was also growing increasingly aware of the corporate concentration and was cognizant of the importance of competition in the industry in order to determine the true value of the forests (Shelford, 1993). This indicated a growing awareness of how fewer companies were controlling a larger volume of the harvest, which was to be a concern that Pearse addressed in his Royal Commission in 1974–1976.

Sloan arguably contributed to the corporate concentration through the creation of the private working circle, to which the larger and integrated companies could contribute their private and old temporary tenure land, which would be combined with additional Crown land to create a sustained yield management unit. The non-integrated mills and independent loggers did not have the land base (either private or through tenure) to be eligible to gain access to this land. Similarly, the appurtenant mill clause, requiring a mill for processing the timber harvested from an FML/TFL, was not something that the independent operator could provide in order to be awarded an FML.

9 The period 1956–1974

In the decade following the Second Sloan Report, the government awarded 20 more tree farm licences. Since that time, a number of them have amalgamated. Initially – in fact for the 20 years following the second Sloan Royal Commission on Forestry – timber from the public working circles (or public sustained yield units (PSYU), as they were later to be called) was accessed by timber sale licences. The timber sale harvesting licence (TSHL), which evolved from the timber sale license, was introduced in the mid-1960s. This was a volume-based licence with a specified annual harvest rate from within a designated PSYU. The first TSHLs carried terms of 21 years, but, after the first 15 TSHLs had been issued, a term of 10 years became the standard (Ministry of Forests, 1991). The TSHL was to evolve into the modern-day forest licence.

In the 1960s another tenure, called pulpwood harvesting area agreements (PHAs), was introduced (Pearse, 1976; Marchak, 1983). The first PHA was signed in 1962 (Pearse, 1976). These were designed to supply pulp mills with pulp logs. PHAs gave the holder the right to harvest pulpwood from within its boundaries, which included several PSYUs. In fact, the right was rarely, if ever, exercised, due to an adequate supply of sawmill residues. There were efficiencies that could be realized by having all logs (saw and pulp) harvested by the same operators, with the saw-logs recovering any dimensional wood and all residue and chips being directed to the pulp mills – the 'chip direction' policy (Ministry of Forests, 1991).

10 The Royal Commission on Forest Resources (Pearse Commission), 1974–1976

Concerns about forest tenures and their policy implications and the ability of the forest service to administer an ambiguous and confusing public policy (Marchak, 1983), along with concerns about the state of the forests and perceived timber shortages, were some of the issues that led to the establishment of the Royal Commission on Forest Resources in 1974. It is noteworthy that the state of the forests and perceived timber shortages were the same concerns that led to the previous three Royal Commissions being held and would indeed be the primary reasons for the future Peel Commission.

Peter Pearse, a University of British Columbia economics professor, was the sole commissioner for what was to be called the Pearse Commission. This 2-year Commission recommended major revisions to the tenure system and addressed a number of issues, including, but not limited to:

- the emergence of problems related to the protection of the natural environment;
- concerns about continuing corporate concentration in the forest industry;

- the need to anticipate the transition from old growth to second growth and deal with the 'fall-down effect';
- the need to address the increasing governmental and public interest in the development of the forest industry;
- the growing need for an expert and efficient public forest administration;
- the need to improve the forest tenures, which had successfully established the forest infrastructure and which now needed to be modified to address more comprehensive forestry and not just the conversion of old growth into managed second growth;
- the need to modify forest tenures to address 'non-industrial values' (Pearse, 1976).

The Pearse Commission made a number of recommendations to address tenure reform and there was obvious interest in and support for smaller tenures and community forestry. Pearse recognized the power of tenures:

> The government's freedom to choose among alternative forms of rights offers one of the most powerful means of shaping the development of the forest industry. Because I want to emphasise the impact of tenure arrangements on the structure of the industry, and hence the need for a deliberate policy to achieve the public objectives. . .
>
> (Pearse, 1976)

Following the submission of the Commission report in 1976, a Forest Policy Advisory Committee was established to deal with implementation of the recommendations. The NDP government lost the election that year and the Social Credit Government failed to implement most of Pearse's recommendations.

11 The 1978 Forest Act

A number of forestry practice recommendations were adopted, leading to new forest and range legislation in 1978, as well as a new Ministry of Forests Act, 1978 (Ministry of Forests, 1991). This Act established explicit goals for the Ministry of Forests and recognized, for the first time, forest values other than timber and range as elements of the ministry's mandate (Haley and Luckert, 1995). There were also changes made that flew in the face of the Pearse recommendations – such as not amalgamating the Wildlife Branch into the Forest Service.

The new legislation brought substantial changes to existing licence agreements and additional new licensing agreements were introduced (Robinson, 1995). The early TFLs lost their perpetual term, and all were replaced without competition (Robinson, 1995). 'Evergreen' replacement provisions were introduced for tree farm licences. This involved updated conditions being offered to the TFL holder every 10 years (later changed to 5 years) and, if they chose to accept the new conditions, the licence was replaced with another 25-year term. If the TFL holder did not accept the new conditions or chose not to apply for replacement, they could retain the TFL under the original terms for the balance of the 15 years remaining in the term, at which point the licence expired.

The PSYUs (originally called public working circles) were reorganized and renamed timber supply areas (TSAs), of which there are now 35 in the province. An annual allowable cut was assigned to each and apportioned to the various types of new licences on a *pro rota* basis determined by the former licence agreements (Ministry of Forests, 1991). A number of new licences were introduced to address harvesting in the TSAs, including the following:

1. Forest licences, which replaced the timber sales harvesting licences and timber sales licences. It should be noted that the forest licence also replaced the timber sale licences which had been earlier awarded as 'third band sales', whose purpose was to improve timber utilization.
2. Pulpwood agreements replaced the pulpwood harvesting area agreements.
3. Timber licences replaced the old temporary tenures, which had remained in good standing, many for over 90 years.
4. Wood-lot licences replaced the farm wood-lot licences and enabled a broader spectrum of entrepreneurs to combine private land with additional Crown land, in order to have a larger and theoretically more economic forest base to manage. Wood-lot licences can be awarded without a private land component. The contribution of private land is one of the three main factors considered in the awarding of a wood-lot licence, the other two being the management intent and the skill and experience of the applicant (Waters, 1997). Restricted to 400 ha on the coast and 600 ha in the interior, the wood-lot licence is essentially a miniature TFL, but with different conditions for award and with different policy objectives.
5. The Small Business Forest Enterprise Program (SBFEP) was established. The objective was that a portion of the allowable cut would be competitively awarded to two classes of non-quota holders (operators with no access to long-term Crown timber supplies), those with sawmills and those without (Robinson, 1995). Today the SBFEP accounts for approximately 15% of the provincial AAC (Haley and Luckert, 1995).

Following the passage of the new Forest Act in 1978, the forest industry in British Columbia continued to expand and the provincial annual allowable cut continued to increase, driven by large increases in the interior cut (Travers, 1993). Poor economic conditions in the early 1980s led to a government decision to introduce 'sympathetic administration', in which standards were relaxed and sometimes ignored in order to foster a more favourable operating climate. This resulted in reduced forest management quality and, from 1982 to 1987, it cost the government $1.1 billion more to administer the forests than it received in revenue (Garner, 1991; Travers, 1993). More specifically, in 1981, 1982 and 1985 the Crown lost money on the timber harvest and in 1983, 1984 and 1986 earned a modest profit. During the 6-year period of 1981–1986, the Crown netted only $55 million (Schwindt and Heaps, 1996).

12 The Forest Resource (Peel) Commission, 1989–1992

Growing public concern over the state of the forests, the concentration of harvesting rights and processing facilities in the hands of the major forest companies and a move by the government in 1989 to convert forest licences into tree farm licences led to a strong public backlash. There was growing discontent with the Ministry of Forests and the Honourable Dave Parker, then Forest Minister, suggested the establishment of a permanent Forest Resource Commission. The Peel Commission was established in 1989 and, while it only lasted 3 years (being dissolved after the NDP were re-elected in 1992), it addressed a number of issues including, but not limited to (Forest Resources Commission, 1991):

- integrated land management for all users;
- land claims from indigenous peoples (it was clear that the people felt there could be no land-use strategies drafted without resolution of these problems);
- the need for public participation and local input in joint management decisions;
- the need for a better inventory for all forest users;
- concern for the environment and wilderness;
- concern for large companies dominating the industry;
- a desire for smaller clear-cuts or none at all;
- restricting the use of herbicides;
- more wood lots, community forests and other new alternative tenures;
- an increased level of coordinated land-use planning;
- more logging determined by silviculture concerns.

The Peel Commission made 108 recommendations (Forest Resources Commission, 1991). The most important of these resulted in:

1. The establishment of the Commission on Resources and Environment (CORE) and its attendant legislation, which addressed coordinated land-use planning;
2. The implementation of the Timber Supply Review (TSR) process, which inventoried all of British Columbia's Crown forest land;
3. The implementation of the Forest Practices Code (FPC) and its attendant legislation;
4. The implementation of the Protected Areas Strategy;

5. The establishment of competitive sort (log) yards to determine the value of logs.

The Peel Commission findings were distributed throughout the province and subjected to review and commentary. The proposals for comprehensive land-use planning and improved inventories were less controversial than the establishment of a Crown corporation to manage the province's commercial forests (Haley and Leitch, 1992). A major concern regarding the Commission and its findings is that it lacked the cohesive force of clearly defined goals for the provincial forests and related industries (Haley and Leitch, 1992).

13 The state of the forest sector in British Columbia today

Forest companies in British Columbia are currently suffering from a number of factors including: low lumber prices; difficulty in exporting to the USA as a result of the softwood tariff dispute; low demand from a weak Asian economy; and high operational costs, which are linked in part to the costs of the Forest Practices Code.

A number of mills have recently shut down permanently, due to timber shortages and high harvesting and processing costs. The current timber harvest is not sustainable and there are increasing incidents of conflict in the forests.

There is concern over the state of the forests and the health of the forest sector and, once again, there has been mention of the need for a forestry commission in the provincial papers.

References

Ainscough, G.L. (1976) The British Columbia forest land tenure system. In: McKillon, W. and Mead, W.J. (eds) *Timber Policy Issues in British Columbia*. UBC Press, Vancouver, pp. 33–55.

Drushka, K. (1995) *HR. A Biography of H.R. MacMillan*. Harbour Publishing, Madeira Park.

Forest Resources Commission (1991) *The Future of Our Forests, Executive Summary*. Forest Resources Commission, Victoria.

Garner, J. (1991) *Never Under The Table*. Cinnabar Press, Nanaimo.

Godwin, G. (1994) *The Eternal Forest*. Godwin Books, Vancouver.

Haley, D. and Leitch, J. (1992) The future of our forests – Report of the British Columbia Forest Resources Commission: a critique. *Canadian Public Policy* 28, 47–56.

Haley, D. and Luckert, M. (1995) Policy instruments for sustainable development in the British Columbia forestry sector. In: Scott, A., Robinson, J. and Cohen, D. (eds) *Managing Natural Resources in British Columbia, Markets, Regulations and Sustainable Development*. UBC Press, Vancouver, pp. 54–79.

Mahood, I. and Drushka, K. (1990) *Three Men and a Forester.* Harbour Publishing, Madeira Park.

Marchak, P. (1983) *Green Gold, the Forest Industry in British Columbia.* UBC Press, Vancouver.

Ministry of Forests (1991) *Forest Tenures, a Public Discussion Paper – November 1991, Draft 7.* Ministry of Forests, Victoria.

Pearse, P. (1976) *Timber Rights and Forest Policy in British Columbia.* K.M. MacDonald, Queen's Printer, Victoria.

Pearse, P.H. (1989) Evolution of the forest tenure system in British Columbia. Unpublished monograph, UBC, Vancouver, British Columbia.

Robinson, A.B. (1995) *Witch Hunt in the B.C. Woods.* Sagebrush Book Publishing, Kamloops.

Schwindt, R. and Heaps, T. (1996) *Chopping Up the Money Tree, Distributing the Wealth from British Columbia's Forests.* David Suzuki Foundation, Victoria.

Shelford, C. (1993) *Think Wood!* Shelford Publishing, Victoria.

Sloan, G. McG. (1945) *The Forest Resources of British Columbia.* Charles F. Banfield, King's Printer, Victoria.

Travers, O.R. (1993) Forest policy. Rhetoric and reality. In: Drushka, K., Nixon, B. and Travers, R. (eds) *Touch Wood.* Harbour Publishing, Madeira Park, pp. 171–224.

Waters, A. (1997) Ministry of Forests. Interview by the author, 28 November, Vancouver.

A Brief History of Italian Forest Policy

27

A.L. Freschi and L. Hermanin

Institute for Forest Management and Technology, University of Florence, Italy

Italy's forest policy, based on the Royal Decree no. 3267 of 1923, consists mainly of regulatory instruments (hydrogeological constraint, provincial forest regulations, forest management plans compulsory for public forests) aimed at guaranteeing the protective function of the mountain forests, foremost against hydrogeological risk and against natural forces (strong winds, avalanches and rockfall). The state forest policy had been implemented directly by the State Forestry Service (Corpo Forestale dello Stato, CFS) and the State Company for State Forests until 1977, when the state transferred its administrative functions concerning forests to the regions. The regions with an Ordinary Statute started to develop, within the framework of the state forest law, their own forest policies. These were characterized by a hybrid state/region competence and a regional delegation system determining a complex and multilevel administrative structure, involving many public institutions but not efficiently coordinated. In 1985 the Galasso law (a landscape law) extended the landscape constraint to forests, but it resulted to some extent in negative effects because of unclear formulation of its measures. The regions with Ordinary Statute forest administration suffered mainly because of a delay in regionalization, which in fact turned out to be prevalently nominal (Hermanin, 1997), and because of the conflicts in forest land uses, which the forest law of 1923 was not able to predict.

(eds M. Agnoletti and S. Anderson)

1 Italian forest policy

In 1877 the state's first unitary forest law (law no. 3817/1887), following the unification of Italy (1861), was issued. The main objective was hydrogeological protection of forest and mountain land. Actually, the law was inadequate to protect woods from destruction, tilling and irrational exploitation. Between 1878 and 1908 more than 2 Mha, about 7% of national land, was estimated to have been transformed into agricultural and pasture land and 28,000 ha to have been reforested. The hydrogeological emergency attracted public attention and lawmakers passed an organic law in 1910. Law no. 277/1910 established new rules for reforestation and compulsory forest management plans for public forests, and instituted the State Company for the State Forests (Azienda Speciale per le Foreste Demaniali, ASFD) as an instrument of and model for forest management. Other measures were defined and implemented in order to make the structure of the ASFD operative.

The basic Act of the national forest policy, replacing previous forest laws (nos 3817/1887, 535/1901, 379/1908 and 277/1919), was issued in 1923. Royal Decree (RD) no. 3267/1923 'Rearrangement and reform of legislation concerning woods and mountain land', the so-called 'forest law', is still effective. Hydrogeological protection in mountain areas was the main objective implemented regulating forest uses and silvicultural activities and hydrogeological constraint in those soils subjected to hydrogeological risk, in order to guarantee the public interest in hydrogeological security. A strong relationship among forest management practices, forest uses, other interventions inducing a direct impact on soil stability and hydrogeological risk inducing public damage of human activities and life was determined by the law (Abrami, 1996). The text of the law provides for: (i) active measures: hydraulic–forest-setting works, reforestation by state intervention, incentives for public bodies and private owners in favour of mountain silviculture and agriculture, promotion of associations among owners of estates subjected to hydrogeological constraint and associations among municipalities and other bodies aimed at land management, and compulsory forest planning for public estates; and (ii) passive (or negative) measures: imposition of hydrogeological constraint on soils of any nature and destination, sensitive to the loss of stability and disturbance of the water equilibrium in case they are utilized without respecting specific use regulations established by the provincial administration (provincial forest regulations, Prescrizioni di Massima e di Polizia Forestale, PMPF). Hydrogeological constraint was progressively imposed on most mountain land, according to a specific administrative procedure. In hydrogeologically constrained soils, transformation of forest into cultivation and periodic tillage land had to be authorized by the administrative authority responsible for hydrogeological constraint, i.e. the State Forestry Service (Corpo Forestale dello Stato, CFS). Following DPR no. 616/77, the administrative authority was generally the province or the mountain community – i.e.

association of mountain municipalities – delegated by the Region (see DPR no. 616/77).

In the 1950s and 1960s, forest policy had two objectives, related to the institution of the 'Fund for the south' (Cassa per il Mezzogorno, laws no. 646/1959 and no. 647/1950) and law no. 991/1952, 'Measures in favour of the mountain'. These were to create work opportunities in disadvantaged zones and to ensure land protection from hydrogeological risks. This combined objective of hydrogeological policy and employment policy remained characteristic of Italian forest policy up to the mid-1970s. In the period 1952–1970 relevant reforestation programmes were implemented in southern Italy, especially in Calabria, resulting in contradictory and negative effects, such as forest fires caused by the same employed workers in order to increase the need of employment for reforestation programmes.

During the 1970s, diverse new important measures and institutional conditions affected forest administration and programmes:

1. Institution of mountain communities (Comunità Montane, i.e. associations of mountain municipalities, by law no. 1102/1971) as public bodies responsible for implementing the main policy objectives of the same law: to balance social and economic differences between mountain areas and other territories of the country, to protect nature and soil by means of public intervention (mountain land reclamation and engineering works), to provide for incentives for economic activities orientated towards valorization of local resources and to favour the cultural and professional position of the populations.

Mountain communities could achieve these strategic objectives by means of economic–social plans implemented by means of operative programmes. Generally speaking, the objective of the law was not achieved, mainly because of scarcity and uncertainty of state funds, scarcity of technical personnel and incapacity of regional governments to delegate organically administrative functions to mountain communities after 1977, when DPR no. 616/1977 transferred the state administrative functions concerning forestry, agriculture and other sectors to regions with an Ordinary Statute.[1]

2. The state transfers to regions with an Ordinary Statute relevant administrative functions: hydrogeological constraint management, hydrogeological setting, soil conservation, mountain land reclamation and arrangement of mountain basins (approbation of plans and programmes, promotion of associations), landscape constraint[2] management and fire control activities.[3]

3. Suppression of the ASFD and transfer of state goods to the regions; the state keeps the functions of forest police, agricultural and forest research and experimentation, measures of international and national relevance, certification of agricultural and forest products, air service for fire control, valorization of the natural environment and phytosanitation interventions.

4. The regions with an Ordinary Statute start to develop their own regional forest policies within the framework of the principles defined by the State Forest

Law (RD no. 3267/23), and to delegate agriculture and forestry functions mainly to mountain communities constituted after 1971. However, the National Forest Service (Corpo Forestale dello Stato) supports, depending on the region, the region itself, mountain communities and provinces for control and technical duties, so that a hybrid state/region situation characterizes forest administration. This is a suitable substratum for negative interactions and conflicts between the central administration and regional administration.

5. On the other hand, northern autonomous regions, taking advantage of a greater independence from the central government (recognized by the Special Statute) and of a strong forestry tradition, developed efficient forest administration by means of regional forestry services substituting the State Forestry Service.

6. The state forest policy formulated and implemented new objectives about state afforestation programmes: to promote private reforestation and forestry aimed at timber production in order to decrease the internal deficit of timber supply, to improve connections with the wood industry and create opportunities of development and economic management in large agricultural and forest areas in the south (Special Project no. 24, a special action of the 'Fund for the south').

During the 1980s, law no. 431/1985 (the so-called Galasso law), a landscape law, directly affected woodland and forest activities. The landscape constraint that was introduced by law no. 1497/1939 for natural beauties and imposed according to punctual administrative procedures, was extended to 'territories covered by woods and forests (even in cases when they are crossed or damaged by fire and subjected to reforestation constraint) to rivers, lakes, glaciers, parks, wet areas, volcanoes, archaeological areas, sea coasts, and mountains'. In addition, the law mandated regional landscape planning aimed at utilization and valorization of such environmental goods. Landscape constraint implies that landscape authorization of the administrative authority responsible for landscape protection was necessary for all actions, including agricultural and silvicultural activities, altering the permanent status of those goods, by means of building or other engineering works, or altering the hydrogeological setting. Therefore forests subjected to hydrogeological constraint (ex RD no. 3267/1923) needed two different authorizations (hydrogeological authorization and landscape authorization).

In regard to silvicultural operations, the law stated that 'in territories covered by woods and forests 'cultural cutting', afforestation, reforestation, land reclamation, fire control and conservation interventions are allowed without landscape authorization but must respect the regulations in force' (i.e. RD no. 3267/1923 providing for provincial forest regulations, hydrogeological constraint and related authorization procedures). The law had not specified the meaning of the general term 'cultural cutting'. A double authorization regime, combined with uncertainty about the silvicultural activities involved, decreased the private owner's interest in forest activity, which was typically discouraged by other conditions determining low profitability, such as the small-size forest estates (Pettenella *et al.*, 1997), the negative influence of the international timber market on the local market and a weak connection between the wood industry and forestry. Only recently, in 1996, did the decision of the Constitutional Court dated 22–29.1.1996 clarify the meaning of 'cultural cutting' operations not subjected to landscape authorization: tending, thinning, conversion from coppice to high forest system, harvest cutting (Postiglione, 1996a).

The regional landscape plan (Piano Regionale Paesaggistico, PRP), as an ordinary instrument of land planning at the regional scale, is a relevant innovation introduced by the Galasso law. The plan included specific considerations of landscape components and values aimed at planning land resource use according to principles and criteria of landscape protection and environmental compatibility (Greco, 1994). Regions with an Ordinary Statute and also Regions with a Special Statute (since the Galasso law is an economic–social reformation law) were to have compiled the PRP within 16 months, but several structural difficulties caused a prolonged delay. At present some regions still lack a regional landscape plan. The regional landscape plans provide for general directives and rules for forest management and forest uses and principles and measures as defined by the state Forest Law (RD no. 3267/1923). It charges provinces and municipalities to specify landscape regulations and interventions at the local planning scale.

In 1987 the Forest National Plan (Piano Forestale Nazionale, PFN) was approved. It constituted the basis for the national forest policy, focusing on the multiple functionality of forest ecosystems. The main objective was to develop forest quality and an equilibrium between forest products and services obtainable from Italian forests. The following directives were formulated for the first period of validity: (i) development of public forest investments with functions of land protection and natural resource conservation; (ii) development of a closer relationship between agriculture and forestry systems; and (iii) development of a more efficient connection between forests and the wood industry. Reduced availability of financial resources has strongly limited implementation of the PFN, so much so that it seems to have become only a document of general directives for forest planning and programmes (Giordano, 1991).

In the late 1980s and early 1990s, four state laws concerning land planning affect to some extent forest land administration and management:

1. Law no. 142/1990 increased autonomy of local public bodies from regional administration and enlarged local administrative competence.
2. Law no. 183/1989, 'Rules for the new organization and functional arrangement of soil protection', instituted hydrographic basins (national and interregional basins) and the Basin Plan (Piano di Bacino) providing for the conservation and recovery of land by means of hydrogeological, hydraulic, hydraulic–forest, hydraulic–agrarian interventions, afforestation, naturalistic

land recovery and natural resource (water, agricultural and forest land) use and management.

3. Law no. 394/91, 'Basic law for protected areas', aimed at promoting conservation and valorization of the natural Italian patrimony. Protected areas were classified into national and regional areas. The Park Plan (Piano del Parco), formulated and implemented by the park body in cooperation with local public bodies included in the park territory, substitutes land and landscape plans of any scale (provincial, municipal, regional) in force in the park. The law provided for two important measures: (i) financial priority by state and regional funds for those municipalities and provinces located in protected areas for environmental conservation and restoration, agriculture and forest activities; and (ii) restrictions and constraints on agricultural activities and forest activities imposed by the Park Plan could be compensated. It has to be pointed out that forests belonging to state natural reserves and parks are managed by the CFS, but public (municipal, provincial, regional) forests belonging to regional and provincial parks are generally managed by mountain communities, as delegated bodies, or by the CFS through a special convention, region or province–CFS.

4. Law no. 97/1994, 'New rules for mountain areas', constituted a new legislative and political attempt to promote a winning development strategy in mountain areas. The development model proposed was based on protection and valorization of local natural resources and a more active role of mountain communities in the development process (Bagnaresi, 1997). The following directives were formulated for the forest sector: (i) social-economic plans also define priorities of environmental valorization by means of the hydrogeological and hydraulic–forest setting and water management coordinated with the Basin Plan (according to law no. 183/1989); (ii) mountain communities have to promote forest management by means of agreements with public owners (municipalities and other bodies) and to constitute, even compulsorily, forest associations aimed at rational forest management; and (iii) the regional government can assign to these associations, to mountain communities and to municipalities tasks of land maintenance and conservation, not only for forest but also for agricultural and landscape purposes, technical assistance and forest surveillance.

The law instituted the National Fund for the Mountains, sharing resources by means of regional funds to be instituted by the region. The region was charged with giving financial concessions of up to 75% of the costs for environmental maintenance activities in agriculture–pasture–forest estates and providing incentives for persons who want to settle in mountain areas.

The state laws described, including the Galasso law, define a complex land planning system, involving different levels and objectives, including regional landscape plans (law no. 431/85), park plans (law n. 394/91), provincial land coordination plans (law no. 142/90) and basin plans (law no. 183/89). The main challenge was coordination of the planning instruments, in order to achieve positive interactions and an efficient process of land planning, and the adequate consideration of ecological conditions in the process. Actually, many perplexities about coordination occurred (Postiglione, 1996b). Overlapping of public constraints (hydrogeological constraint, landscape constraints), prohibitions, limitations, procedures of authorization derived from the planning system and regulations can discourage any land transformation, mainly because of bureaucratic mechanisms (Postiglione, 1996b).

The new frontier of land planning is represented by sustainable development objectives. Many regions are working on this project. For instance, the Toscana region is one of the first regions with a new regional legislation (1995) for land resource management orientated to sustainable development. The Regional Landscape Plan of the Valle d'Aosta region (north-western autonomous region) is based on sustainable development principles.

Other important documents at a national level, such as the National Forest Inventory (DGEMF/1985), the Report on the State of the Environment (MA/1991), the Environmental Statistics Annual Reports (ISTAT), the Inter-Ministerial Committee for the Economy (CIPE) decisions in 1993 and 1994 on directives for implementation of the UNCED Agenda 21 and adoption of sustainable economic development principles, environment and biodiversity conservation, provide bases for a new policy for the forests and the environment. An important step towards this goal could be the recognition of the social and environmental roles of forests in the Italian Constitution, as recently proposed (Querini, 1994).

2 Concluding considerations

The Italian forest policy situation outlined here is certainly not positive, but it has to be considered incomplete, because it does not take into account regional cases. It is better to consider different regional forest policies, some of which are more efficient because of consolidated tradition and administrative autonomy (Hermanin, 1997).

One could also conclude that some characteristics of national policies on forestry, mountain protected areas and land planning, although constituting a good framework of principles and directives, 'suffer' because of limitation factors, pertaining mainly to the policy process (especially formulation and implementation):

- emergency nature of measures;
- fragmentary and uncoordinated character of natural resources policy (several measures and laws);
- innovative principles but not clearly specified and difficult to interpret; also laws are often subjected to modification. Italian legal tradition seems to prefer the 'formal aspect' of laws rather than concrete measures (Postiglione, 1988);

- inadequate and insufficient financial support of laws and programmes;
- times fixed for administrative adjustment to law's measures are systematically unrealistic;
- formulation of protective measures is generic and reverted to further specifications and actions involving the participation of other public bodies. This leads us to conclude that policy actions are not based on a concrete knowledge of reality;
- inefficient implementation determined by different public bodies and public administration and control forms not coordinated.

Coordination and control of expenses scattered between state, regions and other administrative bodies are highly problematic. Nowadays, the greatest deficiency seems to be scarcity of coordination among institutions involved, the forest–timber chain and policies concerning the forest sector.

Thus, in Italy, a certain difficulty of translating policy objectives into concrete measures occurs. The emerging forest issues concern the following aspects, which must be considered in order to enhance the effectiveness and efficiency of Italian forest policy:

- with special reference to regions with an Ordinary Statute, a more efficient regional forest administration;
- development and enhancement of forest planning and promotion of private forest planning by means of subsidies and tax exemptions;
- a greater consideration of forest resources and the forest sector in the process of land planning and programming aimed at integrating forests in the global planning system;
- promotion of social-economic and ecological sustainability in forest management and utilization;
- management instruments for those areas of forest expansion which are evolving, especially in abandoned mountain territories;
- new models of integrated development in mountain areas based on local natural resources, multiple use of forests according to forest suitability and an efficient system for utilizing forest or forest products.

3 State's legislation concerning and affecting forest and mountain land[4]

3.1 Post-national unification legislation (1877–1943)

- 1877 Forest law no. 3817.
- 1910 Law no. 277, the so-called Luzzatti law, institution of the Azienda di Stato per le foreste demaniali (ASFD, State Company for State Forests).

- 1923 Royal Decree no. 3267, reorganization and reform of legislation concerning woods and mountain land; art. 108: Azienda Speciale del Demanio Forestale dello Stato (Special State Company for State Forests).
- 1931 Law no. 973, measures for protection of chestnut woods.
- 1931 Law no. 99, protection of medicinal plants and Alpine flora.
- 1933 Royal Decree no. 215, so-called Serpieri law for land reclamation.
- 1939 Law no. 1497, measures for protection of natural beauties.
- 1942 Law no. 1150, the law for land planning.
- 1943 Royal Decree no. 16-B, suppression of the Milizia Nazionale Forestale (National Forest Army), restoration of the Real Corpo delle Foreste (Real Service of the Forests).

3.2 Post-Second World War legislation (1948–1987)

- 1948 Legislative Decree no. 804, reform of the Corpo Forestale dello Stato (CFS, State Forestry Service).
- 1950 Law no. 646 and no. 647, institution of the Cassa per il Mezzogiorno (Fund for the South).
- 1950 Law no. 841, land property reform.
- 1952 Law no. 901, measures in favour of mountain land.
- 1955 Law no. 1177, extraordinary measures for Calabria.
- 1955 Decree of the Republic's President (DPR) no. 987, decentralization of the services of the Ministry of Agriculture and Forestry.
- 1955 Law 1318, concerning reconversion of coppice woods and recovery of degenerated coppices.
- 1956 Law no. 759, concerning protection of cork-oak woods.
- 1961 Law no. 454, first green plan.
- 1962 Law no. 1360, aids for purchasing land and for reforestation to municipalities.
- 1966 Law no. 910, second green plan, institution of the Forest Fund.
- 1967 Law no. 950, first law depenalizing forest crimes.
- 1970 Law no. 281, take-off of regions with an Ordinary Statute.
- 1971 Law no. 1101, institution of the Comunità Montane (literally mountain communities, i.e. Associations of Mountain Municipalities).
- 1972 DPR no. 11, transfer of the state forest and agriculture competencies to the regions with an Ordinary Statute.
- 1973 Law no. 269, legislation concerning trade of forest seeds and plants for reforestation.
- 1975 'Cassa per il Mezzogiorno', Fund for the South, Special Project no. 24 for afforestation.
- 1975 Law no. 706, depenalizing forest crimes.
- 1975 Law no. 47, integrative measures for protection of forests from fires.
- 1977 DPR no. 616, completion of decentralization.

- 1977 Law no. 984, the so-called *Quadrifoglio* (four-leaf clover) law.
- 1979 EEC Regulation no. 269/79.
- 1981 Law no. 689, depenalizing forest crimes.
- 1982, extraordinary measures for the Mezzogiorno (the South) in the period 1982–1991.
- 1984 Law no. 424, hardening of administrative sanctions against transgressors of regulations concerning protection of forests against fires.
- 1985 Law no. 431, the so-called Galasso law, protection of forests and woods and other landscape components by means of landscape constraint and regional landscape planning. Each region must formulate its landscape plan as the main instrument of environment–landscape policy.
- 1986 Law no. 349, institution of the Ministry of Environment and measures concerning environmental damage and establishing the 'procedure of environmental compatibility', which reflects only partially the environmental impact assessment (EIA) procedure, introduced by the ECC Directive no. 337/85. Two Government Decrees (DPCM) – DPCM no. 377/88 of 10.8.88 and DPCM of 27.12.88 – provide rules for implementing procedures of environmental compatibility.
- 1986 Law no. 752, law for implementation of planned interventions in agriculture.
- 1986 Government Decree (DPCM) 1.2.86 receiving ECC Regulation no. 2088/85, concerning Mediterranean Integrated Programmes.

3.3 Recent legislation (1987–1997)

- 1989 Law no. 183, rules for new organization and functional arrangement of soil protection.
- 1990 Law no. 149, adjustment of the CFS personnel.
- 1991 Law no. 394, framework law on natural protected areas.
- 1992 Law no. 142, a law reforming the arrangement of municipalities and provinces (public bodies), increasing their autonomy and reciprocal coordination in land planning and economic programming. Each province contributes actively to the region's economic, land and environmental programming by means of the Piano Territoriale di Coordinamento (PTC, Land Coordination Plan) containing general directives for provincial land uses and interventions, in conformity with regional directives defined in the Piano Regionale di Sviluppo (PRS, Plan of Regional Development). Mountain communities and municipalities contribute to the formulation of the PTC.
- 1993 Law no. 491, rearrangement of region's and state's competencies concerning agriculture and forestry, institution of the Ministry of Agricultural, Alimentary and Forest Resources (MIRAAF).

- 1994 Law no. 94, new rules for the mountains, law aimed at promoting integrated development in mountain areas based on local natural and social-economic resources.
- 1997 no. 59, delegation to the government for giving functions and tasks to regions and local public bodies (such as Comunità Montane) in order to reform public administration and simplify administrative procedures.
- 1997 Legislative Decree no. 143, giving further administrative functions concerning agriculture and fishery to regions and reorganization of state administration.

With the suppression of the MIRAAF and the institution of the Ministry of Agriculture policies, all the functions and tasks concerning agriculture, forestry, fishing, hunting, rural development, agrarian tourism and food, previously carried out by the MIRAAF, are carried out by regions directly or by means of delegation or attribution to provinces, municipalities, mountain communities or other local bodies. The new Ministry carries out the following tasks:

1. Formulation and coordination of agriculture, agricultural industry and forestry policies.
2. Proposals concerning the government's functions of coordination and orientation of agriculture and forestry.
3. General legislation and national coordination in the following matters: food supply, protection of the quality of agricultural–food products, food education, experimentation and research carried out by national institutes and laboratories, import and export of agricultural and alimentary products, interventions of market regulation, regulations for seeds, propagation materials, phytosanitation and fertilizers, registers of vegetal varieties, cattle genealogical book, national seed-wood books, protection of vegetal and animal biodiversity and genetic resources, management of sea resources of national relevance, utilization of innovative biotechnology in the agricultural–alimentary sector, species suitable for hunting and irrigation networks of national relevance.

Notes

1 The majority of Italian regions have an Ordinary Statute: the regions Friuli-Venezia Giulia, Trentino Alto Adige and Valle d'Aosta in the north and Sicily and Sardinia in the south. The so-called autonomous regions are regions with a Special Statute, which provides them with greater autonomy from the central government.
2 Landscape constraint for natural beauties was introduced by law no. 1497/1939 'Measures for natural beauties'.
3 Regional fire control plans were instituted by law no. 47/1975; before DPR no. 616/77, these plans were approved by the Ministry of Agriculture and Forestry.
4 Elaborated and updated on the basis of the list proposed by Alessandrini (1993).

References

Abrami, A. (1996) Regime giuridico dei boschi e funzione sociale. *L'Italia Forestale e Montana* 1, 29–34.

Bagnaresi, U. (1997) Dinamica della gestione delle risorse montane nell'ultimo cinquantennio. In: Georgofili, I. (ed.) *Atti della Accademia dei Georgofili anno 1996*, pp. 417–433.

Giordano, E. (1991) La gestione dei boschi nel Piano Forestale Nazionale. *L'Italia Forestale e Montana* 1, 33–43.

Greco, N. (1994) Storicità e attualità della legislazione sulle foreste nel sistema di governo dell'ambiente. *L'Italia Forestale Montana* 2, 134–171.

Hermanin, L. (1985) La problematica dell'assestamento forestale nelle aree protette. *Agricoltura e Ambiente* 28.

Hermanin, L. (1997) Considerations of political components of forest reality in Italy. In: Ciancio, O. (ed.) *Forest and Man*. Accademia Italiana di Scienze Forestali.

Pettenella, D. *et al.* (1997) Il bosco privato in Italia: caratteristiche organizzative, problemi e prospettive. Convegno Boschi privati: tutela e valorizzazione. Florence.

Postiglione, A. (1988) Il diritto dell'ambiente e i suoi aspetti istituzionali. In: Amendola Cervia Ambiente Rimini Maggioli (eds) *Corso di Legislazione Ambientale con Appendice Legislativa*.

Postiglione, A. (1996a) Il 'taglio colturale' Sentenza della Corte Costituzionale. *Monti e Boschi* 6, 23.

Postiglione, A. (1996b) La pianificazione territoriale ovvero l'ingorgo pianificatorio. *L'Italia Forestale e Montana* 2.

Querini, R. (1994) Proposte per una nuova politica italiana delle foreste e dell'ambiente. *L'Italia Forestale e Montana* 6, 619–629.

A History of the French Coniferous Forest

28

Vincent Moriniaux

Institut de Géographie, Université Paris IV-Sorbonne, 191, rue Saint-Jacques, 75005 Paris, France

Writing a history of the French coniferous forest is a challenge. It is, however, necessary to know the status of conifers in France over the centuries in order to understand why they are so strongly rejected now[1] – more strongly, it seems, than in other European countries. In Germany, for example, production forestry is indeed condemned for its excesses but the choice of conifers as the dominant species (71.3% of trees planted as against 36% in France[2]) is not contested. French people seem to condemn conifers more because of the trees themselves than because they represent a certain kind of forestry that is no longer highly regarded. We have already tried to explain this rejection by the power of the imagination: the dark coniferous forest reflecting a vision of the Germanic forest which is incompatible with the vision – no doubt equally imaginary – of the Gallic forest, which is probably seen as more 'hirsute' (or broad-leaved). A historical analysis demonstrates that the rejection of conifers by French people also comes from the fact that the expansion of conifers in France is recent and quite sudden.

Isolated in the mountainous areas by competition after the ice age, the few conifers that would have survived in the plains mixed with deciduous trees lost a lot of ground during the Middle Ages due to their low utility value. Interest in these trees goes back to the Renaissance, but their return to prominence is recent. We will show, using unpublished documents from the French Academy of Agriculture,[3] that the first private forest plantations in the 19th century gave rise to grounds for complaints that are sometimes very close to today's discourse.

(eds M. Agnoletti and S. Anderson)

1 The sudden and recent return of conifers

The history of conifers in France is the history of a recent and fairly sudden return to prominence. Driven out by the post-ice age climatic warming and ignored in the Middle Ages, they began their comeback at the beginning of the 19th century. The graph we present in Fig. 28.1 can only be based on estimates because the data are not reliable before the 20th century.[4]

2 Retreat

Pollen analysis shows that conifers, and especially those we consider today to be 'exotics', were even found in the plains during the cold periods of the ice age. The sequoias and Douglas fir (*Pseudotsuga menziesii*), for example, only disappeared from Europe some 12,000 years ago. (There are traces of their presence in the cave of Lascaux, which was occupied about 13,000 years ago.) In the plains, conifers could only survive as relict vegetation in the face of competition from genetically more dynamic deciduous trees. The fact is that fir seems to be indigenous to Normandy in the region of L'Aigle (Orne) and Lemée suggests that *Pinus sylvestris* could be indigenous to Fontainebleau.

3 The eclipse

It is very likely that some stands of conifers could have been present in the forests of the plains at the beginning of the big medieval land clearances, but no documentary record of them exists. There was a very good reason for this silence: conifers did not have any economic value. They were less useful than deciduous trees, which provided acorns and beech mast for pigs and coppice growth for firewood. So the conifers were inexorably eliminated from the forests of the plains during the Middle Ages, because they did not belong to the 'nurturing forest'.[5]

4 The comeback

The comeback of the conifers began at the end of the 18th and beginning of the 19th centuries as a result of a growing interest in exotic forest species used as ornamental trees in gardens and a growing need for wood: forest cover was at its lowest level after the clearances of the revolutionary period (about 12% of the territory).

In fact, interest in conifers emerged as early as the Renaissance and was pioneered in France by Pierre Belon (1518–1564), who in 1553 published *De Arboribus Coniferis, Resiniferis, Aliis quoque Nonnullis Sempiterna Fronde Virentibus . . .*[6] He coined the word *conifère* (conifer) in the French language.

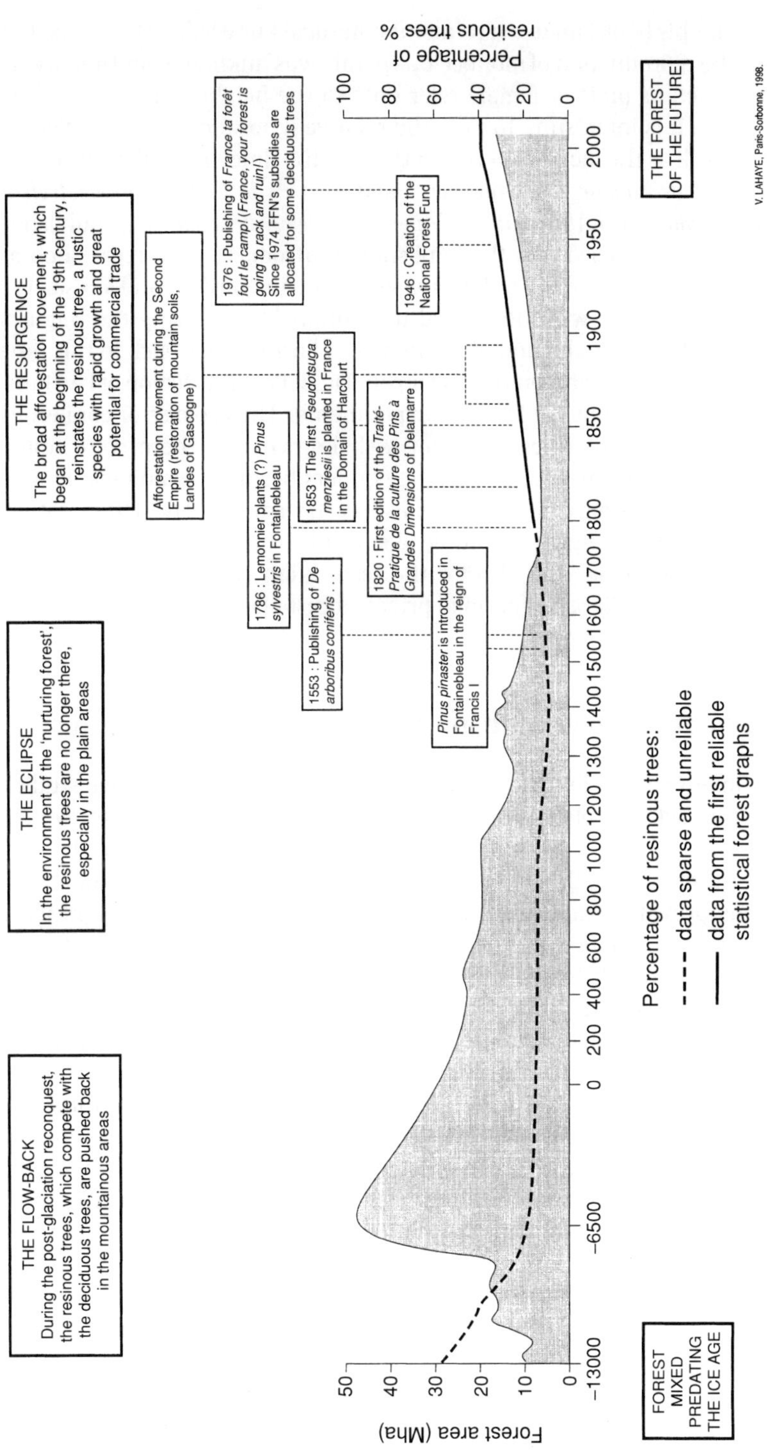

Fig. 28.1. Resinous trees in the French forest. The source for the graph of the forests' surfaces is the *Atlas des Forêts de France*, De Monza Editions, Paris, 1991.

Beyond what his book can teach us about botanical knowledge at the time (for example, the reproduction of monoecious plants was unknown, and qualifying adjectives such as male or female refer only to the height or the strength of the seedling), it is interesting to note the relative ignorance about coniferous species, as well as the lack of interest in them at the time. In his *Remonstrances sur le Défault de Labour et Culture des Plantes, et de la Cognoissance d'Icelles. Contenant la Manière d'Affranchir et d'Apprivoiser les Arbres Sauvages*, published in 1558, Belon proposed that the king should introduce 'wild trees', such as larch (*Larix* spp.), *P. sylvestris* or *Pinus pinaster*, into his gardens and forests. We know that *P. pinaster* was introduced to Fontainebleau during the reign of François I and the introduction of *P. pinaster* in Sarthe dates back to this time too. Belon refers to the gardens of his patron, Du Bellay, the bishop of Le Mans. But there was no follow-up to these experiments. The ecology of coniferous trees, even indigenous ones, was not well known and there were many failures. In modern times, however, the conifers became fashionable first as ornamental trees in gardens and later as forestry species. The 19th century witnessed not only the great plantings decreed by the state (in the Landes of Gascogne and Sologne, and also for the restoration of mountain soils), but there was also a proliferation of small private forest plantations.

5 Speculation on private afforestation in the 19th century

The documents of the French Academy of Agriculture that we have studied attest to the fact that plantations were established in order to make money; speeches about France's self-sufficiency in timber for the navy, about the flow of rivers, the climate, soils and so on are in fact of secondary importance. Files on returns provide a rich mine of information because their authors were obliged, under legislation regulating competition, to declare the costs and benefits they derived from their plantations. The landowners, leading citizens or nobles, claimed that they wanted to and sincerely believed that they would restore soils, make desolate lands fertile, improve climatic conditions, increase the flow of springs and, above all, make large profits. L.G. Delamarre, in his *Historique de la Création d'une Richesse Millionnaire par la Culture des Pins* (1827), promised millions to whoever would plant pines. Documents in the archives of the French Academy of Agriculture prove that Delamarre's promise was not a hollow one, even before the last quarter of the 19th century, when conifers found market outlets in the paper industry. Duchesne-Thoureau, who planted 11,000 ha of pines (mostly *P. sylvestris*) in the middle of the 19th century at Châtillon-sur-Seine (Côte d'Or) in Burgundy, is a good example of one who made a quick fortune. Duchesne-Thoureau was not the first to plant conifers in Burgundy. His inspiration came from the works of an inspector of the Forestry Commission, who introduced *P. sylvestris* and *Pinus nigra* (from Austria) in the clearings and glades of the forests of his district. He in turn obtained followers; he wrote in 1868 that 'today, in this district, there are no more than a few villages which do not have two, three, four or more blocks of fir forest on their territory'. This imitation, which provoked a frenzy of afforestation, is explained by the economic interest in such plantations. One could say that, in a pine tree, everything is good. A light thinning in an 18-year-old *P. sylvestris* forest, for example, produces some stakes, bundles of firewood for bakeries, hop poles and some pine buds for the 'apothecaries of Paris, of the whole of France and Belgium'. A planter from the area of Le Mans named Vétillart notes that every year demand for pine buds was higher than production (15,000–30,000 kg per year, sold at 100–150 francs per 100 kg, for an investment of less than 35 francs per 100 kg). Vétillart boasted about the fact that he was able to make money from his plantations as early as the fifth year. In addition to bundles of firewood for lime kilns, potteries, bakeries and so on, he also sold the cones of his *P. pinaster* for the seeds, which were exported as far as Paris. He leaves us in no doubt about his desire to get rich by planting pines when he writes: 'By management, I mean the way of growing *Pinus pinaster* in order to get the maximum income and the maximum production possible from them every year.'

6 The rejection of conifers

The popular rejection of conifers is very likely to have come from this ostentatious speculation. It is interesting to note that one of the factors in today's rejection by ecological movements is the concept that an individual can make money from the forest. The rejection could also be explained by the rapidity of change: the rate of planting was 100,000 ha a year during the 18 years of the Second Empire, which is more than twice the rate of planting in the best years of the afforestation financed by the Fonds Forestier National (FFN), or the National Forest Fund. The opposition to conifers, then, is old. It was based more on economic considerations (from which jealousy was not absent) and less at the time – at least not clearly – on notions of landscape (except in Fontainebleau, where the Barbizon painters were the first to fight against conifers for this reason) or ecology. This last preoccupation only appears after the Second World War, when the FFN was created. On the contrary, during the 19th century, conifers were considered to be 'ecological' trees because they restored soils, improved climatic conditions and so forth.

In 1854, the Marquis of Vibraie reporting on his experiments grafting different pines on his property in Sologne wrote that about 20 grafts succeeded:

> I have lost a lot of other [grafts] in my first experiments and also unfortunately because of the malevolence of a boorish man who, in a monomaniacal outburst, let it be known that he was an enemy of firs and that he preferred to attack exotic species.

A certain Regnier complained in 1852 to the National and Central Society for Agriculture that his plantations (*Pinus strobus* and *Pinus sapo*) had been disrupted by the hostility of the inhabitants:

> It would have been a pleasure for me to present you with the results of my plantations. Alas! I soon had to give up the satisfaction of doing so. I had hardly settled here when the help of the inhabitants and the local authorities, on which I had imprudently counted, did not materialize. I had to plead, to give up my plantations, to sell my property at a significant loss and finally to escape from an inhospitable region where I, a friend of progress in arboriculture was made ridiculous and called a fool, where chicanery is the dominating activity, where the inhabitants, brandishing the Common Law, help one another in order to despoil a trusting stranger.[7]

Duchesne-Thoureau noted that the population, like his own family, thought he was mad. He declared: 'The public I met was too incredulous . . . they didn't forgive me for having broken away from all customs.' This man, very proud of his fortune, was receiving a rent for the sheep that grazed in his plantations: the owners of the sheep were the same people who, a few years ago, had sold him their 'waste lands' for almost nothing.

7 The present hostility to conifers

The present hostility of French people towards conifers probably has its roots in this bitterness. The government has continued to promote this trend towards plantation forests, which has changed the appearance of French forests, and, since 1815, has continued to expand the land available for forestation. The FFN has highlighted what has come to be called, since the beginning of the 20th century, *enrésinement* (afforestation with coniferous trees and replacement of deciduous trees by planting conifers). The French considered that there was an urgent need to respond to the needs of industry, the paper industry in particular, by increasing the supply of softwoods (cellulose). Even if some conflicts appeared earlier (for example, at the time of the creation of the National Forest Service in 1964), one could date the beginning of the campaign against conifers to the 1970s, when 'ecology' as a political movement first emerged in France. The explanation for this choice of a policy of 'all conifers', against which today's ecological movements are fighting, may be found in the lack of an established common European forestry policy at the time of the establishment of the Common Agricultural Policy (CAP). The six founding states of the European Community agreed to eliminate timber and forests from the Treaty of Rome in order not to establish a European forest policy. 'This was very dramatic,' says M. Cointat.[8]

> If we had had a European forestry policy, we could have decided that France was the only country capable of producing deciduous trees, that is to say oaks and beech, but it would take a long time – 150 years – and it would be more expensive

than to produce resinous trees, and the five other countries would have paid the difference.[9]

8 Conclusion

This short study cannot explain the very intricate history of conifers in France, but it shows the need for historical analysis in order to understand today's situation. The 'natural' forest in France is a mixed one (deciduous and evergreen trees). The rejection of conifers is a consequence of the fact that the French forest has been deliberately managed for 2000 years to become deciduous. The French forest, which has been cultivated for a very long time, is artificial.

Notes

1 Les Français face à l'eurésinement. XVIe–XXe siècles. January 1999, Thesis in geography, directed by Jean-Robert Pitte, Sorbonne-Paris IV, end 1998.
2 Eurostat, 1987–1990.
3 18, rue de Bellechasse, 75007 Paris.
4 Rousseau, P. (1990) L'évolution des forêts françaises métropolitaines d'après les statistiques forestières, *Revue Forestière Française*, 42, 56–68.
5 Nougarède, O. (1997) Un écosystème dominé par les arbres . . . transformé par les hommes, in: Biro, A. (ed.) *Histoire de Forêts*, Centre Historique des Archives Nationales, Paris, pp. 11–24.
6 Pierre Belon is especially known as a zoologist (ornithology and ichthyology).
7 Letter, 26 July 1852, Avignon, about the forest of Joignier (Isère).
8 Minister for Agriculture in 1971–1972.
9 Discussion with M. Cointat, 17 October 1994.

Imagining and Creating Forests in Britain, 1890–1939 29

Judith Tsouvalis and Charles Watkins

School of Geography, University of Nottingham, University Park, Nottingham NG7 2RD, UK

In this chapter we explore the period 1890–1939 as a time of slow transition in the work practices of British forestry and the different types of woodland produced. Drawing on evidence from official publications, forestry manuals, textbooks and scientific papers, we examine the ways in which forestry practices changed over time and space and influenced the form of forests and the shape of trees. Particular attention is given to the decline of traditional woodland practices and the development of new forest techniques, propagated by a burgeoning interest in continental forestry, and the establishment of the Forestry Commission. The consequences of these changes for forest workers and forests alike are critically assessed.

1 Introduction

The practices of forestry and woodland management were transformed in late 19th-century and early 20th-century Britain. On the one hand, there was a long, quiet decline in traditional woodland management techniques, such as coppicing, which went largely unnoticed by contemporary chroniclers. On the other hand, there was an orchestrated campaign to form what was termed a 'new forestry' based on scientific principles developed on the continent (Simpson, 1904). Britain was not, of course, unique in taking on board a more scientific approach to forestry. James (1996) points out that developments in German forestry in the 18th and 19th centuries 'engendered the scientific and intellectual basis for advancements and knowledge in many other countries'. Recent research in Spain and Israel, for example, has shown how the spread of a scientific forestry took place and the implications for traditional land management practices, rural employment, nature conservation and landscape (Boquete, 1998; Liphschitz and Biger, 1998).

In this chapter we consider the context of the introduction of the new forestry in Britain and its implications. Various factors make Britain an unusual case: first of all, Britain at this time was almost wholly dependent on imported rather than home-produced timber and wood products; secondly, only 5% of the area of land was wooded; and thirdly, there was very little publicly owned forest apart from some small remnants of Crown forests, such as parts of the Forest of Dean. This latter factor is enormously important, as it means that, until the establishment of the Forestry Commission in 1919, the espousers of new ideas on scientific forestry had the difficult job of attempting to persuade many hundreds of private landowners of the value of the new approach.

The period 1890–1939 is crucial to an understanding of the overall development of forestry policies and practices in the 20th century. It was a time of slow but irreversible transition (Table 29.1), when the power of the British aristocracy and landowning classes faded and when landed estates and the associated forms of woodland management began to disintegrate; when the state in the form of the Forestry Commission acquired large tracts of land, planted them uniformly with conifers and introduced grant aid for the management of private woodlands; when scientific forestry principles were adopted and technological innovations were made; and when forestry education was formalized and institutionalized and traditional woodland management practices sank into oblivion. The scale of these changes and their consequences

Table 29.1. Broad changes in British forestry, 1890–1939.

1900: traditional woodland management	Theme	1930s: modern woodland management
Complex woods; multiple objectives; varied landscapes	Overall change	Monocultures; strategic timber reserves; modern/national landscape
Traditional knowledge; woodmen learn on the job; local training	Knowledge	Scientific knowledge; formal training; national school
Coppicing, planting, felling, game management; small scale; varied	Work practices	Ploughing of moors; work study; accounting; national standards; efficiency; large scale; uniform
Coppice with standards; 'poor' form; mixed species	Trees	Exotic species; straight trees; monocultures
Scattered; on agricultural land; on private estates	Location	Land classed as 'waste' uplands; coastal areas; heathland
Woods: small; old woodland; game layouts; landscape	Type of woodland	Plantations: large; square layout

varied over time and space, but they were, on the whole, far-reaching for nature and society. In this chapter we shall concentrate on their general effects and provide some examples of their particular manifestations.

2 Changes in woodland management in the late 19th and early 20th centuries

Two principal types of estate forestry existed in the 19th century: traditional woodland management, principally taking the form of coppice woodland or coppice with standards, and plantation forestry, which developed its own strong traditions from the early 18th century onwards. The trajectories of these two types of management followed very different courses.

2.1 Coppiced woodland

The markets for coppice products fluctuated dramatically over the 19th century. The replacement of firewood and charcoal by coal for domestic and industrial energy production brought about massive reductions in two of the principal markets for coppice. However, the population growth enabled by agricultural and industrial development brought about a strong growth in markets for a wide variety of woodland products used in industry, agriculture and the home. Many of these markets were local and regional, and there were considerable regional variations in the strength of the different markets (Collins, 1992).

Standard forestry books of the late 19th century detail the decline in value of coppice woodlands. These texts have to be treated with care because authors were frequently keen to espouse new forestry techniques instead of old established methods. They do, however, provide a broad insight into the views of contemporary foresters. As late as 1880, Ablett in his *English Trees and Tree Planting* was concerned more with the organization of coppice woodland than with its profitability: 'Where copse-wood is cultivated to any considerable extent, it is advantageous so to arrange matters to come on in perpetual rotation, which may be cleared and put to profitable use yearly' (p. 402).

John Nisbet, in his edition of James Brown's *The Forester* of 1894, emphasized the profitability of coppice woodland in areas where there was a strong market for its produce:

> On hillsides of rather shallow description, but otherwise favourable as to soil and situation, Oak coppice-woods are often by far the most remunerative form of silvicultural crop, when there is any favourable market near at hand for the disposal of the bark to tanneries. Again, mixed coppices of Ash, Maple, Sycamore, Sweet-Chestnut, Hazel and the like are also exceedingly remunerative throughout southern England, when near to favourable markets for saplings and poles,

> such as hop districts . . . Alder coppice is likewise often a more remunerative form of crop than almost any other on land that is suited for the Alder, but which cannot be conveniently drained to serve higher purposes.
>
> (Vol. I, p. 44)

Only 9 years later, in 1905, with the publication of Nisbet's new book, also called *The Forester*, but actually a rewritten book with much new material, Nisbet described a completely different and very bleak picture of the state of coppice woodland:

> The Copse-woods or Coppice under Standards, the national form of Arboriculture, and especially typical of the warmer portion of England, have for the most part become practically transformed into game coverts . . . Yet the Copse-woods and Coppices were at one time among the most profitable parts of large estates.
>
> (Vol. I, p. 49)

It was not until the turn of the century, by which time metal implements had almost totally replaced locally produced wooden ones, that the market for coppice finally collapsed in several areas. Many of the characters in Thomas Hardy's novel, *The Woodlanders* (1887), gained their living from coppice trades. In the new introduction to this novel in the Wessex edition written in April 1912, Hardy wrote:

> In respect of the occupations of the characters, the adoption of iron utensils and implements in agriculture, and the discontinuance of thatched roofs for cottages, have almost extinguished the handicrafts classed formerly as 'copsework' and the type of men who engaged in them.

The final collapse of coppicing in the late 19th and early 20th centuries was brought about by the further substitution of wood products by new metal and chemical products and changes in agriculture. This is not to say that the coppice industries did not manage to limp on well into the 20th century in some localities (FitzRandolph and Hay, 1926; Edlin, 1949). On the Eastnor estate in east Herefordshire, for example, it was not until the early 1930s that the demand for coppice wood reached such a low point that auction lots did not attract any purchasers (Watkins, 1993). But by the end of the Second World War many previously coppiced areas had fallen into such decline that they were classed as scrub for the 1947–1949 Forestry Commission census and as such they were threatened with conversion to agricultural land or conversion to high forest (Rackham, 1976; Watkins, 1990). And with the decline in coppicing there was a loss of the traditional working practices and skills which have been termed 'woodmanship'.

2.2 Traditional estate plantations

Plantation forestry became a fashionable and profitable branch of English estate management in the 18th century. The activity was strongly associated with patriotism and improvement (James, 1981; Daniels, 1988; Seymour, 1998) and continued to be a core aspect of rural estate management in the 19th century. Its place was assured less by any inherent profitability, although this was always stressed by forestry publicists and professionals, than by a belief in the seemly nature of tree growing and the clear benefits for landscape, game and hunting. The planting practices developed by professional foresters on Scottish estates in the 18th and early 19th centuries were often used on English ones (Monteath, 1824; Steuart, 1828). In the mid-19th century :

> it was as much the correct thing for an estate to have a Scotch forester as it was for a nobleman's establishment to possess a French *chef*. . . the practice of early and heavy thinning, which prevailed in England for fifty or more years, was introduced by Scotch foresters.
>
> (Forbes, 1904)

One of the most influential forestry texts was James Brown's *The Forester* (1847). Forbes (1904) considered that:

> it is to Brown and his school that we owe the introduction of the mixed plantation – a system of planting that has led to some of the worst results that could possibly be attained . . . Pure plantations, when such were planted, invariably consisted of larch . . . The object in many cases was not so much the ultimate production of first-class timber, as the speedy growth of game cover or screens and belts for landscape effect.
>
> (p. 17)

There was no established school of forestry in England and most foresters were practical men who were trained on the estates themselves. Often the long-term planning of forests was in the hands of land agents or the owners themselves. The situation was criticized by many commentators. A.C. Forbes, who wrote several forestry texts and was lecturer on forestry at Durham College of Science, Newcastle upon Tyne, noted that:

> until quite recently (and to some extent even now) it was no uncommon thing to find all classes of men filling the position of estate forester. Any man with a general knowledge of estate work was considered qualified to manage the woods, more especially on those estates on which the area under wood consisted of coppice with standards. It required no great ability to manage a squad of half a dozen woodmen, to mark and measure the necessary number of trees for estate use or sale, and to see that hedges and fences were more or less in good condition . . . [When] planting was carried on, it was, and still is, usual to get a nurseryman to do it by contract at so much per acre and leave the method of planting and choice of species to him.
>
> (Forbes, 1904, p. 318)

Forbes was especially critical of the many small estates where the 'commercial details and higher branches of the work' were in the hands of land agents, who would have only a broad training in forestry, and the 'practical woodcraft' was the responsibility of a foreman woodman, who was 'little more than a skilled workman at the best, with a rule-of-thumb acquaintance with the elements of planting, thinning, draining, and so forth'. The effects of employing such poorly qualified staff were, in his opinion, 'inevitably bad' (Forbes, 1904).

3 The new forestry: the introduction of scientific forestry in the late 19th century

Modern scientific forestry was introduced to Britain through its Empire. The vestigial Crown forests were managed by a body known as the Office of Woods, but this was small-scale and did not employ trained foresters. The need for specialized personnel for the Empire was first recognized in India. In 1856, the German botanist Dietrich Brandis was appointed Superintendent of Forests in the Pegu Province of Burma and, 6 years later, he became Forestry Adviser to the Government of India. It was Brandis who urged that some form of British school of forestry education should be established and in 1879 the Forestry School of Dehra Dun was opened. Following that first step, a forestry section was established at the school of military engineering at Cooper's Hill in Surrey. The first director was Wilhelm, later Sir William, Schlich, who eventually became director of the Oxford Forestry Institute at the University of Oxford. Schlich had a great influence on the orientation and development of forestry education, which he based mainly on textbooks translated from German. The end result was a first generation of trained foresters whose knowledge had been imported from Germany and who were essentially concerned with plantations rather than with coppice with standards. Just after the turn of the century, in 1904, a School of Forestry for Woodmen was opened at Parkend in the Forest of Dean, and the first readership at Cambridge University was established in 1907 (James, 1990; Gerber, 1997). Table 29.2 shows some of the most important steps in the institutionalization of forestry.

This circulation of ideas, formal and informal, and its implications for British forestry have been described by Forbes in his book *The Development of British Forestry* (1910). He observed that:

> since about 1860, when Cleghom and Brandis inaugurated the Indian Forest service, a small stream of continental trained youths has been going out to India, an equally small stream of retired Indian foresters . . . has been returning from it. Whatever the exact practical results of this inter-mixture of British and Anglo-Indian ideas may have been, there is little doubt that fresh ideas were instilled into British foresters and proprietors, and a wider knowledge of forestry as an industry instead of a hobby resulted.
>
> (p. 252)

Formal education was not, however, the only way in which Continental scientific modes of forestry were popularized in England. Schlich pointed out in the *Quarterly Journal of Forestry* in 1914 that:

Table 29.2. Key dates in the institutionalization of English forestry.

1856	Dietrich Brandis, German botanist, joins the Indian Forestry Service, marking the start of scientific forestry in India
1883	Dietrich Brandis retires; William Schlich takes over his post
1884	The first exhibition solely concerned with forestry – International Forestry Exhibition – is held in Edinburgh
1885	As a result of the 1884 exhibition, a select committee is set up to consider the value of establishing forest schools
1885	The Royal Engineering College at Coopers Hill, Egham, Surrey, is established by William Schlich for people wanting to enter the Indian Forest Service
1886	The final report of the 1885 select committee recommends the establishment of a Board of Forestry
1889	An Act is passed which establishes a Board of Agriculture, which does little to meet the recommendations of the 1887 report. However, it grants £250 towards the salary of a Professor of Agriculture and Forestry at the Durham College of Science, 1891
1897	The English Arboricultural Society holds the first examination in forestry
1889	William Schlich's *Manual of Forestry*, Vol. 1, is published
1902	A Departmental Committee is appointed by the Board of Agriculture to enquire into British forestry. Its report proposes the establishment of educational opportunities for foresters at Oxford, Cambridge, and other places
1903	Lectureships in forestry are established at the University College of North Wales at Bangor and at the Armstrong College at Newcastle upon Tyne
1904	The School of Forestry for Woodmen at Parkend, Forest of Dean, is opened
1905	The School of Forestry at Oxford University is founded by Schlich
1906	The Royal Engineering College at Coopers Hill closes, its forestry section being transferred to Oxford University, where Schlich lectures
1907	Establishment of a readership in forestry at Cambridge University
1907	The first technically trained forester, E.P. Popert, a former conservator of forests in India, is appointed by the Crown for the Forest of Dean

Sources: James, 1981; Gerber, 1997.

> another form of education, totally unconnected with the State, has also had a very great and probably lasting effect upon British forestry during the last fifty years. This consists in the constant visits made by the British landowning class to the Continent in search of pleasure, sport or health. These visits, although totally unconnected with forestry, perhaps, can scarcely have failed to open the eyes of landowners to the possibilities of scientific forestry.
>
> (p. 80)

The establishment in 1882 and increasing popularity of the Royal English Arboricultural Society (later Royal Forestry Society) (James, 1982) was a clear indication of the rising importance given by many foresters and landowners to scientific aspects of plantation forestry. Many English foresters, however, remained uncertain of some of the benefits of German forestry. A debate in the *Quarterly Journal of Forestry* in 1914 highlights this conflict. Thomas Bewick, writing on a successful visit of British foresters to Germany in 1913, noted that 'we cannot under present conditions adopt the German system of forest management in extenso, but I am sure that some thing could be done in a modified way' (Bewick, 1914). Schlich did not allow Bewick to get away with this rather lukewarm appreciation of the merits of German forestry. In the following number of the journal he summarized his view by noting that:

> it has often been said that the continental systems of forest management are not much use in these islands, because here entirely different conditions have to be dealt with. I do not think any sensible person has ever suggested that we should adopt the continental systems en bloc, but [we should] . . . profit by experience in other countries.
>
> (p. 75)

His main contention was that British foresters must stop their tradition of 'haphazard procedure' and start making use of long-term working plans. From such plans, he noted, 'no deviation' is allowed 'except with the sanction of higher authorities'. Moreover, 'in every case it is strictly laid down what is to be done during the next ten, or in some cases twenty, years'. His conclusion was that the 'preparation of well considered working plans is a crying necessity in all British forests, even if they be of moderate size' (Schlich, 1914).

N.D.G. James (1981) has noted how difficult it is to 'assess what impact these continental ideas and opinions made on English forestry at the time' (p. 196). In the early years of the 20th century many of the larger estates would employ forestry specialists to assess their woodlands and possibly draw up plans. In practice, however, the foresters and woodmen employed by estates were still untrained in the new methods and ideas of forestry. James (1981) observes that:

> when Schlich visited private estates in England in order to provide working plans for them, he tried to encourage the idea of systematic forest management which he had been taught in Germany. [Later] . . . he regarded [this work] as some of the least successful of his career and he attributed this to the frequent changes in the ownership of private estates and to the absence of any tradition of forestry in [England].
>
> (p. 196)

The lack of any significant 'higher authorities' who could lay down rules that private landowners would have to follow was, perhaps, the main reason why modern forestry had not gained a significant foothold in Britain before the First World War. The tremendous social and political changes associated with the onset of that war, however, allowed the idea of a state forest service to become credible. The formation of a state forestry service was by no means welcomed by all forestry specialists. A debate held by the Royal English Arboricultural Society in September 1916 demonstrated the divisions between those members who saw a state forest service as essential to the development of British forestry and those who saw it as an attack on the freedom of private estate owners (Anon., 1917). By the end of the war, however, the establishment of a state forestry service was generally welcomed as a necessity. The luxurious forestry traditionally practised by the landed estates was seen as anachronistic, and the need for a strong state forest organization to produce large quantities of timber and a modern forest landscape was seen as essential.

4 The rise of state forestry

The First World War shattered the confidence of the landowning classes in Britain. In the Edwardian period just before the war, private landed estates and the rural way of life associated with them had symbolized the power of private property and aristocratic values. The country-house weekend and the associated shooting parties epitomized social éclat; wealthy industrialists wished to buy into that way of life. In practice, however, the power of the landowning classes had already begun to decline in the late 19th century (Beckett, 1986; Cannadine, 1996). The electorate was enlarged from a little more than 3 million men in 1883 to nearly 6 million in 1888, and in 1918 universal adult suffrage for men was introduced. This erosion of the political power of the aristocracy coincided with the agricultural depression, whose most intense period lasted from the late 1870s to the late 1890s. This depression led to a dramatic fall in estate rentals and, correspondingly, land values. Landed wealth could no longer compete with industrial wealth, and the introduction of estate taxation reduced what wealth remained. Increasingly landowners began to sell off their land. The final blow was brought directly by the war as hundreds of heirs to estates were killed. Many estates were put on the market, and in 1919 there were some half a million acres of English land up for sale, with twice as much being actually sold (Cannadine, 1996).

The direct impetus for state forestry in the Edwardian period came from an unexpected source. In 1909 the Royal Commission on Coastal Erosion reported. One of its conclusions was 'whether in connection with reclaimed lands or otherwise, it is desirable to make an experiment in afforestation as a means of increasing employment' (James, 1990). In the same year, Lloyd George in his budget speech announced that money was to be made available for schools of forestry, the acquisition of land for planting and the creation of experimental forests. An advisory committee on forestry was appointed in 1912 by the Board of Agriculture and its report, published in the same year, pursued this last idea and recommended that a forest survey should be carried out and that 5000 acres of land should be acquired for the establishment of such an experimental forest.

These endeavours were interrupted by the First World War. Enormous pressures were placed on existing woodlands and plantations and approximately 450,000 acres of woodland were felled. The urgency of the timber requirements resulted in the imposition of state control of timber felling and supply (Stewart, 1989). A Home-grown Timber Committee, chaired by F.D. Acland, was established in November 1915. This worked until March 1917, when its duties were transferred to the Directorate of Timber Supplies set up by the War Office. When postwar policy was being considered, the Reconstruction Committee set up a subcommittee on forestry in 1916, also chaired by Acland. The secretary to the committee was R.L. Robinson, a former pupil of Schlich.

This committee had to put forward the case for an adequate forest policy in the absence of any comprehensive data concerning the amount of trees and timber sources left. The main arguments for a positive and expanding policy made by this committee were, first, the great dependence of Great Britain on imported timber (over 90% at the time, i.e. 400 million cubic feet in 1914 alone), which was a great weakness in times of war. Secondly, the great reliance on imported softwood, largely from extra-Imperial countries, was a cause for concern for balance-of-trade considerations. Thirdly, large areas of land were considered as 'waste', and to make them grow trees would increase their productivity as well as bringing employment to rural populations. Fourthly, economic gains and wealth creation were stressed (Ryle, 1969).

The Acland Committee recommended the establishment of a Forestry Authority, consisting of three full-time paid commissioners, an unpaid commissioner appointed by the government and responsible to Parliament, as well as two further unpaid commissioners, one of whom should be nominated and appointed as chairman. Next in the hierarchy were to come the assistant commissioners for England and Wales, Scotland and Ireland, each provided with the necessary indoor and outdoor staff. These three assistant commissioners were to be advised by consultative committees, of which Wales would have its own, which would consist of representatives from land-using interests, local authorities and forestry societies. This hierarchy was implemented and remained until the Second World War.

The overall aim of the new commission was to afforest 1,770,000 acres of land over the next 80 years in order to establish what was referred to as a 'strategic reserve of timber'. An Interim Forest Authority was appointed in 1918 under the chairmanship of Acland. The Forestry Act of 1919 came into force on 1 September 1919, and the Forestry Commission came into being. With the Forestry Commission, modern forestry was introduced in Britain on a large scale and, together with the private sector, which the Forestry Commission would increasingly come to regulate, the forestry industry was born.

The first annual report of the Forestry Commission (1920) emphasized the fact that, unlike many Continental countries, British forestry at this date had:

- no organization of higher executive staff with British experience;
- no forest officer personnel;
- no body of foresters and foremen with state forestry experience or customs;

- no forestry code;
- no national statistics and survey data of woodlands.

Modern forestry was, so to speak, institutionalized with the Forestry Commission, and it is the practices and their implications that ensued for the trees and the workers alike that we shall turn to in the last section of our chapter.

5 Modern forestry and its implications

The transformation of British forestry heralded by the establishment of the Forestry Commission was not to take place through the introduction of revolutionary new techniques. Rather, it took place through the instigation of well-tried scientific forestry techniques introduced from the Continent, combined with the following institutional and policy developments:

- the establishment of a corps of trained foresters;
- the state purchase of a reserve of land suitable for afforestation;
- the state afforestation of land;
- the state management of woods according to working plans;
- the encouragement of planting by private landowners;
- the provision of scientific forestry advice.

The 'strategic reserve of timber' had to be established as quickly as possible and the form of management chosen was to use even-aged, high forest plantations. Usually a single species was planted and the crop was clear-felled at the end of the rotation. George Peterken (1993) has characterized the practices associated with this form of management:

- plantations are established on bare ground;
- seedling trees are raised in nurseries and planted when 2–3 years old;
- species selected according to commercial judgement and site conditions (mostly coniferous species);
- most planting is arranged in regularly spaced rows (efficient use of plants; care less expensive);
- ground vegetation ('weeds') is removed ('cleaned') so as not to destroy seedlings (woody crop from previous woodland or self-sown trees are known as 'scrub');
- 'thinning' is undertaken at intervals to clear out the surplus stock (those trees exceeding the stock of mature trees ('final crop') required);
- mature plantations are allowed to stand until growth rates decline, cash is needed by the owner, a good market for timber prevails;
- the final crop is clear-felled, leaving the ground available for the cycle to be repeated.

The implications of these practices were far-reaching for the British woodland landscape and for work practices. In terms of species choice, primarily non-native conifer trees were used. From 1827 onwards, the introduction of conifers from western North America had begun with the Douglas fir, followed by the Sitka spruce in 1831, the western hemlock in 1851, the western red cedar in 1853, the lodgepole pine in 1853/54 and the Lawson's cypress in 1854 (James, 1981). These exotic species, especially Sitka spruce, were to become of major importance to forestry in Great Britain. The rise of the popularity of Sitka spruce is already detectable in the *Forestry Practice Bulletin* of 1933:

> Conditions justifying selection [of Sitka spruce]: In general a substitute for Norway spruce at high altitudes or wherever volume production is more important than quality of timber . . . General remarks: Produces timber much more rapidly than Norway spruce and under a wider range of conditions.
>
> (Forestry Commission, 1933)

Most of the land purchased or leased by the Forestry Commission was marginal land of relatively little importance for agriculture. Land not under economic production was considered as 'waste', as George Ryle's observations make clear:

> Land transactions during that period [the early years of the Forestry Commission] were by no means limited to the poor-grade grazings: there were heathlands and other really unused wastelands such as those in Hampshire, Dorset and more particularly the Brecklands of East Anglia. There were sand-dunes on the coasts at Culbin, Pembrey and Newborough. There were the bracken and heath wastes of the Sherwood Forest. There were, too, the large expanses of derelict or nearly derelict woodlands.
>
> (Ryle, 1969)

Once a piece of land had been acquired, several operations had to be carried out to prepare it for the trees that were finally to grow on it. Most work before the Second World War was manual, requiring large numbers of people as well as horses. At this time the Forestry Commission drew most of its employees from the armed forces that had fought in the war and now had no other work. As soon as the Forestry Commission was set up, a coordination officer was employed to 'study methods and procedures in planting operations with special reference to securing economy in the employment of labour' (Forestry Commission, 1921).

In 1929, the method of time-study was extensively used and it was found that an unsatisfactory gang of planters planting steadily dealt with the plants at a rate of three plants in 2 minutes, or 90 per hour per man, and that each man planted approximately 450 trees in an 8-hour working day. The measures taken to eliminate 'wasted' time included portable shelters and training workers to work without pickets to set lines straight. In that way it was hoped that the 'maximum time spent on useful work' could be secured (Forestry Commission, 1929).

Furthermore, before a good road network was established and social conditions improved, woodmen often had to walk long distances to reach their working site; a walk of 2 h in the morning to get there and 2 h in the evening to return home was not uncommon (Ryle, 1969). Afforestation often took place in remote regions, and the journey to work proved a great cost and loss of time not only to the workers but also to the Forestry Commission. It was in 1924, under Lord Lovat's chairmanship, that the idea of smallholdings for forest workers was first conceived. With work being so hard to find at the time and due to the fact that employment in the forests was secured for certain times while agricultural activities could be pursued at others, these smallholdings were greatly sought after. After 1936, when industrial depression led to the designation of 'Special Areas', the pressure on the Forestry Commission by the government to create more smallholdings increased, and the attempt was made to settle unemployed miners in the forests. This, however, was largely a failure. Some families returned to their valleys within a few weeks. The idea of special forest villages designed to build new forest communities was not developed until after the Second World War.

The major attempts to improve the pace at which forest workers could work can thus be summarized as follows:

- Overcome workers' physical limitations via science and technology, replacing people with machines if necessary.
- Rationalize physical movements by standardizing them.
- Identify time losses and spatial barriers and eliminate them as far as possible (e.g. portable lunch shelters, keep planting gang supplied with plants, forest smallholdings, etc.).
- Offer financial incentives (piece-rate additions or threats like 'no good work, no pay') to overcome 'motivational' barriers.

About 162,000 ha of new forests had been planted and purchased by the Forestry Commission in Britain in the interwar period and over 4000 people were employed (Ryle, 1969). The new plantations were celebrated by many as the efficient conversion of wasteland into modern forests which would provide rural employment and a strategic reserve of timber. The principal criticism of the new forests was not the adverse effect of such plantations on the nature conservation value of semi-natural habitats. This was not a significant criticism until the 1960s. Rather, the new landscapes of forests were criticized for their artificiality. They were seen as monotonous intrusions into the 'natural' landscapes of moor and heath (Revill and Watkins, 1996).

Richard St Barbe Baker, the forest commentator and founder of the organization, Men of the Trees, commented in his 1944 autobiography that, when he had been in Germany in the late 19th century, the foresters were by then laying great stress on the importance of 'planting mixed woods' and that the German foresters were 'trying to remedy their past mistakes and the short-sighted policy of planting pure coniferous woods'. He considered it ironic that the Forestry Commission in Britain had 'slavishly' followed 'outworn' policies which the German foresters had 'themselves revised nearly fifty years ago' (St Barbe Baker, 1944).

6 Conclusions

The period 1890–1939 was one in which the concept of scientific forestry finally gained a foothold on British soil (Table 29.1). The declining power of the landed estates and the broad social and political changes associated with the First World War were the factors which allowed the establishment of a state forestry service. The thousands of acres of coniferous trees that were planted not only formed the kernel of a strategic reserve of timber but also challenged traditional landscape values, leading to considerable protest in some areas, notably the Lake District. The geometric form of the new, extensive plantations marched starkly across the semi-natural moors and heaths. These new forests were the physical manifestation of newly established institutional structures and working practices. The forest workers who established the plantations were increasingly monitored in terms of the speed and quality of their work. In later years, they were to be replaced by machines, their numbers dwindling as science and technology advanced. The rise of modern forestry obscured the decline and finally collapse of traditional woodland practices, which would not be revived until the interest in conservation management at the end of the 20th century.

References

Anon. (1917) Report on the REAS meeting in September 1916 on the present position and future development of forestry in England and Wales. *Quarterly Journal of Forestry* 11, 20–58.

Beckett, J.V. (1986) *The Aristocracy in England 1660–1914*. Basil Blackwell, Oxford.

Bewick, T. (1914) Report on the visit of Royal English Arboricultural Society to German forests (1913). *Quarterly Journal of Forestry* 8, 1–15.

Boquete, E.R. (1998) The expansion of the forest and the defence of nature: the work of forest engineers in Spain 1900–1936. In: Watkins, C. (ed.) *European Woods and Forests. Studies in Cultural History*. CAB International, Wallingford, pp. 181–190.

Cannadine, D. (1996) *The Decline and Fall of the British Aristocracy*. Macmillan, London.

Collins, E.J.T. (1992) Woodlands and woodland industries in Great Britain during and after the charcoal iron era. In: Metaille, J.P. (ed.) *Protoindustries et histoire des forêts*. University of Toulouse, Toulouse, pp. 109–120.

Daniels, S. (1988) The political iconography of woodland in later Georgian Britain. In: Daniels, S. and Cosgrove, D. (eds) *The Iconography of Landscape*. Cambridge University Press, pp. 43–82.

Edlin, H. (1949) *Woodland Crafts in Britain*. Batsford, London.

FitzRandolph, H. and Hay, M. (1926) *The Rural Industries of England and Wales.* Oxford University Press.

Forbes, A.C. (1904) *English Estate Forestry.* Arnold, London.

Forbes, A.C. (1910) *The Development of British Forestry.* Arnold, London.

Forestry Commission (1920) *First Annual Report.* HMSO, London.

Forestry Commission (1921) *Second Annual Report.* HMSO, London.

Forestry Commission (1933) *Forestry Practice – a Summary of Methods of Establishing Forest Nurseries and Plantations with Advice on Other Forestry Questions for Owners and Agents.* Forestry Commission Bulletin No. 14, HMSO, London.

Forestry Commission (1974) *Fifty Years of Forestry Research – a Review of Work Conducted and Supported by the Forestry Commission, 1920–1970.* By R.F. Wood. Forestry Commission Bulletin No. 50, HMSO, London.

Gerber, J. (1997) The social construction of nature: the case of forestry in Great Britain since the turn of the twentieth century. D.Phil. thesis, Oxford.

James, N.D.G. (1981) *A Forestry Centenary. The History of the Royal Forestry Society of England, Wales and Northern Ireland.* Blackwell, Oxford.

James, N.D.G. (1990) *A History of English Forestry.* Blackwell, Oxford.

James, N.D.G. (1996) A history of forestry and monographic forestry literature in Germany, France and the United Kingdom. In: McDonald, P. and Lassoie, J. (eds) *The Literature of Forestry and Agroforestry.* Cornell University Press, Ithaca, New York, pp. 15–44.

Liphschitz, N. and Biger, G. (1998) Afforestation policy of the Zionist Movement in Palestine 1895–1948. In: Watkins, C. (ed.) *European Woods and Forests. Studies in Cultural History.* CAB International, Wallingford, pp. 165–180.

Monteath, R. (1824) *The Forester's Guide and Profitable Planter,* 2nd edn. Stirling and Kenney, Edinburgh.

Nisbet, J. (1894) *The Forester,* 6th edn. Blackwood, Edinburgh and London.

Nisbet, J. (1905) *The Forester,* 2 vols. Blackwood, Edinburgh and London.

Peterken, G. (1993) *Woodland Conservation and Management,* 2nd edn. Chapman and Hall, London.

Rackham, O. (1976) *Trees and Woodland in the British Landscape. The Complete History of Britain's Trees, Woods and Hedgerows.* Dent, London.

Revill, G. and Watkins, C. (1996) Educated access: interpreting Forestry Commission Forest Park Guides. In: Watkins, C. (ed.) *Rights of Way: Policy, Culture, and Management.* Pinter, London, pp. 100–128.

Ryle, G.B. (1969) *Forest Service – the First Forty-five Years of the Forestry Commission of Great Britain.* David and Charles, Newton Abbot.

St Barbe Baker, R. (1944) *I Planted Trees.* Lutterworth Press, London.

Schlich, W. (1914) Report on the visit of Royal English Arboricultural Society to German forests (1913). *Quarterly Journal of Forestry* 8, 75–81.

Seymour, S. (1998) Landed estates, the 'spirit of planting' and woodland management in later Georgian Britain: a case study from the Dukeries, Nottinghamshire. In: Watkins, C. (ed.) *European Woods and Forests. Studies in Cultural History.* CAB International, Wallingford, pp. 115–134.

Simpson, J. (1904) *The New Forestry, or the Continental System Adapted to British Woodlands and Game Preservation.* Pawson and Brailsford, Sheffield.

Steuart, H. (1828) *The Planter's Guide.* Murray, Edinburgh.

Stewart, P.J. (1989) *Growing Against the Grain. United Kingdom Forestry Policy*, 2nd edn. A report commissioned by the CPRE, London.

Watkins, C. (1990) *Woodland Management and Conservation.* David and Charles, Newton Abbot.

Watkins, C. (1993) Coppicing: the decline of a traditional woodland management technique. In: Lightbown, R. and Searle, A. (eds) *Proceedings of the Coppice Restoration Seminar.* Wessex Group, Institute of Chartered Foresters, Cowes, pp. 3–14.

Watkins, C. (ed.) (1998) *European Woods and Forests. Studies in Cultural History.* CAB International, Wallingford.

Some Aspects of Forest Management in Spain from the 16th to the 19th Century

30

Carlos M. Manuel

Departamento de Silvopascicultura, Escuela Técnica Superior de Ingenieros de Montes de Madrid, Ciudad Universitaria s/n 28040 – Madrid, Spain

Some aspects of historical forest management (both coppice and high forest) are analysed from local by-laws of several Spanish councils and leasing documents. Ancient management practices in stands of different species are evaluated, comparing them with some forest conservation already undertaken in the modern age. Strong demand for forest outputs implied negative consequences, especially in high forest. New technical forest knowledge meant, in some cases, sustainable exploitation of resources compatible with long-term conservation.

The implications of the phytosociologists' interpretation of traditional forest management in Spain are also analysed. From a historical point of view, some aspects of these interpretations are not really accurate, as is the case with the secondary role given to pine formations, and with the very widespread assumption of important reforestation activity with pines from ancient times. The case of Valsaín forest shows us how historical records provide useful information for understanding the natural vegetation distribution and the role of human beings in changing it.

1 Introduction

Despite the great number of studies related to the management and the ancient exploitation of Spanish forests published recently, very little is known about the main silvicultural practices applied in the different historical periods. Further investigations on this subject may allow us to discover how a particular

(eds M. Agnoletti and S. Anderson)

kind of silvicultural practice could imply positive consequences from the point of view of forest stand quality or, alternatively, how some practices could have a significant role in some forested areas' deterioration.

This chapter focuses on two main subjects, both related to the different branches of forest traditional practices. First, some characteristics of the silvicultural rules in early modern Spain, such as local by-laws, leasing documents, etc., are discussed, and will be compared with some of the new concepts brought by foresters in the 19th century, especially those related to forest management. Secondly, the dynamic vegetation model most widely assumed in Spain will be discussed, taking into account its strong connection with a particular view of historical forest management.

2 Silvicultural rules in early modern Spain (16th–19th century)

Historical research from different approaches has led to different interpretations of ancient forest management. Some authors share the opinion that traditional management (before the creation of the Spanish forest administration in the second half of the 19th century) was adequate for sustainable forest exploitation and conservation. This idea comes from the historical trend that strongly defends an appropriate and harmonious relationship between human beings and nature in traditional communities (before the Industrial Revolution), especially in the case of communal forests. Consequently, new conditions in the exploitation of public forests, with the intervention of foresters, change for the worse the relationship between the local population and forest resources (Cobo Romero *et al.*, 1992; Moreno Fernández, 1994).

Therefore, local by-laws and ordinances from the 15th to the 18th century are described as 'ecological rules' in some recent studies (Rubio Recio, 1985; Ostos *et al.*, 1994; Araque Jiménez and Gallego Simón, 1995). But in my opinion this is not an adequate interpretation. These ordinances were made to protect economic resources, not ecosystems or high valued natural areas. Indeed, forest products had an important function in local economies. The ancient exploitation was the result of an eminently productive and economic vision of forest land. Rules promoting care actions for the forests must be understood as regulations to avoid the extinction of some of the forest resources. In this sense, the establishment of different kinds of punishment was very frequent, depending on potential profit, on the forest type or on the forest location.

This kind of normative trend always tended to favour one species above the others, depending on the particular products demanded by the population at the time. Therefore, even in the same region, one species could be favoured and decades later be damaged. Due to the great role of cattle raising, trees with edible seeds were commonly favoured. Among these trees we can mention *Quercus ilex, Quercus pyrenaica, Quercus faginea, Quercus suber* and *Castanea sativa*. For instance, it was forbidden to remove any tree by the Guadalhorce banks (Antequera's by-laws, 1531), except *Rubus* sp. and *Tamarix gallica*. Furthermore, the different value given to the species can also be inferred from the different amount of fines for cutting trees in this local by-law: *Q. ilex, Populus* sp. or *Fraxinus angustifolia* was 600 maravedis, but *Q. faginea* was only 100 maravedis, and in some areas its cutting was even allowed without punishment (Alijo Hidalgo, 1979). There were similar rules in many other regions involving other species. In La Alberca's by-laws (1515) the fine for cutting *Q. ilex* or *Q. suber* in the *dehesa* was 500 maravedis, but for *Q. pyrenaica*, the amount was only 100 maravedís (Berrogain, 1931).

Additionally, the quantity of fines was also variable depending on the location where the infringements occurred. In Baños de la Encina's ordinances (1742) the fine for cutting *Q. ilex* in the *dehesas* was four times higher than that for the *baldíos* (Araque Jiménez and Gallego Simón, 1995), due to the different economic significance of the two systems.

Apart from fines, the exclusive socio-economic aim of the ordinances is evident. In Niebla's county ordinances (1504), Medina Sidonia's duke only allowed cork extraction by the common population in very distant areas, keeping for himself the nearby cork oak forests. Shepherds and farriers could take dead fuel wood, but only out of the *dehesas* and other preserved areas (Ladero Quesada and Galán Parra, 1984).

These normative documents did not ensure at all the conservation of the regulated resources in the long term. In my view, the kind of rules contained in ordinances and leasing documents were not always positive for forest conservation.

Basically, traditional local by-laws are punishment rules trying to avoid the negative consequences of uncontrolled exploitation. We can mention the prohibition of tree logging without a licence, the feeding of livestock in logged or burnt forests, taking products out of the jurisdiction the forest belonged to or the creation of forest guards. These rules are very common in such documents.

Ancient practices of tree logging contained in ordinances and leasing documents show negative consequences from the point of view of forest conservation. Local by-laws and other documents give us a very complete list of negative common practices: logging small trees; logging trees with the aim of getting fruits or torches; barking living trees; obtaining pitch from healthy trees instead of from damaged ones; using green wood instead of dead wood.

Frequently, the kind of management was subordinated to the special needs of the owner or the community. It is usual in the case of logging in coppice forests, used to obtain charcoal or fuel wood. We can hardly speak of rotation in a strict sense. The rotation depended on the requirements for fuel or money. In Navalquejigo, logging of evergreen oaks (*Q. ilex*) was done, by the middle of the 18th century, every 8, 10 or 12 years. Generally, a 'rotation' was very short: from 7 to 15 years for *Q. pyrenaica*. Nowadays for this species a rotation period of 20 or more years is recommended. In some other places, no system at all was applied to manage coppice, as was described by Pedro Rodríguez Campomanes at the end of the 18th century (Artola Gallego, 1986).

Concerning high forests, a very widespread custom was high grading, or creaming. It consists of logging just the better trees of a particular forest. This practice is documented for Valsaín forest (Segovia) from the 16th to the 19th century, and also in other pine forests from Segovia, Burgos, Soria and other provinces. In these and other forests it was common to leave pumped and warped trees as seed trees. This practice has very negative effects on tree stands in terms of wood quality, especially if it is repeatedly applied. Perhaps this could be one of the reasons to explain the curved shape of the boles in some *Pinus pinaster* forests (as in Talayuela, in Cáceres, or Oña, in Burgos).

Another negative consequence comes from spatially irregular logging. Logging was normally done in areas near roads or in easy logging places, while far or difficult places to work in were not affected by woodcutters. This practice is well documented in Valsaín pine forest from the 16th to the 19th century. As in coppice forests, we cannot speak of rotation: the different needs for timber or money were the main factor determining the logging.

Possibly one interesting argument supporting the inadequate forest practices for its own conservation is the large number of areas in Spain with no trees in the 19th century, but also in the 18th century and even before. This argument can be illustrated with some examples. We certainly know that a number of industries had to stop their activities for the extreme lack of fuel wood in great areas near these factories, as happened at the end of the 18th century with blast furnaces in Liérganes and La Cavada, in Cantabria (Alcalá-Zamora and Queipo del Llano, 1974), or at the beginning of the 19th century for the weapon factory of Marbella, in Málaga (Nadal, 1970).

Significant periods of price rise of fuel and timber are also well documented in different areas; for Cataluña, Pierre Vilar estimates a 300% increase in wood and firewood prices from 1740 to the end of the 18th century (Vilar, 1962, 1987). Besides, people in some regions had to use different organic remains instead of fuel wood (thistles, straw, vine shoots or even dung) due to the lack of wood in the 16th century. In Cabañas de la Sagra (Toledo) and Fuenlabrada (Madrid), at the end of the 16th century straw and thistles were used as fuel. In Toro (Zamora), a town with different communal forests in the past, a lack of forested areas is detected around 1542; for this reason roots and vine shoots were used as fuel (Vassberg, 1986). In the middle of the 19th century in different villages of La Mancha, rubbish and straw were used, and the stems of *Euphorbia* were sold as small fuel wood and *Onopordum* as heavy fuel (Laguna, 1868). The great number of references showing these facts indicates that forest land use was changed into other uses and that forest exploitation practices were not always adequate to ensure its conservation.

3 A new forest management (last quarter of the 19th century)

The forest management introduced in Spain in the last quarter of the 19th century had some positive effects on forest conservation. Although its application does not mean that natural resources were guaranteed as a whole, it is true that forest management (especially high forest management) introduced new ways of harvesting. Knowing the stocks and growing the stand made possible the continuity of harvests without compromising the future of the stand. It implied the substitution of an irregular harvest by a regular and constant one. We can possibly consider this 'new' forest management as the first application of sustainable production. In some forests where new management has been applied (for 100 years), records show us the benefit from a forestry point of view: increase in the number of trees; increase in the allowable cut; increase in harvested trees; better stand structure; and maintenance or even increase of environmental, landscape and recreational values (Rojo y Alboreca and Manuel Valdés, 1992; Hernández Fernández de Rojas and Montero González, 1993; Manuel Valdés *et al.*, 1993; Montero González *et al.*, 1993; González *et al.*, 1996).

However, this kind of silviculture has some problems from the point of view of natural areas management. Possibly one of the most censurable aspects of forest management was the usual practice of considering a principal species. This is normally negative for the rest of the species or resources of a determinate forest, especially in the case of mixed stands. There are references in the last century that show the decrease in the number of *Q. faginea* in managed forests where the principal species is *Q. suber*, the decrease of *Pinus pinea* or *Pinus nigra* in those with *P. pinaster* as principal species, or *Fagus sylvatica* or *Q. pyrenaica* in mixed stands with *Pinus sylvestris*. In general, we can speak of a negative practice for biodiversity, as happened with traditional management. On the other hand, some kinds of management systems have negative consequences from the point of view of landscape or soil erosion, as is especially the case with clear-cutting. We can also point out the usual practice of cutting down those old trees living in inaccessible places, trying to get even-aged stands.

4 About the peculiar interpretation of traditional practices included in a more peculiar and extended dynamic vegetation model

The second aspect concerning traditional forest management comes from some particular botanical interpretations. In Spain, most botanists follow the International Mediterranean and Alpine Geobotanical Society (SIGMA) phytosociological interpretation of ecological dynamics. This school, which arose in Spain in the mid-1960s, supports (in the case of Spain) the concept that pine forests have a secondary role in Spanish vegetation: generally, pine forests are not considered as potential vegetation, so pine forests are very seldom included in the most advanced stages of the ecological dynamic (Rivas Martínez *et al.*, 1987).

But the fact is that pine forests are widely spread across Spain (from before the massive pine plantations of the second half of the 20th century, which amounted to more than 2.5 Mha from 1941 to 1995), covering large areas with very different ecological conditions. So how do phytosociologists explain this evidence when they consider that pine formations are not potential or, still less, natural vegetation?

The argument always given by this school is quite simple: human beings had a major historical role in the spread of pine forests, since pine timber was a useful construction material. In this way, we should accept that traditional forest management was responsible for pine forests in areas for which the phytosociological school does not admit its natural condition. Therefore, pine reforestation should have been a very common practice. But this opinion is not borne out by historical evidence.

Even in areas where the natural presence of pine forests is broadly accepted (as with *P. sylvestris* on Guadarrama ridge and *Pinus uncinata* and *P. sylvestris* in the Pirineos mountains), the phytosociological view and reality do not coincide. Let us look at the case of Guadarrama ridge. The Sierra de Guadarrama is located in the centre of the Iberian Peninsula. Lower heights are around 800 m and the highest reaches 2430 m. In this area, phytosociological studies consider that pine forests (mainly Scotch pine) have a potential distribution between 1600–1800 m and 2000–2200 m (Rivas Martínez *et al.*, 1987; Rivas Martínez, 1992).

However, Scotch pine forests can really be found from the upper limit established down to, in some places, 1200 m – that is, 400 m below the mentioned lower limit. The explanation in these cases is the same: human beings had favoured pine populations. Is it possible to admit this explanation, even if these arguments are never supported by historical records? In my opinion, it doesn't seem to be an appropriate scientific approach.

There is an interesting example that stresses the discordance. Valsaín forest is a wooded area located on the northern slope of the Guadarrama mountains. It is formed by a Scotch pine forest in the upper part (about 7000 ha) and an oak forest (*Q. pyrenaica*) in the lower part (about 3000 ha); there is also grazing land, with no trees, at lower levels. Today pine forest stretches in Valsaín from 2100 m to 1400 m, and even down to 1200 m in some areas near the Eresma River. This range is not admitted as natural by phytosociologists, who fixed the pine forest lower limit at 1600 m. Some authors have stated that the greater part of the Valsaín pine forest is a result of human management and reforestation (Izco, 1984).

But historical records show that pine forest existed (as a mixed stand with oaks) at 1200 m by 1560. Documents mention that some pines were almost dying, due to their great age (AGS, Leg. 267/1, fo. 101). Many historical records from the 15th century suggest a distribution very similar to the one we can see today. On the other hand, we have not found any reference to any kind of management favouring pine forests. There are several conclusions in my own research about the historical management behaviour in the case of Valsaín forest:

- The first protected area was oak (*Q. pyrenaica*) forest, dedicated to fuel production as coppice forest, at the beginning of the 16th century (1539, AMS, Leg. LVI-2). The main reason for that protection was the construction of a royal residence for hunting nearby, and also the progressive diminution of oak coppice forests.
- We know about reforestation activities. Two of them took place also in the 16th century (AGP, RC 3, fo. 104–105; AGS, C. y S. R., Leg. 267/2, fo. 138). Both *Quercus* and *Pinus* were used in these small-scale experiments. We know that at least one of them failed: wild fauna (probably deer) ate the seed previously sown in the reforested area.
- In the second half of the 18th century, some other references to reforestation are found again. In these cases, oak, pine, chestnut tree and cork-oak were used (AGP, FSI, Ca. 13,602, 13,607, 13,609 and 13,629). However, these activities were not very effective: only a few of them were successful. In one particular case the success affected a pine reforestation in a very small part of the forest (about 5 ha), and it was related to an enclosed area to avoid livestock action (AGP, FSI, Ca. 13,654).
- During the whole of the 19th century and even in the first third of the 20th century, a meaningful activity was frequently developed: pine trees from the administrative area of oak forest were cut down. Due to the degradation of some oak areas, pine trees spread into the area assigned to the oaks. But the personnel in charge worked to avoid this 'semi-natural' dynamic, trying to keep pine trees out of the oak area (AGP, FSI, Ca. 13,756 – 1835, 13,763 – 1838, 13,801 – 1855, 13,846 – 1881 and so on).
- Also reforestation activities in the oak area were achieved in the 19th century and at the beginning of the 20th century; only oaks were used in all those cases of reforestation (AGP, FSI, Ca. 13,713 – 1806, 13,791 – 1851, 13,807 – 1861, 13,843 – 1881, and others). It is from the last quarter of the 19th century that we can find practices focused on favouring pine regeneration, but always in the pine environment (AGP, FSI, Ca. 13,846 – 1877, 13,940 – 1900, 13,963 – 1904, etc.). Therefore, at least for this particular forest, during 400 years there had been no management to increase the pine range by decreasing oaks.

Finally, I would like to add some general ideas on reforestation activities before the 20th century in Spain. The most surprising aspect is that phytosociological theories, especially concerning reforestation, have been assumed with no discussion by many historians. We can read – even in recent historical textbooks – that pine reforestation was a common practice in several long historical periods, and that these actions affected 'natural vegetation', in which surprisingly pines are not included (Sanz Fernández, 1986; García de Cortázar and González Vesga, 1994). It can be inferred that reforestation was a common practice in all times, but in my opinion this is not correct at all for the following reasons:

- First, we must admit that the species used in historical reforestation were generally found near the reforested area. Only kings, the nobility and monasteries could afford the costs of collecting seed or plants from distant places.
- Additionally, it is difficult to assume that these actions could strongly change forest landscapes, due to their consistently small size. In all governmental dispositions related to reforestation, we can verify the delegation of these works, both the financing and the execution, to landowners and the local population. We can even find, in the middle of the 19th century, some decrees (20 November 1841; 14 October 1851) that entrust councils and local populations with these works, in a very similar way to the norm in the 18th century. It will be necessary to wait until the end of the 19th century for the direct implication of forestry administration in reforestation labour.
- Ancient recorded reforestation does not mention the exclusive use of pine. The most important species used were mainly fruit species: some *Quercus*, chestnut and cherry trees, and also riparian trees. Many of these species had several uses, including livestock feeding. We even know of reforestation in a noble estate where it was explicitly ordered not to use pine seed, trying to eliminate it from that area due to low interest in its products (Yun Casalilla, 1991).
- Even in those areas affected by special administration (such as the whole coastal region, which was administered by the Navy from 1748, or the area surrounding Madrid, managed by the king's administration from the same date), the consequences of reforestation were not very relevant. For instance, in some northern areas, instead of planting oaks – one of the more important species for shipbuilding – the local population planted chestnut, cherry trees and apple trees (Barreiro Mallón, 1992). But, in most regions, reforestation was almost never undertaken.
- Ancient species replacement is difficult to understand, taking into account the technical capabilities of the local population in the pre-industrial period. Elimination of coppice *Quercus* forests implies hard work that only could be understandable for new cultivation or grasslands, not to replace them with pines.

Abbreviations

AGP: Archivo General de Palacio.
AGS: Archivo General de Simancas.
AMS: Archivo Municipal de Segovia.
Ca.: Caja.
C. y S. R.: Casa y Sitios Reales.
fo.: Folio.
FSI: Fondo San Ildefonso.
Leg.: Legajo.
RC: Reales Cédulas.

References

Alcalá-Zamora and Queipo del Llano, J. (1974) *Historia de una empresa siderúrgica española. Los altos hornos de Liérganes y La Cavada, 1622–1834.* Diputación Provincial de Santander/Institución Cultural de Cantabria del Consejo Superior de Investigaciones Científicas/Centro de Estudios Montañeses, Santander, 260 pp.

Alijo Hidalgo, F. (1979) *Ordenanzas de Antequera (1531).* Universidad de Málaga, Málaga, 123 pp.

Araque Jiménez, E. and Gallego Simón, V.J. (1995) *Regulación ecológica en Sierra Morena. Las Ordenanzas Municipales de Baños de la Encina y Villanueva de la Reina. Segunda mitad del siglo XVIII.* Diputación Provincial de Jaén, Jaén, 85 pp.

Artola Gallego, M. (1986) Pedro Rodríguez Campomanes: Viaje a las Sierras y Castilla la Vieja (1–30 octubre, 1779). *Estudios de Historia Social* 12–13, 325–411.

Barreiro Mallón, B. (1992) Masa arbórea y su producto en Asturias durante la Edad Moderna. In: Cabero Diéguez, V. *et al.* (eds) *El Medio Rural Español. Cultura, Paisaje y Naturaleza,* Vol. 1. Universidad de Salamanca/Ministerio de Agricultura, Pesca y Alimentación, Salamanca/Madrid, 663 pp.

Berrogain, G. (1931) *Ordenanzas de La Alberca y sus Términos Las Hurdes y Las Batuecas.* Tipografía de Archivos, Madrid, 65 pp. (originally published in *Anuario de Historia del Derecho Español*).

Cobo Romero, F., Cruz Artacho, S. and Gonzalez de Molina Navarro, M. (1992) Privatización del monte y protesta campesina en Andalucía Oriental (1836–1920). *Agricultura y Sociedad* 65, 253–302.

García de Cortázar, F. and González Vesga, J.M. (1994) *Breve Historia de España.* Alianza Editorial, Madrid, 740 pp.

González, A., Torres, E., Montero, G. and Vazquez, J. (1996) Resultados de cien años de aplicación de la selvicultura y la ordenación en los montes alcornocales de Cortes de la Frontera (Málaga), 1890–1990. *Montes* 43, 12–22.

Hernández Fernández de Rojas, A. and Montero González, G. (1993) Evolución de la selvicultura en los montes de pino silvestre de Soria: métodos de cortas y regeneración aplicados, *Actas del Congreso Forestal Español (Lourizán), Pontevedra* 2, 511–516.

Izco, J. (1984) *Madrid Verde,* Ministerio de Agricultura, Pesca y Alimentación/ Comunidad de Madrid, Madrid, 517 pp.

Laguna, M. (1868) Un pedazo de Sierra Morena, Revista Forestal. *Económica y Agrícola*, 437–450.

Manuel Valdés, C.M., Rojo y Alboreca, A. and Montero Gonzalez, G. (1993) Intervención dasocrática en los pinares de Cercedilla y Navacerrada. *Actas del Congreso Forestal Español (Lourizán), Pontevedra* 2, 693–698.

Montero González, G., Rojo y Alboreca, A. and Hernandez Fernandez de Rojas, A. (1993) Teoría y práctica de la Selvicultura. *Actas del Congreso Forestal Español (Lourizán), Pontevedra* 2, 433–447.

Moreno Fernández, J.R. (1994) *El Monte Público en La Rioja durante los Siglos XVIII y XIX: Aproximación a la Desarticulación del Régimen Comunal*. Gobierno de La Rioja, Logroño, 295 pp.

Nadal, J. (1970) Los comienzos de la industrialización española (1832–1868): la industria siderúrgica. In: Schwartz Girón, P. (ed.) *Ensayos sobre la Economía Española a Mediados del Siglo XIX*. Servicio de Estudios del Banco de España, Madrid, pp. 203–233.

Ostos, P., Romero, M. *et al.* (1994) *Las Ordenanzas de Zalamea la Real, 1535, una Normativa Ecológica del Siglo XVI*. Ayuntamiento de Zalamea la Real.

Rivas Martínez, S. (1992) La vegetación de la Sierra de Guadarrama. In: Sáenz de Miera, A. (ed.) *La Sierra de Guadarrama. Naturaleza, paisaje y aire de Madrid*. Agencia de Medio Ambiente de la Comunidad de Madrid, Madrid, pp. 167–195.

Rivas Martínez, S. *et al.* (1987) *Mapa de Series de Vegetación de España, 1:400.000*. ICONA, Madrid.

Rojo y Alboreca, A. and Manuel Valdés, C.M. (1992) La intervención dasocrática en los montes públicos españoles. El caso del Pinar y Agregados de Cercedilla, Madrid, *Agricultura y Sociedad* 65, 415–452.

Rubio Recio, J.M. (1985) El sentido ecológico de quienes redactaron unas Ordenanzas Municipales en 1535. *Paralelo* 37, 507–519.

Sanz Fernández, J. (1986) La historia contemporánea de los montes públicos españoles, 1812–1930. Notas y reflexiones (II). In: Garrabou, R., Barciela, C. and Jiménez Blanco, J.I. (eds) *Historia Agraria de la España Contemporánea. 3. El Fin de la Agricultura Tradicional (1900–1960)*. Crítica, Barcelona, pp. 142–170.

Vassberg, D.E. (1986) *Tierra y Sociedad en Castilla. Señores, 'poderosos' y Campesinos en la España del Siglo XVI*. Crítica, Barcelona, 331 pp.

Vilar, P. (1962) *Cataluña en la España Moderna*, 3 vols, Crítica, Barcelona.

Yun Casalilla, B. (1991) Vasallos y señores en el marquesado de Cuéllar: relaciones sociales, aprovechamiento de recursos y gestión señorial a fines del Antiguo Régimen. In: Saavedra, P. and Villares, R. (eds) *Señores y Campesinos en la Península Ibérica, siglos XVIII-XX*, Vol. 1. Crítica/Consello da Cultura Gallega, Barcelona, pp. 239–260.

The Impact of Economic, Social and Political Factors on the Ecology of Small English Woodlands: a Case Study of the Ancient Woods in South Yorkshire, England

31

Ian D. Rotherham and Melvyn Jones

Countryside Research Unit, School of Leisure and Food Management, Sheffield Hallam University, City Campus, Sheffield S1 1WB, UK

Most English woodlands are small, which has made them vulnerable to clearance and neglect at various periods in their history. In South Yorkshire in the northern Midlands region of England, there are 333 documented ancient woodland sites. Only one exceeds 200 ha and 157 are less than 5 ha in size. The ecology of these woodlands has been affected by a succession of woodland management regimes – wood pasture, coppice with standards, high forest – reflecting changes of function, ownership and perceived economic, cultural and social importance. Urban influences have also been profound, leading to significant disturbance. This chapter examines a range of economic, social and political factors which have had an impact on woodland structure, woodland composition, herbaceous vegetation and breeding birds and mammals.

1 Introduction

England is not only one of the least wooded countries in Europe, but its surviving semi-natural woodlands are small. Woods over 50 ha are uncommon and those over 100 ha quite rare. In South Yorkshire there are 333 ancient woodland sites covering 4451 ha or just 2.8% of the land surface (Eccles, 1986). Only one woodland exceeds 200 ha, four are in the 100–200 ha range and five in the 50–100 ha range. Forty-seven per cent of the ancient woodland sites (157) are less than 5 ha in size. The small size of most of these woodlands has led to their vulnerability to clearance and neglect at various periods. The ecology of these South Yorkshire woodlands, like those in the rest of the country, has been affected by a succession of active and passive management regimes reflecting changes of function, ownership and perceived economic, cultural and social importance. Urban influences have also been profound, leading to significant disturbance.

Most of the ancient woodlands in South Yorkshire were managed intensively as coppice with standards from the late medieval period until at least 1850. In the second half of the 19th century most of the woods were gradually converted to high forest through singling of the coppice, planting and natural regeneration. Woodlands were also affected by mining for ironstone and coal. A small core of woods are still managed as commercial timber woods or are used as game preserves (for pheasant shooting) or both. Some of the largest woods have been converted to commercial plantations. Most of the woods, however, have been sold during this century to public authorities and these are now recreational woodlands and nature reserves open to the public.

Changing function and ownership, with the associated changes of management regime, have had – and continue to have – a profound effect on woodland ecology. This chapter pays particular attention to the impact of a range of economic, social and political factors on tree and shrub communities, woodland structure, herbaceous vegetation and breeding birds and mammals.

2 Changing woodland management systems during the last millennium and their ecological impacts

2.1 Wood pasture, AD 1000–1450

Woodland cover had already been drastically reduced in South Yorkshire by the beginning of the 11th century. In some districts the typical English countryside of hedged or walled fields, winding lanes, isolated farms, hamlets and villages and scattered woods and commons must have already existed for centuries.

Oliver Rackham has calculated that the Domesday survey of England in 1086 covered 11 Mha of land, of which 4.1 million were wooded, that is 15% of the surveyed area. His figure for the West Riding of Yorkshire is 16% (Rackham, 1980). Our own calculation for South Yorkshire is just under 13%. By way of comparison, woods today, including plantations, cover just over 6% of South Yorkshire. This means that by the 11th century, South Yorkshire was relatively sparsely wooded even by 20th-century standards.

In 1086, in the western half of South Yorkshire in the millstone-grit country and on the coal measures (Fig. 31.1), woodland was relatively

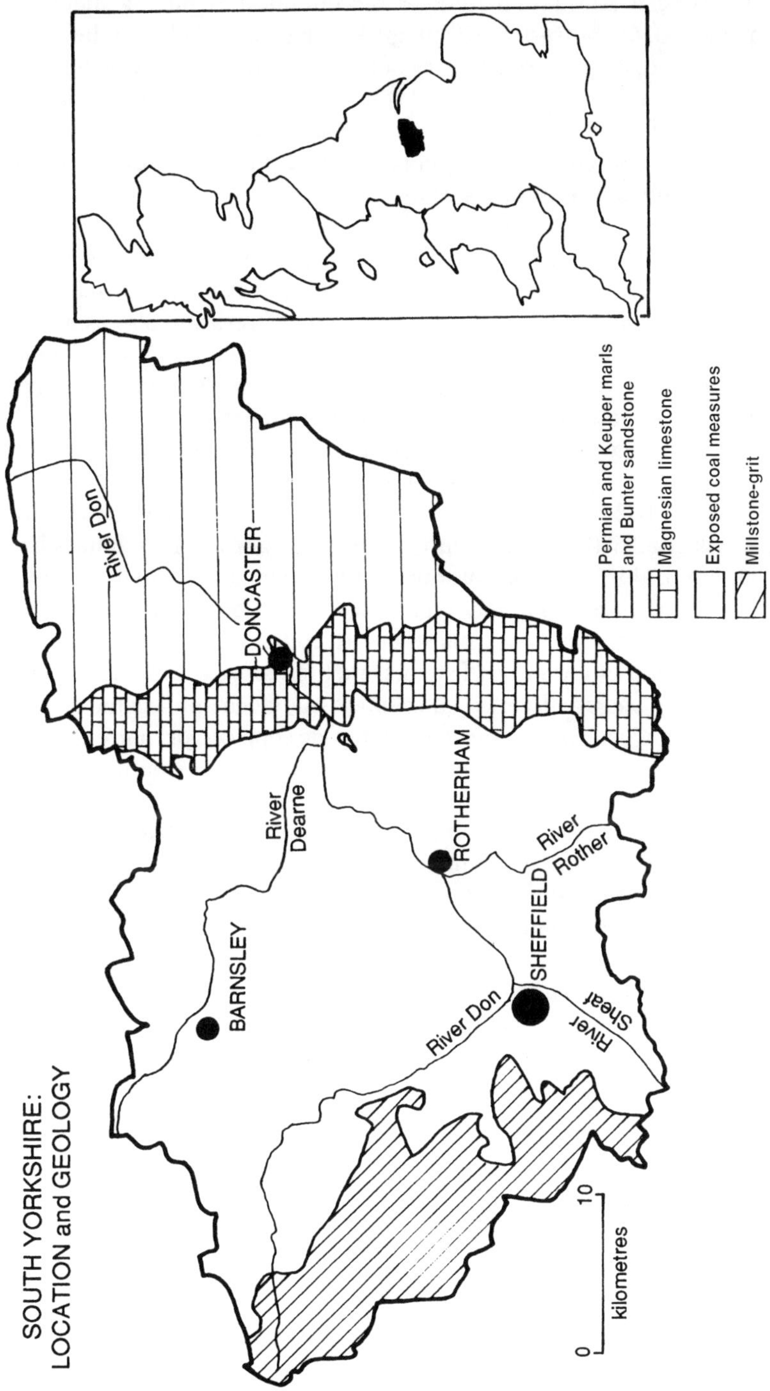

Fig. 31.1. South Yorkshire: location and geology.

extensive, with a substantial number of communities having more than 400 ha of wood. In contrast, in the magnesian limestone belt and the eastern lowlands beyond, the picture was very different. In those areas woodland was more scattered, amounts in individual communities were much smaller than to the west, and in 19 places no woodland was recorded at all. This suggests early clearance and continuous occupation and cultivation by a relatively dense population for thousands of years.

The types of woodland in South Yorkshire at Domesday also suggest a shortage of woodland in the east of the county and a relative abundance on the coal measures. When woods were widespread and populations sparse and scattered, they would have been exploited for their trees and also as pastures for cattle, pigs and sheep, i.e. as wood pasture. As populations grew, demands on the timber increased and pressure from grazing animals prevented regeneration, a type of management which gave a continuous and self-renewing supply of trees had to be introduced. This was achieved by the more widespread use of coppicing. Only seven manors had coppice woods in South Yorkshire in 1086, while 102 had wood pastures. All seven areas of coppice wood were in the eastern half of the county, with five in the magnesian limestone belt. On the other hand, although wood pasture was found in all parts of the county, it was very extensive and the only type of woodland found on the lower and middle coal measures and in the millstone-grit country.

As populations grew and settlements and therefore clearance of woodland increased, wood pasture declined and coppice management became predominant. However, wood pasture survived alongside coppices for many centuries. In the late medieval period and beyond, wood pastures were found on wooded commons, in deer parks and on chases.

In the manor of Sheffield in 1637, for example, there were more than 21,000 acres of common, much of it wooded (Ronksley, 1908). Almost all of this disappeared at the time of the Parliamentary enclosures between 1750 and 1830 and only fragments remain, one of the largest stretches being Loxley Common, referred to about 1650 as 'one Great Wood called Loxley the herbage common and consisteth of great Oake timber'.

Deer parks, large and small, sometimes well wooded and sometimes only moderately so, studded South Yorkshire in the medieval period (Jones, 1996). Unlike commons, deer parks were areas of private land, bounded by a wall or bank with a cleft-oak paling fence, in which the owner kept deer, hares and rabbits in order to provide his family with a reliable source of meat, and grew timber and underwood. There were more than 200 deer in Tankersley Park in 1653 and 1000 in Sheffield Park in 1637. In the 18th century many medieval deer parks were landscaped but frequently retained their herds of deer, often joined by sheep and cattle. Parks contained coppice woods, holly hags (see below) and great timber trees. Sheffield Park, its site now covered by housing developments at Park Hill, Norfolk Park, the Manor and Arbourthorne, was famous in its day for the size of its trees. John Evelyn, the 17th-century writer, in his book, *Silva* (1706 edition), described a massive oak tree growing below the Manor which provided 1400 'wairs', which were planks 2 yards long and 1 yard wide, and 20 cords (30 tons) of wood from the branches. The local woodward described to Evelyn another oak tree in the park that was so big that, when it was felled and lying on its side, two men on opposite sides of it on horseback could not see each other's hats.

Wood pasture was also a feature of chases, the private equivalent of the royal forests. Chases were areas of land, usually a mixture of heath, moor and woods of various kinds, on manorial wastes and, unlike parks, chases had no fences and were often intermingled with settlements and farms. Excluding Hatfield, which was primarily a wetland, there were two wooded chases in South Yorkshire, Rivelin Chase, which belonged to the lords of Hallamshire, and Wharncliffe Chase, which belonged to the Wortley family. Grazing and tree growth do not go well together and on many commons and in parks and chases pollarding was practised in order to produce poles and rods beyond the reach of grazing animals.

The decline of wood pasture management began in the Middle Ages with the expansion of coppicing, although it was still widespread on wooded commons and in deer parks in the 17th century, as Harrison's 1637 survey shows. However, it was already in decline. A substantial part of Sheffield Park was let to tenants by 1637 and eventually the whole park disappeared, first through its subdivision into separate farms, then through the spread of manufacturing and mining activity and finally through the spread of urban Sheffield. Wood pasture on commons declined steeply in the second half of the 18th and the first third of the 19th century as commons were enclosed under the Parliamentary Enclosure Acts. These 'new' enclosures on old commons are easily recognized by their square and rectangular fields.

2.2 Coppice management, AD 1450–1850

In the centuries following the Domesday survey, although the wood pasture tradition in South Yorkshire continued to be strong, in the form of wooded commons, deer parks and chases, as described above, coppice management gradually became dominant. This was in order to conserve wood supplies, which were becoming depleted as the population grew and as more and more woodland was cleared for agriculture. However, evidence available at the present time suggests that coppice management, which had probably been practised within wood pastures since neolithic times, did not replace wood pasture as the dominant form of woodland management in South Yorkshire until a comparatively late date. Coppice with standards had emerged as the most important form of woodland management in economic terms, if not in physical extent, by the late Middle Ages and continued to be so until at least the middle of the 19th century. The underwood in the coppice woods, though

Fig. 31.2. (a) An ancient wood on the coal measures of South Yorkshire in 1650. It occupies a typically steeply sloping site on a parish boundary marked by a stream (A). It is being actively managed as a coppice with standards. The main standards are oak (*Quercus petraea*), wych elm (*Ulmus glabra*), ash (*Fraxinus excelsior*) and alder (*Alnus glutinosa*). The underwood (coppice) is mixed but dominated by oak, birch (*Betula pendula* and *Betula pubescens*), hazel (*Corylus avellana*), hawthorn (*Crataegus monogyna*) and holly (*Ilex aquifolium*). Typically there would be 150–200 young standards under 20 years old (called wavers) and 25–50 older standards, 40–100 years old, called black barks or lordings, depending on their ages, per hectare. The main products would be timber for building projects, poles for a myriad of woodland crafts, charcoal for iron furnaces and forges, whitecoal for lead smelting and oak bark for leather tanning. The wood contains charcoal platforms (B) and whitecoal kilns (C). The woods are grazed in the later stages of the coppice cycle (about 20 years). Where not bounded by a watercourse, the wood is protected by a bank or wall (D) or both. Turf stripping, associated with charcoal making, removes the topsoil and leads to loss of many plants from ground flora. (b) The same wood today. In the period 1850–1890 the wood was converted

predominantly of oak, was mixed (ash, maple, hawthorn, crab apple, elm, alder, birch, hazel, holly, willow, rowan), but the standards were overwhelmingly of oak (*Quercus petraea*), though ash and alder were locally important (Fig. 31.2a).

By the beginning of the 17th century, coppice management in South Yorkshire was entering a golden age. The manor of Sheffield was particularly well wooded. In an undated document written for the seventh Earl of Shrewsbury, the major landowner in Sheffield, who succeeded to the title in 1590 and died in 1616, 49 coppice woods were listed. In 1637, John Harrison in his *Exact and Perfect Survey and View of the Manor of Sheffield* (Ronksley, 1908) listed 36 separate coppice woods, in which the underwood varied in age from 4 years to 40. The coppice woods supported the local metalworking industries by providing the fuel for iron smelting (charcoal) and lead smelting (whitecoal) and through the provision of building material for the millwrights, who built and maintained the many water-powered industrial sites, weirs and river banks (Jones, 1997). The timber and underwood also supported a myriad of local crafts and industries, of which leather tanning was the most important (Jones, 1993).

One of the features that differentiated coppice woods from other woods was the care and vigilance with which young coppice growth was protected. The proceedings of local manorial courts and woodwards' accounts are full of instances of waiting for and catching offenders, accusations of theft and trespass, appearances in court, fines and payments to woodwards' helpers for tracking down suspects and repossessing stolen wood, timber and bark. The straying of animals into the coppice woods in the early stages of coppice growth was guarded against by building and keeping in good repair stock-proof fences. These could be in the form of stone walls or banks and ditches, with the bank nearest the wood and the ditch on the outside. The banks were surmounted by hedges or walls. When the coppice was well grown, tenants' animals were allowed access to the woods on payment of a due. This practice was called herbage or agistment.

Fig. 31.2. *(Continued)* to high forest by singling the coppice, by natural regeneration and by extensive planting. Many coppice stools were grubbed up or poisoned. Broad-leaved species not native to the site are common, such as planted sweet chestnut (*Castanea sativa*), beech (*Fagus sylvatica*), lime (*Tilia* × *europaea*) and hornbeam (*Carpinus betulus*) and the invasive sycamore (*Acer pseudoplatanus*). A small number of conifers, such as Scotch pine (*Pinus sylvestris*) and larch (*Larix decidua*), originally planted as a nurse crop for the broad-leaved species, may also survive. Standing and fallen dead elms may also be notable features. Many of the trees are of roughly the same age and form a dense canopy. As a result, the shrub layer is sparse and the ground flora is impoverished. The wood is now not likely to be managed as a commercial wood; it is more likely to be in public ownership and used for recreation. Bluebells (*Endymion non-scriptus*) spread and dominate large areas of the ground flora following the cessation of coppicing.

Occurring side by side with the coppice wood system throughout the late Middle Ages and into the 18th century was the management of holly trees (*Ilex aquifolium*) as winter fodder for farm stock, especially sheep, and for deer. The practice was not restricted to South Yorkshire but seems to have been generally associated with upland farming areas in northern England (Spray and Smith, 1977; Spray, 1981). The holly woods, which were either small, separate woods or compartments within woods or on wooded commons, were called holly hags. The holly hags were cut on a rotation, like coppice woods.

2.3 The decline of coppicing and conversion of coppice woods to high forest since AD 1850

Coppicing itself went into steady decline during the 19th century and had virtually disappeared in South Yorkshire by the beginning of the First World War. This decline was partly the result of nationwide changes but there were also important local factors that accelerated the process in South Yorkshire. The adoption in the 18th century by lead smelters, iron makers and some steel makers of coke instead of charcoal and whitecoal, meant the loss of major markets. This, coupled with the presence of valuable coal and ironstone seams under many local woods and rapid urban growth, led to the eventual destruction of a number of woods and the further contraction of the coppicing tradition. The increasing amount of trespass from the growing urban populations seeking recreation in the surrounding countryside induced woodland owners to sell or donate a number of woods to local authorities. For example, Roe Wood was donated to Sheffield Borough Council by the Duke of Norfolk in 1897 and Wincobank Wood in 1904. Sheffield Council also purchased Bowden Housteads Wood from the Duke in 1914.

By the 1890s, coppicing was nearing its end. Income from local coppices had declined sharply and management problems were increasing. As a result, more and more woods were gradually converted into canopy woods or 'high forests'. In essence they were becoming plantations, and forestry was replacing woodmanship. This was achieved by singling the multi-stemmed coppice stools in order to allow the best stem to grow into a standard tree, by clearing away altogether the oldest stools and sickly trees, and in their place planting young trees to be grown for timber on a long cycle (Jones and Walker, 1997; Jones, 1998). Many of the newly planted trees were not native to South Yorkshire, trees such as beech, sweet chestnut, common lime and sycamore. Amongst the broad-leaved trees, conifers were often planted as a 'nurse crop' for the slower-growing broad-leaved species (Fig. 31.2b).

Forestry management, including coniferous plantings, has continued to be important right up to the 1990s in some parts of the area, and large woods such as Wharncliffe (Forest Enterprise) and Greno (Fountain Forestry) have been almost completely planted to conifers.

A large number of former coppice woods have come into local authority ownership and are now mostly amenity woods. Since their purchase, almost all of them have been managed only on a 'care and maintenance' basis, and in consequence they have grown overmature and their shrub layers have greatly diminished or have disappeared almost completely.

None of the known late medieval coppice woods in South Yorkshire have disappeared through the wanton behaviour of woodmen and other exploiters of wood and timber. Woods have disappeared or have been reduced in size over a long period as a result of the clearance for farmland and mining operations, through the spread of settlement and through road building. Nevertheless, the survival rate is surprisingly high, a testament to the central importance of woods to the local economy in the past and, more recently, to local conservation groups, who have assured their survival in the face of heavy pressure from developers. However, with the exception of the woods on private estates still managed for timber or for game and those managed as commercial 'forests', the surviving woods, particularly those owned by the local authorities, have been neglected. By the beginning of the 1980s, many of South Yorkshire's surviving ancient woods were even-aged, with dense canopies, poorly developed shrub layers and impoverished ground layers. The more accessible woods were sometimes heavily vandalized.

The benign neglect of the amenity woods made them much less attractive than in the past, in that they contained much poorer displays of spring flowers, a significant number of breeding birds and butterflies had been reduced in number or were no longer found in woods where they were once common (Medforth and Rotherham, 1997), and local residents were increasingly afraid of walking in the woods because they were dark and gloomy and engendered a fear of personal attack. After many centuries of intensive management and careful protection, the 20th-century attitude seemed, at best, to leave them to their own devices and, at worst, to abuse them unmercifully.

During the last two decades there have been some interesting developments. These have been particularly prolific in Sheffield, where the City Council has responsibility for more than 60 woodlands, including nearly 40 ancient woods. In the early 1970s, the city's Recreation Department created an Amenity Woodland Advisory Group, made up of council officers and representatives of environmental organizations (Gilbert, 1982). In 1986, the group commissioned a study to determine the status and management history of the major woods in the city (Jones, 1986) and this was closely followed by a woodland policy document, which was approved by the City Council in 1987. In 1991, this was incorporated into the Sheffield Nature Conservation Strategy (Bownes *et al.*, 1991). Meanwhile work had begun in the winter of 1988 on the implementation of a management plan for Bowden Housteads Wood, an important inner-city woodland, which had remained unmanaged since its purchase from the Duke of Norfolk in 1914. This was the first example of active woodland management in a non-commercial wood in South Yorkshire in the whole of the 20th century! Other management plans followed, prepared not only by City Council officers but also by local environmental groups. Most recently, a series of carefully monitored coppicing experiments have been set up in four different woods in the city to assess the ecological and economic impact of the coppiced coupes and to evaluate visitor reaction to the reintroduction of intensive management in well-visited woods (Jones and Talbot, 1995; Rotherham, 1996a).

Similar developments have taken place in Rotherham Metropolitan Borough, where there are more than 20 ancient woods in public ownership. In 1988 the Planning Department commissioned detailed studies of all the surviving woods and in many cases their neglected state (Jones, 1989–1992). A woodland officer has been in post for a number of years and there is now a wider appreciation of the value of the borough's woodlands and the need to actively manage them. Detailed management plans have been drawn up for half a dozen woods and others are in preparation.

A major influence on local attitudes toward woodland management in the last few years has been the South Yorkshire Forest Project. This project, established in 1991, is a partnership between Barnsley, Rotherham and Sheffield Councils, the Countryside Commission and the Forestry Commission. Its aim is to 'develop multi-purpose forests which will create better environments for people to use, cherish and enjoy'. It is not just concerned with ancient woodlands. Among its objectives are commitments to protect areas of historical, archaeological and ecological interest (i.e. the existing ancient woodlands), to increase opportunities for access and recreation and to encourage the development of timber-based industries, employment opportunities and woodland products. Following a year of public consultation, the South Yorkshire Forest Plan was published in August 1994 (South Yorkshire Forest Team, 1994). The plan established a policy framework and a strategic approach to woodland management throughout the South Yorkshire forest area – for private as well as publicly owned woods – and will guide developments well into the next century.

One particular issue that has emerged in recent years has been the increase in deer population and its implications for woodland management (McCarthy *et al.*, 1996; Rotherham, 1996b). Since the decline of the medieval deer parks, deer populations in South Yorkshire have been rather low and geographically very restricted. However, in recent decades populations of the native roe deer and the alien muntjac deer have increased dramatically. The implications of these changes, including colonization right into the urban heartlands, are potentially very significant.

3 Summary and conclusion

Woodland cover in South Yorkshire has never been extensive in the last 1000 years. In 1086, at the time of the Domesday survey, woodlands covered only 13% of the county. Since then, population growth, the extension of cultivated and grazed land and, after 1840, rapid industrialization have continued to reduce woodland cover. Today, ancient woodlands cover less than 3% of the area.

Despite their small extent, woodlands, particularly in the western half of the area, have played a vital part in the initial development of industry in the early Middle Ages and its subsequent expansion into the modern period.

A number of economic, political and social factors (Table 31.1) have had a profound impact on the ecology of the ancient woodlands in South Yorkshire.

1. Changing patterns of ownership, from a concentration in the hands of a few large estate owners to a mixed pattern of ownership, including a range of smaller private owners and public bodies, such as the Forestry Commission and city, town and parish councils.
2. There has been a change of function since about 1850 from the role of suppliers of timber, wood and bark to the role of suppliers of timber only and then, in a substantial number of cases, to the role of recreational woods/nature reserves.
3. Change of function has been accompanied by a change of structure from coppice with standards to high forest and by changes in composition, with the introduction of many trees not native to South Yorkshire.
4. The onset of large-scale industrialization from about 1840, with the proliferation of 'smokestack' industries, had a negative impact on ancient woods in South Yorkshire, causing the loss of a number of woods and the reduction in size of many others, together with widespread disturbance and pollution. Only since about 1980 has there been a rapid decline in coalmining and heavy industry and a shift to a postindustrial, service economy.
5. Since about 1970, public pressure, organized conservation campaigns and increased research activity (Rotherham, 1996c) have resulted in the widespread recognition of the heritage value of the county's ancient woods and the resumption of active management.
6. The future of South Yorkshire's ancient woods looks much better than it did three decades ago. Awareness of their cultural and ecological importance has been raised to a much higher level than at any other time this century and interest in their economic as well as recreational potential has been reawakened. However, the renewed interest in the sustainable management of these ancient woodlands is just the beginning; the challenge is to sustain it in the medium and long term.

Table 31.1. One thousand years of woodland change in South Yorkshire resulting from a combination of economic (Ec), political (Po) and social (So) factors.

Chronology	Ec	Po	So	Ecological implications
1086 At this early date, according to the Domesday Book, only about 15% of the area was wooded through clearance for arable land, meadow and grazing. More than 95% of woodland described as wood pasture	•			Woods regularly but not intensively grazed by deer, cattle, sheep and horses
1200–1450 Forty-four grants of free warren in period 1250–1325, and more than 70 between 1200 and 1441	•	•	•	Woodland (in the form of canopy woods and later coppices) and old trees preserved within demesne land, mainly in the form of deer parks and chases (Rivelin Chase, Hatfield Chase, Wharncliffe Chase)
1421 Earliest-known surviving record of coppice-with-standards management	•	•	•	
1550–1800 Golden age of coppice-with-standards management, associated with production of charcoal for iron furnaces and forges, whitecoal for lead smelters, bark for leather tanning, oak timber for building and poles for a myriad of coppice crafts	•			Surviving woods protected by banks and/or walls; grazing animals only allowed into woods in middle and later stages of the coppice cycle; certain species favoured (oak, hazel, holly, alder); canopy opened periodically allowing high light levels on the woodland floor, resulting in development of distinctive ancient woodland ground flora; ground flora and soils extensively disturbed and impoverished by deturfing in course of charcoal and whitecoal production
From *c.* 1600 disparking of deer parks and gradual extinction of chases through intensification of agriculture (Rivelin Chase) or drainage (Hatfield Chase)	•	•		Deer increasingly maintained in a few landscaped parks attached to large country houses. Red deer survive on one chase till late 1940s
From 1775 lead produced in coal-fired cupola furnaces and iron in coke-fired blast furnaces: market for whitecoal disappears; and market for charcoal much reduced	•			Coppicing begins steady decline after 1800 and accelerates after *c.* 1825. Gradual conversion of coppice woods to high forest by natural regeneration, singling of coppice and planting.
1750–1830 Parliamentary Acts of Enclosure eradicate all but a few tracts of wooded common	•			Non-native broad-leaved and coniferous species widely planted in ancient woods and new plantations

continued

Table 31.1. *Continued.*

Chronology	Ec	Po	So	Ecological implications
By *c.* 1900 coppicing virtually extinct	•			In many woods, in public and private ownership, a long period of absence of active management begins, which lasts until the end of the 1980s. Woods become increasingly even-aged, shrub layers are greatly depleted and, in woods under beech, sycamore and sweet chestnut, the ground floras are severely impoverished and blubells become dominant. Birds typical of coppice woods decline; birds typical of high forest increase. Sensitive lichens affected by air pollution
From 1900 many woods acquired by local authorities for use as recreational woodlands			•	
1850–1970 Urban woods in particular affected by air pollution. Some woods cleared for urban expansion, others reduced in size	•		•	
From 1920 some large woods converted to commercial coniferous plantations	•	•		Recovery of lichens. Populations of red and fallow deer increase; muntjac begins to colonize the region
From 1970s increasing public pressure to conserve old woodlands and manage them 'actively'			•	Increasing number of public woods actively managed for first time in 100 years. Renewed broad-leaved planting
1991 South Yorkshire Forest Project established	•	•	•	

References

Bownes, J.S., Riley, T.H., Rotherham, I.D., and Vincent, S.M. (1991) *Sheffield Nature Conservation Strategy*. Sheffield City Council, Sheffield.

Eccles, C. (1986) *South Yorkshire: Inventory of Ancient Woodland.* Nature Conservancy Council, Peterborough.

Evelyn, J. (1706) *Silva or a Discourse of Forest-Trees*, 4th edn. Scott, Chiswell, Sawbridge and Tooke, London.

Gilbert, O.L. (1982) The management of urban woodland in Sheffield. *Sorby Record* 20, 47–48.

Jones, M. (1986) *Sheffield's Ancient Woods, Past and Present.* Report for Sheffield City Council. Sheffield City Polytechnic, Sheffield.

Jones, M. (1989–1992) *Inventory Survey of Ancient Woods in Rotherham Metropolitan Borough*, 3 vols. Report for Rotherham MBC Planning Department. Sheffield City Polytechnic, Sheffield.

Jones, M. (1993) *Sheffield's Woodland Heritage*, 2nd edn. Green Tree Publications, Rotherham.

Jones, M. (1996) Deer in South Yorkshire: an historical perspective. *Journal of Practical Ecology and Conservation*, Special Publication No. 1, 11–26.

Jones, M. (1997) Woodland management on the Duke of Norfolk's Sheffield estate in the early eighteenth century. In: Jones, M. (ed.) *Aspects of Sheffield: Discovering Local History*, Vol. 1. Wharncliffe Publishing, Barnsley, pp. 48–69.

Jones, M. (1998) The rise, decline and extinction of spring wood management in south-west Yorkshire. In: Watkins, C. (ed.) *European Woods and Forests: Studies in Cultural History*. CAB International, Wallingford, pp. 55–72.

Jones, M. and Talbot, E. (1995) Coppicing in urban woodlands: a progress report on a multi-purpose feasibility study in the City of Sheffield. *Journal of Practical Ecology and Conservation* 1, 48–54.

Jones, M. and Walker, P. (1997) From coppice-with-standards to high forest: the management of Ecclesall Woods 1715–1901. In: Rotherham, I.D. and Jones, M. (eds) The Natural History of Ecclesall Woods, Part 1. *Peak District Journal of Natural History and Archaeology*, Special Publication No. 1, 11–20.

McCarthy, A.J., Baker, A. and Rotherham, I.D. (1996) Urban-fringe deer management issues – a South Yorkshire case study. *British Wildlife* 8, 12–19.

Medforth, P. and Rotherham, I.D. (1997) The birds of Ecclesall Woods. In: Rotherham, I.D. and Jones, M. (eds) The Natural History of Ecclesall Woods, Part 1. Peak District. *Journal of Natural History and Archaeology*, Special Publication No. 1, 21–33.

Rackham, O. (1980) *Ancient Woodland: Its History, Vegetation and Uses in England*. Edward Arnold, London.

Ronksley, J.G. (ed.) (1908) *An Exact and Perfect Survey and View of the Manor of Sheffield with other Lands by John Harrison, 1637*. Robert White, Worksop.

Rotherham, I.D. (1996a) Woods in the heart of a city. *Yorkshire Wildlife*, Autumn, 10–12.

Rotherham, I.D. (1996b) Deer or the new woodlands? A local authority perspective. In: Jones, M., Rotherham, I.D. and McCarthy, A.J. (eds) Deer or the New Woodlands? *Journal of Practical Ecology and Conservation*, Special Publication No. 1, 59–63.

Rotherham, I.D. (1996c) The sustainable management of urban-fringe woodlands for amenity and conservation objectives. Proceedings of the symposium on vegetation management in forestry, amenity and conservation areas: managing for multiple objectives. Association of Applied Biologists, York, 1996. *Aspects of Applied Biology* 44, 33–38.

South Yorkshire Forest Team (1994) *South Yorkshire Forest Plan*. Sheffield: South Yorkshire Forest Project, Sheffield.

Spray, M. (1981) Holly as a fodder in England. *Agricultural History Review* 29, 97–110.

Spray, M. and Smith, D.J. (1977) The rise and fall of holly in the Sheffield region. *Transactions of the Hunter Archaeological Society* 10, 239–251.

Index

Page ranges in **bold** indicate a whole chapter in which the subject is a major topic. Page locators in *italics* refer to figures and tables. References to end notes are indicated by (n).